Praise for K. C. Cole's *Something Incredibly*

"Go to the Exploratorium. Read first K. C. Cole's elegant new book . . . a biography of that museum's founder and a portrait of the world he created, which spawned a revolution in science museums. Oppenheimer created a palace of experiments . . . in which it is impossible not to be engaged, amazed, frustrated and amused."

EDWARD ROTHSTEIN, *New York Times*

"Oppenheimer's charisma, on full display in anecdotes, reminiscences and bon mots, wins us over, just as it did Cole and countless others. Like one of Oppenheimer's science classes or a visit to the Exploratorium, this joyful and loving portrait has the potential to change minds and make explorers of us all."

JESSE COHEN, *Los Angeles Times*

"Something incredibly wonderful will happen when you open this book. You'll come face to face with a man who had an uncanny knack for making the wonders of nature available to the rest of us, and you'll get to look inside his extraordinary mind. . . . As always, K. C. Cole delivers science to us as a thrilling ride, a deeply human story, and a gallery of unimaginable beauty."

ALAN ALDA

"With vivid insight, Cole relates the fascinating story of Oppenheimer's life, and the confrontations of his career. She came to know his ability to charm everyone he met, from rambunctious kids in this city's projects to the wealthiest Silicon Valley pioneers. . . . Her story is at once an absorbing—even dramatic—biography of Cole's hero, and a wonderfully anecdotal history of [the Exploratorium] itself."

San Francisco Gate

"It is unlikely that anyone will ever write a more perceptive biography of Frank Oppenheimer."

TIMOTHY FERRIS, *New York Review of Books*

"A deeply personal and moving narrative . . . [Cole] sweeps readers through the events that led Oppenheimer to ultimately create a world of his own: The Exploratorium, a revolutionary science museum . . . dedicated to thinking, tinkering and rumpuses of all sorts."

LAURA SANDERS, *Science News*

"Cole's real subjects are the early years of the Exploratorium and its charismatic, maddening creator . . . Exhibits were donated and jerry-built . . . a traffic light became an optics lesson. . . . Children played truant to hang out there . . . and adults found it exotic and magical."

ROBERT CREASE, *Nature*

"'No one flunks a museum,' Frank repeatedly said. . . . Once Frank explained all this to his biographer, . . . she realized that, 'at its core, the Exploratorium was a political institution.' Frank really believed that scientific curiosity . . . unleashed and satisfied would have enormous moral consequences. . . . The Exploratorium was an institution made in Frank Oppenheimer's image—childlike, silly, naïve and, ultimately, quite wonderful."

STEVEN SHAPIN, *London Review of Books*

"K. C. Cole's life-size portrait of Frank Oppenheimer depicts a complex character who sowed inspiration in his wake. The book made me laugh and cry—and wish that Oppenheimer were alive to read it."

DAVA SOBEL

"This is a gorgeous story about that rare leader who lived a life with his eyebrows continually raised in uncontaminated wonder and passed that wonder on to thousands of others. It is the book I've been waiting to read for over fifty years and should be required reading for every thoughtful world citizen."

WARREN BENNIS, author of *On Becoming a Leader*

Something Incredibly Wonderful Happens

Books by K. C. Cole

The Universe and the Teacup: The Mathematics of Truth and Beauty

First You Build a Cloud ... and Other Reflections on Physics as a Way of Life

The Hole in the Universe: How Scientists Peered Over the Edge of Emptiness and Found Everything

Mind Over Matter: Conversations with the Cosmos

Something Incredibly Wonderful Happens: Frank Oppenheimer and His Astonishing Exploratorium

SOMETHING INCREDIBLY

FRANK OPPENHEIMER AND

WONDERFUL HAPPENS

HIS ASTONISHING EXPLORATORIUM

K. C. Cole

Foreword by Murray Gell-Mann

The University of Chicago Press

The University of Chicago Press, Chicago 60637

University of Chicago Press edition 2012
Printed in the United States of America

SPECIAL EDITION FOR STRIPE, CONVERGENCE 2019

ISBN-13: 978-0-226-11347-0 (PAPER)
ISBN-13: 978-0-226-00936-0 (E-BOOK)
ISBN-10: 0-226-11347-7 (PAPER)

Library of Congress Cataloging-in-Publication Data
Cole, K. C.
Something incredibly wonderful happens : Frank Oppenheimer and his astonishing Exploratorium / K. C. Cole ; foreword by Murray Gell-Mann. — University of Chicago Press edition.
pages. cm.
Includes bibliographical references and index.
ISBN 978-0-226-11347-0 (pbk. : alk. paper) — ISBN 978-0-226-00936-0 (e-book) — ISBN 0-226-11347-7 (pbk. : alk. paper) 1. Oppenheimer, Frank, 1912–1985. 2. Physicists—United States—Biography. I. Title.

QC16.O618C65 2012
530.092—dc23 [B]

2012025914

♾ This paper meets the requirements of ANSI/NISO Z39.48-1992 (Permanence of Paper).

For the Dinosaurs (you know who you are) and all the other deeply committed people who helped Frank realize his dream, keep it alive, and continue to spread his passion and vision in a multitude of most delicious ways . . .

CONTENTS

FOREWORD

As children in blank darkness tremble and start at everything, so we in broad daylight are oppressed at times by fears as baseless as those horrors which children imagine coming upon them in the dark. This dread and darkness of mind cannot be dispelled by sunbeams, the shining shaft of day, but only by an understanding of the outward form and inner workings of nature.

—LUCRETIUS, *De Rerum Natura,* 55 B.C.E.

SOON AFTER I HELPED Frank Oppenheimer obtain a critical million-dollar grant from the J.D. and C.T. MacArthur Foundation for his fledging Exploratorium, I took my son there for a visit. Frank welcomed us, of course, and wandered around the five hundred or so exhibits with us, pointing out some particularly interesting ones. Then he said it was a pity that he was accompanying us because we would have enjoyed the place much more by ourselves, with no guidance.

Frank liked to say he lived his life on the edge of chaos. That seems entirely appropriate, given that the edge of chaos is a name that some have given to the magic zone between order and disorder where life flourishes and complex adaptive systems evolve. He created his "museum of awareness" as a "woods of natural phenomena" where people could play with science, art, and technology much as one could explore a wilderness teeming with life. Toddlers and physicists play with pendulums; little old ladies and teenagers wonder at the magic of mirrors. One finds real science there—important concepts related to momentum, inertia, wave motion, optics, animal behavior, human perception, language, and many other subjects. There is also mathematics.

Because his forest is so fertile, people can make personal discoveries about nature, giving them a feeling of comfort in the often intimidating world of science. The point of the place, he often said, is to get people to notice things, to make connections, to see that science is not a collection of facts, but rather a way of knowing, a crucial part of human culture. Nothing is more important, Frank thought, than the realization that most things are understandable. If we stop trying to understand things, "we'll all be sunk," he said.

And he was right. Humans now have the ability to destroy a great deal of the rich complexity of life created over the eons through evolution, and along with it much of the culture, including science, that humans have built. We are already altering the climate and degrading the biosphere, with its ecosystems on which we depend for survival. People have always had weapons, but their scale today is unprecedented: thermonuclear weapons can wipe out a significant portion of life on the planet.

The only headlight we have is usually rather feeble and flickering, but it can be powerful. It is that "human intelligence" of which we are often proud. Through science and related disciplines, we have learned to think both rationally and creatively, to examine ideas critically, and to see ahead a bit into the darkness.

Yet many people involved with the arts and the humanities admit (some even brag) that they know almost nothing about science. By contrast, an English-speaking scientist might be ignorant of most of the works of Shakespeare but would never brag about it.

Frank's genius was to create a place where these worlds are intermingled, where the workings of science are transparent, accessible, playful, and clearly related to art, technology, politics, and social institutions. In "the world he made up," ordinary people learn to feel more at home in the universe. They may gain a sense that their ideas can make a difference and that they have a stake in the future.

Frank Oppenheimer was an often underrated physicist who made important discoveries concerning the nature of cosmic rays, only to be banished from physics during the McCarthy years. (He didn't handle wisely questions about his former Communist associations.) During World War II, he worked with Ernest Lawrence to develop the Calutrons used for separating the rare, easily fissionable U-235 nuclei from the common U-238 nuclei. At Los Alamos, he was a safety inspector for the Trinity test of the first nuclear weapon, and he watched the explosion that changed the world alongside his brother, Robert—the "father" of the atomic bomb.

Like his best friend, the physicist Robert Wilson, Frank was essentially a pacifist helping to build a weapon of enormous destructive power—a tragic irony that never left him. His horror at the dropping of nuclear weapons on Hiroshima and Nagasaki was one of the seeds from which the Exploratorium grew. Science, he thought, had to be steered by the collective wisdom of the people, not by the judgments of "experts" alone. What he built was, as an early Chinese delegation put it, a "people's science museum."

Frank knew that without a sense of planetary community we are unable to meet the challenges that face us. That sense of community is often blocked by divisions between groups of people based on differences in religion or other aspects of culture that sometimes can be almost imperceptible to those outside. Even when the differences are striking, do they really need to divide people so sharply into "us" and "them"? Yet in general humans of all kinds tend to share important commonalities; most people aspire to be safe, to be free to express themselves without fear, to hope for a better future with adequate food and shelter and protection from disease.

By basing his museum around human perception, Frank showed how easy it is for the mind to jump to incorrect conclusions based on unexamined assumptions. He loved to point out how lessons from physics could help. At heart, Frank built the Exploratorium as a political institution. Its ultimate goal was to get people so addicted to understanding that they would somehow become inoculated against the clever deceptions of some advertisers and politicians. He would persuade them to use those brains of theirs to get involved, to add to the collective wisdom—the only true way, he thought, to solve our pressing global problems.

The author of this book, K. C. Cole, is a distinguished science writer who met Frank when she was twenty-six. She was almost completely ignorant of science but fascinated by his work. Frank saw in her a talent for writing about physics and math. (She thought he was nuts to believe in her.) She spent the next thirteen years, until his death in 1985, working with him on one project or another.

K. C. thinks it is largely because of his guidance and inspiration that she won a number of prizes (from the American Institute of Physics, for example). In my opinion she would have been a prizewinner no matter what. She has written almost a dozen books, translated into as many languages. She has written about science in publications ranging from *The New Yorker* to *Seventeen,* as well as in major newspapers.

She also does radio commentaries on science. I can't imagine why she doesn't do more television. She and I once appeared together on a program on art and science at a wonderful, funky museum in Santa Fe, and I was greatly impressed with her performance, as was the audience.

K. C. has written a splendid account of Frank Oppenheimer's life, his achievements, his mistakes and difficulties, and especially his way of looking at the world. She has restored him to life.

MURRAY GELL-MANN

PART I

THE WORLD HE CAME INTO

1

PALACE OF DELIGHTS*

I MET FRANK OPPENHEIMER soon after I got my first real writing job with the venerable *Saturday Review,* which had just moved from New York to San Francisco. It was the early 1970s, and San Francisco was, in the lingo of the day, happening — the place where you wore flowers in your hair.

For an East Coast kid, everything about it was magic. There were golden hills and wildflowers; rows of pastel wedding-cake houses set shoulder to shoulder on impossibly steep streets; a tunnel you entered through a rainbow. You could take a ferry boat to Angel Island and commune with seals. The stealth fog spilled over the hills fast and fierce, snuffing out the red towers of the Golden Gate Bridge and engulfing you in a silent, splendid isolation. Nothing seemed real. But the air of possibility was palpable.

I had no interest in science whatsoever, having taken the usual dull courses in school and gone on to more interesting and relevant things, such as politics and culture. But one of the editors at the magazine had heard about a strange new science museum in the city, and I was sent to write about it. I had no idea what to expect.

* "Palace of Delights" was the title given to a *Nova* documentary about the Exploratorium. Frank thought it made the place sound like a brothel, but loved the film nevertheless.

This so-called Exploratorium was housed in the Palace of Fine Arts, at the foot of the Golden Gate Bridge — a huge "Roman ruin" created by Bernard R. Maybeck for the 1915 Panama-Pacific International Exposition. Maybeck designed it to crumble under the weight of time. In the interim, the city had put it to use as a fire station, a warehouse for telephone books, a garage for limousines, an airplane hangar, tennis courts. It almost did become a ruin, but by the 1960s, sufficient public support had rallied to restore it — and by that time it had become enough of a landmark that even Maybeck went along. It had just reopened in 1968.

My taxi pulled up beside a quiet lagoon in a sunny park with swans and ducks and old folks on benches feeding the pigeons. Behind rose a massive salmon-pink colonnade and Romanesque rotunda; maidens leaned over giant cisterns draped with garlands. The air was sweet with eucalyptus — a scent that can still take me back to that day.

Nearly hidden behind all this architectural pageantry was an enormous semicircular building. I walked under an archway at one end and opened a door into a dark cavern that seemed to pulse with dim lights and eerie sounds. It was spooky and inviting.

Although I couldn't see it all from where I stood, the space was vast — ninety thousand square feet. The concrete supports and steel girders looked like the bare ribs of a giant fish, and I felt as if I'd entered the belly of a whale. The floor was rough and ugly, the walls concrete. Wires and cables dangled from the rafters. The only light came from rows of fluorescent lamps forty feet above; despite the presence of four enormous fireplaces, there was no heat, and it was cold. An early 1970 article said the place "has all the charm of a blimp hangar."

It took a while for the eyes to adjust — and not only in a literal sense. The strange hums and flickers gave the place an ethereal, otherworldly feel. There was a big yellow tent on my left that housed an electronics shop. On my right was a noisy machine shop, all sparks and whines of saws and welding and lathes. Hanging above the shop, a handmade sign announced: HERE IS BE-

ING CREATED THE EXPLORATORIUM, A COMMUNITY MUSEUM DEDICATED TO AWARENESS.*

I dropped a quarter into a donation barrel and jumped back as 300,000 volts of purple lightning from a tesla coil zapped up a tall pole. Then I did it again.

I walked into a swirling wash of color where fine webs of sensuous red and blue and yellow created huge sheets of undulating light forms. A very tall man was twirling Mylar lassos in the light, spinning swirls of colors. Several children — and adults — joined him. When I traced the colors back, I could see that they were extracted from an ordinary white sunbeam coming through a hole in the ceiling; the beam spread out into a spectrum as it pirouetted through a rack of prisms; the prisms sent the whole palette of sunlight to a stand of thin vertical mirrors which then sliced it into thin ribbons of color — each uncannily pure. The tall man was Bob Miller, and this was his "Sun Painting," honored (along with Bob) in Muriel Rukeyser's poem "The Sun Painter."

I watched disembodied legs wearing patent-leather Mary Janes chase each other around in a circle — a shoe tester used by the National Bureau of Standards in the 1930s. Frank loved cool machinery.

Someone had made a dance floor of color-coded patches that played music as you stepped on them (the idea was later made into a toy sold at FAO Schwarz that was used to great effect in the movie *Big*). One day, I was told, a would-be John Travolta decked out in full *Saturday Night Fever* regalia came and played his heart out. In truth, it was almost impossible not to perform. (For amateurs, cards spelled out how to play "The Old Gray Mare" and the theme from *Close Encounters of the Third Kind*.)

Farther on, two small children were playing with a giant checkerboard of colored lights, "shooting" squares of different colors on and off with red and green light guns. "Bang, I got the red

* The sign was hand made by the wife of George Gamow — the late physicist who figured out that the universe began with a Big Bang (though he hated that term) and author of the popular Mr. Tompkins series, which inspired many a physics career.

one," a little girl said. She didn't know that the colors turned on when the beam was pointed vertically, off when it was horizontal, but her older brother quickly figured it out. When I came back, the brother was engrossed in conversation with a red-jacketed "Explainer" — a high school student — over the rudiments of polarization.

A boy and an old man were trying to get a four-hundred-pound cylinder of cement moving in a circle by throwing tiny magnets at a metal girdle encircling the center. Even the gentlest tugs, given the right timing, could get the cement weight swinging using the same principle of resonance that "pumps" a child on a swing. This "accelerator," I later learned, was built by Robert Wilson, a physicist and Frank's close friend from Los Alamos; it was, in essence, a miniature version of the four-mile-circumference particle accelerator Wilson had built at Fermi National Laboratory in Illinois.*

I played an old-time electronic instrument called a theremin, where you make music merely by waving your hands in the air to modulate a current. (Talk about air guitar!) I spun around on a "Momentum Machine," toyed with all kinds of gyros, sang to a TV screen that responded by weaving colored strands of electronic spaghetti into beautiful, complex patterns. I moved around large red, blue, and green blobs on a color-TV screen with a powerful magnet — entranced as they "condensed" at the poles just like iron filings.

There were "glass catfish" that spun polarized light, which made them look transparent; a big red face called Albert that followed you when you passed; sculptures made of bathroom window glass, Christmas tree balls, and Polaroid filters; glass rods that disappeared when dipped in clear liquid; a "Magic Wand" that could pluck an image out of thin air; swirling soap bubbles

* Just to clarify how dumb about science I was: The first time Frank talked to me about accelerators, I assumed he was referring to the gas pedals in cars. Actually, particle accelerators push subatomic particles to speeds approaching that of light, then bash them into each other (or some target). The energy of the collision condenses into all manner of exotic particles — allowing physicists to get a glimpse, for example, of the state of matter just after the Big Bang.

that made storms of color, then faded into invisibility as the film stretched so thin that it shrank to less than a single wavelength of light.

I watched a father and son discover that it takes a lot of pedal power to keep just three light bulbs burning; the "Pedal Generator" was fashioned from a nineteenth-century lathe from the Oppenheimer ranch. A little girl talked to an electronic tree that talked back in colored electronic twinkles. "I'm going to stay here all day!" the girl announced.

The place was full of stuff that played on my own perceptions, making the mind's eye the real object of investigation. I tried to thread a giant needle with one eye; watched shadows on a wall pop into 3-D as I looked through smoky glass; reversed the images from my right and left eyes — in the process reversing depth perception and making near far and far near. People danced behind a white screen illuminated with red and green lights; if you looked at them through red/green glasses, the shadows came toward you or moved away.

There was a full-sized distorted room where people shrank and grew as you watched; I saw the blood cells in my eye — tiny corpuscles coursing through capillaries, pulsing with each heartbeat; I saw colors where there was only black and white; illusions that made the walls — and my head — spin. I had never realized how much of perception is guesswork — that the brain has a mind of its own and uses it to sometimes downright creepy (even fatal) effect.

The far end of the cavern took on a different cast, brightly lit by skylights and strewn with mirrors: not just kaleidoscopes, but mirrors that focused heat as well as light, for example, or allowed you to shake your own ghostly hand or grasp for a spring that you could shine a light on, yet still not touch. A woman in a sari emerged laughing from a three-sided wooden structure with a sign reading DUCK INTO KALEIDOSCOPE on the side. "Come on in! It's a party!" she invited, after seeing herself and her husband reflected into an infinite expanse of images of themselves.

There were giant lenses, a prism tree, a blower that balanced a ball in an airstream. A Montgomery glider hung from the rafters. I climbed and crawled and slid my way through a giant pitch-black igloo, created for the place by the artist August Coppola, brother of Francis Ford Coppola.

The whole place seemed to have a heartbeat, a breath, a voice — as if it were whispering to itself. There were hums and whines and bells and whistles and hisses and drips and splashes and echoes and pings. There were also oohs and aahs and shouts and giggles and song and screams and running feet.

It was, in effect, a playground. But in place of jungle gyms and slides were nifty gadgets and natural phenomena — rainbows and magnetic fields and electric oscillations. It was not so much a place as a way of being in the world — a verb rather than a noun: spin, blow, reach, vary, strum, look, throw, fiddle, watch, wonder.

I found it hard to get over the feeling of playing hooky, or getting away with something. As if I'd broken into the zoo at night to pet the animals, or stayed in the art museum after closing time and climbed on the sculptures. There were no guides and no path and no right way to go through — no more than you'd expect a guide in a park or at the beach. Stuff was simply there to mess with. And what stuff! I thought there was nothing like it in the world, and I was right.

"It was like, Man, what *is* this place?" one staff member remembers thinking the first time she saw it. "I couldn't believe it. It was like no other place I had been to in my life . . . It was like, Oh, God, this is heaven!"

One thing I knew for sure: it wasn't science, and it wasn't a museum.

"It did not look like a museum," recalled Alan Friedman, a physicist and now the director of the New York Hall of Science. "The look of the exhibits was right off the lab bench. Rough wood. Things nailed into the table. This looked just like my low-temperature physics lab when I was a graduate student. It looked really friendly. It looked like home."

But the big shocker, Friedman said, was that the people he met really seemed to care whether visitors were having a good time. "My prior museum-going experience was that the people who worked in the museum's job was to *keep* you from having a good time."

I always had a hard time convincing people that the Exploratorium wasn't a children's museum — just because people were having so much fun.

Caneology

"He rapped me on the ankles with his cane a couple of times. 'I don't want it that way.' It was very gentle. It was a point of emphasis."

Frank's office was inside a small trailer that had been pulled into the building. I walked up some stairs and into a tiny backroom, rank with tobacco and occupied by an enormous black dog tied to a small sofa. (Whenever another dog came near, Frank's Labrador Orestes would drag the sofa to the doorway. The only thing that kept Orestes from escaping was that the sofa was too big to go through the door.)

Inside was a machinist's tool chest, a steel desk, and an old wooden desk, its surface scarred with parallel lines of finger-sized brown-edged indentations — burns from forgotten cigarettes left to smolder to their last. A bottle of whiskey, I later found out, could always be found in the right-hand drawer.

Of course, Frank wasn't there; he rarely was. I found him instead on the floor of the machine shop, rooting around for a two-by-four, encased in a cloud of cigarette smoke and sawdust. He wore a dark suit, a belt with a big silver buckle, a white shirt, and a thin tie knotted as if to strangle his skinny neck. The tie was tucked into his shirt, but often it hung down precariously, in imminent danger of getting caught in a table saw or lathe.

Tufts of wild gray hair poked up at odd angles, like unkempt hedges, on either side of his shiny head. His ears were big, his chin small, and his nose a little bulbous. He had a crooked stance, an awkward gait, a childlike giggle, and a harsh smoker's cough. Tinkering intently over some contraption of his own creation, he looked like something out of a cartoon.

The most remarkable thing about him was his eyes: intense, translucent, twinkly, powder blue. They looked different from other people's eyes, as if they saw through things, beyond things.

Frank paced while he spoke, and if he was speaking with you, you didn't know at first whether to walk along with him or wait for him to wander back – which he might, or might not. He fidgeted endlessly, fiddling with small objects he kept in his desk or his pockets: a slide rule, a top, a magnifying glass, a pocket spectroscope. He smoked nonstop, and on more than one occasion set himself on fire by putting out butts in his pockets.

"He had all these unnerving mannerisms," one staff member said, recalling her job interview. "He had this smelly dog in his office. He's conducting the interview and he's walking in and out of the room in the middle of the conversation so that I couldn't even hear half of what he was saying."

He used his whole body to speak, or think – rubbing his forehead with the heel of his hand as if to push something in (or out of) his brain, bobbing his head back and forth like Howdy Doody, giving a quick nod and a shy smile to punctuate a point. When he played the flute, people said you could tell the notes even if you couldn't hear them because of the way his eyebrows rose with the pitch. Sometimes you'd see cigarette smoke wafting through the holes.

Frank laughed a lot, and he hummed while he worked, like a motor running. His intensity could scare people. But to me, he was like Tom Sawyer in a business suit.

In later years, Frank carried a cane, which became an extension of himself – a prop he used for almost everything but walking (sometimes he carried it on the wrong side). He had a bad hip, and was told to have it replaced, but he figured it would outlast him (he was right). One time when his wife Jackie was in the hospital, Frank ran down the corridor after a doctor. A nurse shouted to Frank: "You could run a lot faster without the cane!"

The cane was so much a part of Frank's persona that you could read it like a weathervane to see which way the winds were blowing inside his head. Staffers called this game "caneology." When he was agitated or impatient, he'd tap the cane in a sustained staccato. When he was angry, he'd slam it down. If he was in a good mood, he'd swing it around in circles, not paying much attention to who happened to be inside the perimeter, so he was always hitting and poking people. When you walked behind him, you had to watch out.

Like his tie — which he'd put to use erasing blackboards and securing trailers — the cane was an all-purpose tool. He used it to demonstrate physical principles, such as phases or reaction time or center of gravity. He also used it to stop traffic when he crossed the street against the light. He used it to manage the staff as well. When he got annoyed with one staffer for doing something not to his liking, he rapped her gently on the ankle.

He even used it as a "gun." When a contractor continued to pound away on the building's rotting ceiling even after Frank yelled at him several times to stop (he was afraid a piece of cement might fall on someone's head), Frank aimed his cane at the man and shouted, "Stop or I'll shoot!" The worker was far enough away that the true nature of the object Frank held was ambiguous; he took one look, and stopped.

Sightseeing

"Misbehavior is as important in the study of nature as in people."

I introduced myself, and Frank took me on what amounted to a sightseeing trip around the museum. In fact, a place for sightseeing was exactly what Frank had in mind when he built the Exploratorium. Sightseeing, he liked to say, is the basis for all discovery. Marco Polo and Charles Darwin were both sightseers. "Individual sights combine to form patterns," he said, "which constitute a simple form of understanding." The patterns that Darwin noticed changed the way people see themselves, their origins, their relationship with other living beings.

But the kind of sightseeing available to most people in a classroom or on television is like sightseeing from the window of a train that is unstoppable, irreversible, and dominated by the smells, sounds, and motions of the train rather than the landscape outside. The people and towns along the way never become part of your experience. The train is always rushing toward its next destination.

Real sightseeing requires you to get out of the train, wander at will, get lost, get dirty, linger as long as you like, and try things just for the heck of it. You can't be guided, and you can't have an agenda.

Frank's deep regard for undirected sightseeing grew out of the connection he saw between his experiences exploring mountains and teaching physics, which he wrote about in a piece he called "Everyone Is You and Me":

> In my youth I used to wander in the mountains. I would gain a "feel" of the terrain and gradually build up a reliable intuition of how to get from here to there and back again. Always, on these expeditions, I would discover special places—a tiny area, the only one, where fairy slippers grew; a pool in a rushing stream that was deep enough to swim in. Invariably I would find myself excitedly climbing some promising knoll or uppermost peak. Suddenly a whole new vista would open up, showing a great expanse of prairie, a hidden lake in some inaccessible canyon, an entire new ridge of peaks.
>
> As a result of these solitary expeditions, I would tell friends what I had found and would want to show it all to them. Somehow it was especially important that they see the view, often at a special time of day—perhaps precisely when the sun was setting. But we would start late or they would be unable to walk as fast as I. I would point to the place where I had found the fairy slippers, but we would walk on by. We would reach the top, and the view would be well worth the effort and the hurry. But gradually I began to realize that there was something wrong with these revisiting expeditions. Although the view far outweighed anything along the way in wondrous and memorable ex-

perience, the events along the way had been an integral part of the trip for me, and would also have to be so for the people I wanted to bring pleasure to. If the trip was spoiled through hurry or painful effort, then no one was moved to go searching for views on his own.

When I was teaching physical science to high school students, I felt the same kind of thing happening. The course was certainly an improvement on my mountain expeditions. I cleared the trail, mapped out switchbacks when the grade was too steep, and built bridges or steps when the terrain was impassable. But for those who built the course and came to know it well, it was crucial to reach the panorama of the final chapters, which put together everything that had gone before and opened up grand new vistas . . . But in order to reach the vantage point soon enough, the trip had been spoiled. There was no opportunity to explore unexpected and pleasant nooks along the way.

The Exploratorium was Frank's own antidote to the rushed sort of schooling most people experienced. Over the years, I've tried to take many friends on tours of the Exploratorium – guiding them toward my favorite sights, rushing past the other sights that inevitably draw them. It never works. Willy-nilly, people make their own paths, finding their own nooks and taking time to enjoy whatever form of fairy slippers capture their imaginations.

To Frank, there was no sightseeing destination more exotic and magical than the world of science itself – and he intended to make it accessible to anyone inclined to wander in. "The explorations of science have revealed some of the most amazing sights and undreamed-of novelty," he wrote. "What sights we have uncovered in the field of atomic physics alone! They are even more unbelievable than the bulging lips of the Zulu and are as beautiful and replete with form and symmetry as the Taj Mahal . . . Perhaps you think that I exaggerate . . . but really, some of the things one sees are so beautiful."

As we wandered about, I began to notice more and more decidedly un-museum-like features of the place. For example, people rarely enjoyed the sights in solitude. They seem compelled to share, spontaneously interacting with strangers.

"What are we supposed to do?" I'd hear someone ask to no one in particular. Or: "Do it to me!" Or: "I figured it out!" Or: "Wait. Not yet. Let it build up." Or: "We're running out of sand." Or: "No, you have to sit here; you have to look just right." Or: "Whoa!"

Clots of excited children egged each other on, egged on their parents, egged on the blue-haired ladies and the teenage lovers and janitor who put down his mop to play. "Hey, look at this!" "What about this one?" "No, put it this way." "I can do it." "I know what it is." "Look at that!" Everyone was explaining, showing things off, asking questions – it didn't matter to whom.

There was a huge range of activity. Some people cruised by, giving things a cursory look, running a hand over exhibits as if feeling the fabric on the clothes rack at the store, or strumming their fingers on the bark of a tree.

Others became riveted – crouching, or standing on toes, the better to see or touch or hear; flicking, fingering, reaching, ringing, pumping, rearranging. But mostly just watching, listening, puzzling things out. Who says people have short attention spans? I thought.

It was a wonderful irony that, in the days when the Exploratorium was free, children used to play hooky from school to go there – an educational institution. It was a place where babies rarely cried, where teenagers went to get high.*

I was also immediately struck by the fact that everything seemed so *honest,* nothing hidden up the magician's sleeve, no fancy veneers or gilded edges. A paper handout explained: "We do not want people to leave with the implied feeling: 'Isn't somebody else clever?' Our exhibits are honest and simple so that no one feels they must be on guard against being fooled or misled."

---

* Frank loved telling the story about the teenage hitchhikers he picked up who told him they liked to come to the Exploratorium to get high because it was "a place of such intense perceptual stimulation."

"The Exploratorium is not designed to glorify anything," Frank liked to say. "We have not built exhibits whose primary message is 'Hasn't someone done a great service to the American way of life?'"

In truth, many exhibits looked like train wrecks. If something was built from junk or army surplus, well, then Frank thought it should look like junk or army surplus. He hated fakes and secrets. In the Exploratorium, every sight and sound had to be authentic, every innard exposed. For a long time, Frank wouldn't allow computers in the place at all because the insides of a computer aren't transparent (although the Exploratorium did have the very first "Pong"—the ancestor of all of today's computer games).

There was a reason that the first thing you saw when you entered was the machine shop—a large open area where young men (and women!) in safety goggles cut and drilled and welded and the smell of sawdust and oil from the lathes wafted out onto the exhibit floor. The shop was enclosed by a low fence, so that people would stop by to ask questions or just chat "like neighbors," Frank said. He wanted there to be no "behind the scenes," because what would be the point of having a place dedicated to demystifying how things work if the process of creating the place is not visible?

"This show of reality represents a basic honesty that has a surprisingly important effect on learning," he said. "A reproduction of a painting does not stir the imagination or lie as deep in memory as does the original. Re-created gold jewelry of ancient Egypt or modern versions of a suit of armor do not have nearly the stirring effect on the museum visitor evoked by objects that actually were used in the past. It is the same with scientific phenomena."

The difference was that Frank wanted people to be able to hold the reality in their hands, play with prisms and lenses and pieces of Polaroid, turning them over, this way and that, to see what would happen. People had become information rich but experience poor, with very little access to many aspects of nature or technology. Frank wanted a place where people could get back in touch with what he called the "three-dimensional neurophysiological world." He liked to repeat the old saw that you couldn't learn how to swim without going near the water.

So when people asked Frank why he built the Exploratorium, he often told them: "It's like asking, Why did one build a park? It's because there are no trees around." The exhibits were the trees that you could sit on or climb or rest beside or find shelter underneath; pick at with a stick to root out insects or tear up a leaf or carve your initials in the bark. He called his Exploratorium a "woods of natural phenomena."

Of course, true exploration sometimes winds up in dead ends. Things break, or don't work. In fact, without seeing what made things *not* work, you couldn't find out what elements were essential. So Frank got angry if there was no way to make an exhibit "misbehave." And if one of the exhibits broke while someone was playing with it, well, that was considered the museum's fault.

A Decent Respect for Play

"In our era of incessant and rapid cultural change, legitimizing play in the appropriately structured educational institutions is essential for adults as well as children."

The longer I wandered Frank's woods, the more I realized how apt the metaphor was. For example, there wasn't a guard to be seen anywhere. Frank didn't believe people would destroy things if they weren't bored, weren't talked down to, were trusted and respected. And, as it turned out, there was very little vandalism.

This respect was conveyed in a multitude of large and small ways. Things weren't tied down, so many objects used in the exhibits "floated." Balls from the "Gravity Well" or color filters from Bob Miller's "Aurora" might wind up at the other end of the museum. "We discovered if you tie things down, they get stolen," one staffer explained. "If you don't tie things down, they wander, but they don't get stolen. By having pieces that can float, we're saying to them, in effect, 'We trust you.'"

Not only were there no guards, there were no rules.* Frank thought rules showed a disrespect for people — an unwillingness to give each individual case the merit it deserved. "We resist making rules whose sole purpose is to reduce the amount of work or decision-making required of the staff," he said. Besides, he warned, once a rule is made, people have to live with the often unintended consequences. When a young woman came in to interview for the job of receptionist, Frank's first question was "Can you break rules?"

Frank never bought the popular notion that the general public was irresponsible, uninterested, unreachable. He thought those oft-cited short attention spans were a product of the thin gruel of stuff people were often expected to pay attention to.

The general unruliness of the place was amplified by the presence of a great number of dogs. Dogs slept next to staff members, followed people around the museum, or waited for their walks tied to pieces of furniture.† Staffers were prepared to go to the mat for the dogs because it seemed that dogs were part of the real world. "So if you're trying to rip open the world to peek inside," as one put it, "the kids and dogs and everything are part of that."

The king of the pack, of course, was Orestes — a dog Frank and Jackie got for their "old age" who grew to be one hundred pounds of pure, thick stubbornness. "That dog embodied the barely controlled chaos," a staff member recalled, "but it was an important part of Frank's life to have this huge, ridiculous creature under his roof — a strange balance between propriety and wildness."

The dogs were part of the matter-of-factness of the place. You never felt judged or uncomfortable — that you were being controlled or had to control yourself. Frank didn't think people should have to decide whether they were supposed to learn something or just enjoy themselves. "No one flunks a museum," he often said.

* Actually, there was one: no bicycle riding among the exhibits when it was open to the public.

† When I returned to the Exploratorium for a visit in 2005, I knocked on the door of a trailer only to have it opened by a dog who had been trained to pull on a rope and open the door when someone wanted entry.

People don't list the museums they've attended on job applications. Museums are free of the tensions that "make education unbearable and ineffective in school."

The bare girders and concrete walls and rough floors — though initially serendipitous — made it easier for people to feel there was nothing they couldn't touch, or do. The lack of walls made it easy to wander.

When architects were hired to build a mezzanine, they tried to engineer things so that children wouldn't run inside the museum. Frank didn't like that idea at all. "Why should we stop children from running?" he said. "They hardly ever ran into anybody, but they appeared to be a little bit out of control and it worried the architects to see children behaving naturally. I think it's quite wonderful that we don't mind losing some control."

So the playfulness of the place served multiple purposes. For one thing, it allowed people to feel comfortable in what might seem like forbidding and alien territory (science). Play, by definition, is noncoercive. Visitors could make genuine discoveries — not the kinds of discoveries students are urged to make in the "discovery method" of teaching, where they can discover only what the teacher had in mind. Instead, all kinds of unexpected things were discovered in the exhibits, even by the people who built them.

Frank thought play was serious business. "So much time is spent just playing around with no particular end in mind," he wrote. "One sort of mindlessly observes how something works or doesn't work or what its features are, much as I did when, as a child, I used to go around the house with an empty milk bottle pouring a little bit of every chemical, every drug, every spice into the bottle to see what would happen. Of course, nothing happened. I ended up with a sticky grey-brown mess, which I threw out in disgust. Much research ends up with the same amorphous mess and is or should be thrown out only to then start playing around in some other way. But a research physicist gets paid for this 'waste of time' and so do the people who develop exhibits in the Exploratorium. Occasionally though, something incredibly wonderful happens."

When Frank looked into the subject of play, he came to believe that it was much harder than it seemed, especially for adults. He asked his friend Bob Karplus, a physicist at the University of California at Berkeley, if he thought there was anything a young person must learn before it is too late, and Karplus's answer was "play." Frank found this fascinating: if play was so integral to art and science, and so many people couldn't or didn't do it, surely it was worth finding out why. "I can only conclude that it must require an inordinate amount of self-discipline for adults to remain playful in their work," he said.

For Frank, play was never off limits, so he was usually a hoot to have around. He might invite you to a meal that consisted entirely of experimenting with different ways of eating ears of buttered corn (if you slice each row down the middle with a sharp knife, for example, you can eliminate the crunch).

He might demonstrate, while driving, how you could keep time to radio music with the accelerator and brake to produce "quite a remarkable motion of the car." This demonstration of the phenomenon of resonance might lead him to ask you whether you'd ever tried to get the water in the bathtub sloshing back and forth in a resonant mode so that you could eventually empty the whole tub in a single swoosh – and if not, why not?

During dinner at a restaurant, he'd fill the beer glasses just so, to get a triple harmonic series. "Of course, we had to drink a couple of beers apiece before we could get it exactly right," a staff member recalled. "And he played incredibly high, screeching third harmonics that had the entire restaurant rushing in there to see what was happening."

And when an earthquake struck the Exploratorium, sending everyone else running out the doors of the 1915 Palace of Fine Arts, Frank ran *inside* to see what would happen. He stood there looking up at the swaying ribs of the cavernous structure with a goofy smile on his face, fascinated. Such passion was infectious – and ultimately irresistible.

Naturally enough, people liked to play *with* Frank as well;

Frank's frenetic pacing during conversations, often out of hearing range, tended to drive people nuts. One time, a physicist colleague decided to see what would happen if he stood in the doorway of Frank's tiny office to block his exit, bottling him in. "It was like his kinetic energy increased until he blew past the door," he said. "Popped out of the door like a champagne cork and off he went."

After my day of "sightseeing" with Frank, I went back to my office at the magazine and told the science editor about my miraculous day. Did this Oppenheimer have "twinkly blue eyes"? the science editor asked. In fact, I said, he did. "Then he must be Robert Oppenheimer's younger brother," he said. Since J. Robert was the "father of the atomic bomb," Frank joked that he must be the "uncle" of the atomic bomb.

I began to realize that my "wizard" had depths I'd no idea of. Where had these ideas come from? What set of circumstances had produced such a fascinating mind? Who was this guy Frank Oppenheimer anyway? As I was soon to learn, Frank's Exploratorium embodied virtually every aspect of his former life – from his Manhattan childhood to the Manhattan Project; from his frequently wild adventures in science to his hard Colorado exile and hounding by the FBI; from his experiences teaching high school to the influence of his ethereal older brother; and most of all, his passion for art, music, and mischief. Somehow or other, it was all there.

A well-intentioned colleague once tried to get Frank to face up to the limits posed by the so-called real world. He responded in words that are now almost legendary among those who knew him (and many who did not). "It's *not* the real world," he insisted. "It's a world we made up." We could make it up any way we liked.

The world Frank made up was the Exploratorium, and the Exploratorium, in turn, was Frank's life story writ large.

The point of view in this book is mine alone, of course – much of it based on my own flawed remembrances of times spent working with Frank between 1972 and 1985, either at the Exploratorium, on work of my own, or on a book we were writing together at the time of his death. My relationship with Frank was decidedly erratic. We floated in and out of contact. Some years, we saw or talked to or wrote each other nearly every day. At other times, we went off in different directions, and communication was sporadic.

I decided that to the extent the book was based on my personal experience, it would have to be like John Barth's novel *The Floating Opera*. In that book, Barth imagines a showboat that drifts up and down a river; one play is performed continuously on deck; the audience sits on the banks. "They could catch whatever part of the plot happened to unfold as the boat floated past," Barth's narrator explains, "and then they'd have to wait until the ride ran back again to catch another snatch of it, if they still happened to be sitting there. To fill in the gaps they'd have to use their imaginations, or ask more attentive neighbors, or hear the word passed along from upriver or downriver . . . I needn't explain that's how much of life works."

Luckily, I had lots of "attentive neighbors." I was able to interview more than a hundred people who knew Frank, both intimately and in passing. I had dozens of hours of audiotapes that Frank and I made together while discussing "our" book – during those years I was sitting on the banks. I had access to interviews with Frank by a half-dozen scholars and journalists. Most important, I had his papers, talks, personal letters.

Like all such larger-than-life figures, Frank was a man of extremes and contradictions. In a fit of rage, he could pick up a dog by the tail and throw it against a barn door. Then he could somehow stir up the sentiment to feel sorry for a noxious pot of boiling tar left on the street by roofers. "Oh, that poor thing," he once said out of the blue. "It looks so lonesome there sitting all by itself."

One morning, he came into the Exploratorium complaining that there was garbage in front of the museum; he'd been out there picking it up and was annoyed that no one else had thought to do so. Throughout this harangue, he'd been peeling an orange on the counter in the reception area. When he was done, he left the peels there for someone else to deal with.

Most people remember Frank as courtly, gentle, "profoundly humanistic"; a kind and patient man who never raised his voice; unfailingly understanding and perceptive. Others wanted to knock his block off, or avoided him because too many people who went into Frank's office came out crying. He had a bad temper, and could be cruel; he cheated on his wife. As one colleague put it: "First I met Frank the Physicist, then I met Frank the Humanitarian. And then later I met Frank the Flawed."

On the one hand, Frank was a jolly old soul with a gift for serious play. He found pleasure in everything from the smell of dog shit to the patterns waves make on a pond, and especially in art, music, and all human relationships. He was an optimist who believed that people could change the world — as he did. One of his favorite expressions was "Whoopee!" At the same time, he carried around a heavy weight. He never got over "all those flattened people" in Hiroshima and Nagasaki. He was acutely aware of how profoundly the world had changed because of the atom bomb he'd helped to build — and also how little progress humankind had made toward peace in spite of (or because of) it.

Frank was my friend and mentor, and mostly I saw the good side (though there were times I couldn't wait to get away from him). Still, I think it's fair to say that while the unpleasant side of Frank's nature is a part of his legacy — mostly affecting his family — even those who saw his dark side tend to remember him with admiration and affection. What's more, his negative qualities emerged only rarely in the course of hundreds of interviews I conducted.

So to write this book, I decided to take a cue from Frank himself. After the death of his dear friend the poet Muriel Rukeyser, a biographer sent Frank a long questionnaire about her life, with spaces to check off those personality traits that applied to her. Was she kind? mean? cheerful? gruff? brilliant? slow? stingy? generous? attentive? distracted? And so on, for many pages. Frank checked "yes" to everything.

Making a Difference

The last time I saw Frank was a few months before he died of cancer in 1985. He was weak and dizzy but still stayed up to see a concert after I had gone to bed with the flu. He went to the Exploratorium every day until his final week, and he never stopped playing. When he had to have a bone scan, which involved getting injected with a radioactive tracer, he persuaded some staffers to carry his wheelchair to an exhibit containing a Geiger counter — he wanted to see just how radioactive he was. (The answer was *very.*) He rolled himself back and forth, back and forth, enjoying the whole thing immensely.

The legacy Frank left to me is both enormous and personal. When I started writing about science for the *New York Times, Discover,* and the *Washington Post,* he was "my friend the physicist" who appeared in my essays, which were often informed by his ideas.

My friends and acquaintances still benefit (or suffer) from the curious addictions I picked up from Frank. At a friend's birthday dinner in a restaurant, I couldn't help showing Frank's method for finding the center of gravity of a serving spoon. I drop dinner rolls and paper napkins at family dinners to show the equivalence of gravity and inertia. At a New Year's Eve party, I demonstrated how we could use the blind spots in our eyes to decapitate people. (Off with their heads!)

Last fall, I found myself sneaking out of a journalism meeting, hiking past the No Trespassing signs to perch on a rock and watch the river. I loved noticing the interference patterns created as currents collide, knowing why the foam was white, understanding the mechanism that turned the leaves to brilliant reds and yellows.

Later, during an agonized conversation about how people could still *believe* that George W. Bush had made us safer despite everything they *knew* to be true, I demonstrated how it is quite easy to believe contradictory, even impossible, things — which I proved by using a penknife to cut a hotel map into a "corner" illusion that I held together (absent glue) with a bit of mashed banana.*

And I'm not even embarrassed to talk about it.

Frank's legacy to the world at large is both expansive and deep. Versions of the Exploratorium can be found in hundreds of countries; hundreds of thousands of teachers have been touched by his vision; artists influenced by Frank and his ideas have won numerous MacArthur Foundation "genius" grants. Most of all, people who knew Frank — even those who met him for the briefest encounters, or merely encountered his writings and creations (including human creations) — feel profoundly changed. They find themselves infected with a passion for sightseeing, serious play, and breaking rules; they display a sometimes obnoxious insistence on transparency, respect for ordinary people, and tolerance for chaos; they have a taste for big questions and believe in the power of individuals to effect change.

At Frank's memorial service, the most moving tribute (and certainly the most tearful) was delivered by his best friend and soul mate, Robert Wilson, creator of the world's most powerful particle accelerator and, not incidentally, a work of art (Wilson was a sculptor as well as a physicist). The two young men — both committed pacifists — bonded when they found themselves building an atomic bomb, the first and only true weapon of mass destruction. Wilson understood Frank better than anyone, and he left us with an image of Frank that captures his essence almost perfectly:

> . . . most of all we are talking, talking and walking, talking intensely — talking about a world of reason, a world without the follies of bigotry, without the tyranny of war. Talking about painting and sculpture deep into the night. Talking about philosophy. Frank is always pacing back and forth — that typical posture — walking away with his hand on his hip, looking back to say just the quintessential thing of importance about whatever problem he was gnawing away

* This illusion was created by Bob Miller, who turned it into an Exploratorium piece called "Far Out Corners."

deep into the night trying to get somehow to the truth. He used to wear me out. He somehow felt that if we just hung on to the next morning we would understand it. But we seldom did. Frank was a romantic. But he wasn't Don Quixote. When Frank rode forth, it was not to charge windmills, but to wrestle with real, hard, obdurate problems. Sometimes he lost. Sometimes, as we know, he won. In his romantic way, Frank believed that one man could make a difference.

Well, my friend Frank Oppenheimer did make a difference. He Did, He Did.

2

A LITTLE ROYAL FAMILY

The most beautiful youngster he ever saw, "out of an Italian painting."

The world Frank made up had deep roots, some sunk well before he was born. His family, his schooling, his profession, his dip into the political waters that would eventually (if temporarily) swamp him — all shaped his inner world and eventually the concrete outer world he created.

He was, in other words, a man with a past. On August 6 of each year, he would sink into a profound funk, drinking whiskey from the bottle he kept in a drawer of the desk in his smoke-filled office. He rubbed his forehead, hard, as if he were trying to erase something deep in his mind. He was grim, tense, quiet; he looked as though he'd been up all night. August 6 was the day the first atomic bomb was dropped, on the Japanese city of Hiroshima.

The more I found out about Frank's former life, the more clear it became that he wore his history on his sleeve not just in this, but in almost every respect: his kindness, his insatiable curiosity, his love of junk and art, his passionate distaste for dishonesty, his bottomless appetite for everything the human and physical world had to offer — all were transparently tied to his previous experience. And the more I learned, the more it seemed as if everything he'd done had led inexorably (if somewhat erratically and eccentrically) to the Exploratorium.

It was no surprise to learn that Frank grew up in a home full of art, books, conversation, music, and commitment to moral causes. Frank's father, Julius, had emigrated from Germany to the United States in 1888, at the age of seventeen, with little money and no command of English, but by the time Frank was born, in 1912, he had become wealthy as an importer of fabric for men's suit linings (notably linings for army jackets during World War I). Julius, Frank recalled, was very proud of his ability to judge and match colors and his eye for fine fabrics.

Julius was lively, sociable, kind, well liked, self-taught, and well versed in history, art, and politics. He had a hand in setting up New York's Museum of Modern Art and was an early member of the Society for Ethical Culture, which was dedicated to "love of the right." Founded by Felix Adler in the late 1800s as a kind of secular Judaism devoted to good works ("deed, not creed"), it was committed to arousing "the conscience of the wealthy, the advantaged, the educated classes" – moral teachings that deeply influenced Frank and Robert alike.

Frank's mother, Ella Friedman, was born in Baltimore to a Bavarian family that immigrated to the United States in the mid-1880s. A serious painter, she studied for a year in Paris, taught art at Columbia and Barnard in New York, and had her own studio and students until she was married. After that, she returned to her painting only sporadically. "Every once in a while she would set up an easel and do a still life or a landscape," Frank said.

An "exquisitely beautiful" woman with "expressive gray-blue eyes and long black lashes," Ella had a congenitally malformed right hand which was always hidden in a prosthetic glove – and apparently never mentioned. She was quiet, formal, sensitive, considerate. Just a year after her marriage to Julius, in 1903, she gave birth to Robert, and four years later to another boy, who died in infancy.

Frank Friedman Oppenheimer was born four years after that –

a frequently smiling child with very blue eyes and very black hair. The physicist I. I. Rabi, a friend of Robert's and later a Nobel laureate, described Frank as the most beautiful youngster he ever saw, "out of an Italian painting." Perhaps his future mistresses saw some of that beauty in Frank, though for me, his physical presence was closer to how I imagine Merlin.

From birth until college, Frank's home was an expansive eleventh-floor apartment on Eighty-eighth Street and Riverside Drive, with great views of Manhattan and the Hudson River. Books were stacked everywhere: in the halls as well as in the library, spilling even into the bathrooms.

The Oppenheimers' art collection grew along with Frank; eventually it included works by Rembrandt, van Gogh, Vuillard, Derain, Renoir, Picasso, which hung on the walls alongside Ella's own art. He got to know painters, including Max Weber, and art dealers. He considered the family paintings "old friends."

Frank never learned to draw or paint very well, but his experiences trying stayed with him. He was especially inspired by a second-grade art teacher who pointed out that a tree he'd drawn had a cavity where the branches parted. A real tree constructed like that would collect water, rot, and die, the teacher told him. After that, Frank began to look at trees much more carefully.

When Frank was about twelve, he started going to Greenwich Village once a week to learn how to do charcoal drawings – most of them of fire escapes and the "toits de New York." "To this day, I remain highly sensitive to and moved by the views from the back windows of city houses," he recalled. When he was fifteen, the family vacationed on Nantucket, and Frank and his brother spent several weeks learning how to reproduce the colors of the island with oils on canvas. The marvelous colors and textures of the island "sank deeply into us," Frank later said.

Ella and Julius took Frank to Zurich to see an extensive collection of van Gogh's paintings; they took him to Philadelphia to see the Barnes collection by special arrangement. On Saturday afternoons, Frank would "do" the galleries on Fifty-seventh Street with his dates.

Hours spent exploring New York's museums were among Frank's fondest memories. A school trip to the Metropolitan Museum of Art while he was studying the knights of yore, for example, impressed him with the importance of having "real" objects on display, not simulations. "The opportunity to see those beautiful and ornamented steel vestments enabled me to realize their scale, to appreciate their articulation, and to imagine how strange it must have felt to be inside one of them," he wrote. He was so inspired that he made a coat of mail using the key-ring scouring pads that were popular at the time.

Such experiences led him to believe that a cultivated aesthetic sense was as central to decent behavior as it was to art; he believed that art influenced his emotional makeup, the way he felt about things, the kinds of decisions he made about his own life and also about the needs of society.

In addition to the immediate family, the Oppenheimer household included Frank and Robert's nurses, several live-in maids, a chauffeur, and Frank's grandmother. His parents hosted frequent formal dinner parties where the guests discussed ideas, current events, "political things." Like Frank, Julius loved to argue. "The family had a feeling of friendliness and warmth and gentleness, and quite a lot of conversation," one guest remembered. They also liked to have fun. On the occasion of Julius's fiftieth birthday, in 1920, eight-year-old Frank dressed up as Einstein, in a dark suit and mustache, and recited a "pompous polysyllabic congratulatory speech" written for him by Robert.

At the same time, the household must have felt intensely claustrophobic. Ella was terrified of germs, and the boys grew up largely confined within familial walls. Tutors came to them, as did the barber; Frank even had his tonsils out in his bedroom.

So it must have been a great relief when the family closed up the New York apartment in the summers and traveled with their chauffeur and possessions to their waterfront estate in Bay Shore, Long Island. Here, the forty-four-foot yacht, the *Lorelei,* and her captain awaited. For Frank's birthday in August (his mother's birthday was also in August), the family cut flowers from the gardens, so that when they went downstairs for breakfast, the chairs and tables would be strewn with petals. Friends stayed for extended periods. One guest who visited when Frank was a young teenager remembered "high-spirited goings-on all the time."

Ella and Julius were indulgent parents who seemed to cater to their sons' every enthusiasm. When Frank got interested in Chaucer in high school, they gave him a 1721 edition of Chaucer's works. When he fell in love with the flute at age fourteen, they brought him a wooden flageolet. He learned to play it, and wanted to move on, so his parents bought him a well-made flute and found one of the greatest flutists in the country, George Barere, to teach him. Frank invited Barere to come to the apartment and play for his parents' anniversary, but the musician's agent wouldn't let him, so Barere instead gave a lesson that was, in essence, a private concert. "And my parents sat there very formally," Frank recalled, "and he said he felt he was playing for a little royal family."

Later, Frank joined the New York Flute Club. During college at Johns Hopkins, he took lessons at the Peabody Conservatory of Music and joined the Baltimore Bach Club. For a time, he even played for money as a street musician at tube stops in London.

For all his parents' indulgences, their relationship with Frank seems more protective than particularly close. He never spoke about them much, and when he did, it was about the trips the family took together rather than about his mother and father as living, breathing people.

Two European trips in particular were to have a considerable impact on Frank and Robert both. The first, in the summer of 1921, included a visit to Frank's grandfather in Hanau, Germany. Frank was only nine years old, but even so, what he saw greatly upset him. At the time, France occupied the Rhineland. "An army of occupation is such a horrible thing that even to a nine-year-old it was very clear the way the soldiers treated the Germans at that time," he said. The French soldiers pushed the Germans around, acting rudely, behavior Frank attributed to the fact that "they were an army that had been shoved into another country." He concluded that the Germans' thirst for power and their own ability to be nasty came from the nasty treatment they'd received from the French occupying army.

During this trip, the brothers spent many days in the mountains gathering minerals for Robert's collection. Robert got dysentery, becoming so sick that he had to be carried on a stretcher onto the ship home, and had to postpone going off to Harvard for a full year. However, his illness turned out to be an important turning point for both boys, because as part of Robert's convalescence, he was sent with his English teacher to the mountains of New Mexico and Colorado—places that would soon become the brothers' second (and frequently first) home.

The family took another trip to Europe in 1924, and Frank remembered hiking, first by himself and later with Robert, all through the Swiss Alps. He loved to tell the story about how he'd mistaken the little basins he'd found along the path for urinals. Only later did Robert explain that they were receptacles for holy water.

As for Frank's formal education, it was nearly as magical and sheltered as his home life. Like Robert before him, Frank attended the Ethical Culture School on Sixty-third Street and Central Park West. Originally set up as the Workingman's School, offering free education to children of the working poor, it began admitting paying students in 1890. The curriculum was focused on moral education, service to community, and justice, with the avowed aim of creating responsible, caring citizens.

In elementary school, Frank remembered only a general feeling of not being very good at anything: arithmetic, writing, spelling. Starting at around age eight, he got a tutor, a premed student named Micky, who became a beloved friend. When Frank was in the sixth or seventh grade, Micky died of blood poisoning, a personal tragedy that Frank, decades later, still spoke of often. Micky's family donated money to the school to set up a science library, and Frank picked out the books and decorated the small space with paisley rugs and "funny stirrups and brass things brought from home." Frank continued to give gifts to the library until at least 1982.

When Frank was in seventh grade, the Ethical Culture School began plans for expansion, building a new upper school—called the Fieldston School—in the Riverdale section of the Bronx. The new school would have a science laboratory, and Frank's remarkable science teacher Augustus Klock (who also taught Robert) let the students help design it. Klock put Frank in charge of the electrical system in the lab. The young teenager spent much of the year planning the most efficient way to arrange the wiring—even making paper replicas of the switchboard.

When Frank entered high school, Klock asked him if he'd tutor the younger students, so Frank formed a science club and taught kids in the lower grades on a regular basis. In 1956, just as Frank was starting to teach high school in Colorado, Klock wrote him a letter remembering that experience. "You are a natural born teacher," Klock said. "The job you did as leader of that club was so outstanding that I have referred to it scores of times during the last 25 years."

All the Junk He Gave Me

"To try to be happy is to try to build a machine with no other specifications than it shall run noiselessly."

Robert was away at college or graduate school for most of Frank's childhood, but his presence in many ways remained on Riverside Drive. Frank grew up around Robert's prized mineral collection—a collection of such high quality (and on which Robert became such an expert) that he was asked to give a talk at the New York Mineralogical Club when he was only twelve. Pieces of the collection were scattered everywhere in the Oppenheimer home, and Frank was entranced by their beauty.

Once Frank decided to surprise his older brother by cleaning all his minerals. He cleaned one so well it dissolved.

Robert would send Frank books on physics and chemistry, and

also a variety of scientific instruments, which Frank later referred to as "all the junk he gave me." There was a sextant, a barometer that also measured altitude, compasses, a metronome. Frank drove his teacher crazy by bringing new gadgets to school every day and distracting the class, so the teacher suggested that he bring them in all at once. The next day, the family chauffeur helped Frank pack all the junk into the car, and they hauled it off to school.

The microscope in particular had a deep impact on Frank. He told the story again and again of how he'd discovered his own sperm. "It was the most wonderful scientific discovery that I made," he said, "so full of emotion."

Robert was, by his own admission, "an unctuous, repulsively good little boy." Frank was anything but. By four or five, he was already breaking out of the family cocoon, taking apart coaster brakes on bikes and putting them back together. He learned to ride a two-wheeler when he was still too small to reach the pedals except at the top of the stroke, and he soon became the bicycle instructor for his playmates.

At age eight or nine, he started climbing to the tops of trees to watch lightning. "We had trees for every age around our place," he remembered. "Little crab apple trees for six-year-olds, spruce trees for ten-year-olds, and then maple trees for the fifteen-year-olds." All the kids climbed, but no one else was so foolhardy as to do it during an electrical storm.

Frank's longtime friend Philip Morrison of MIT recalled how in his day kids would put tinfoil across the tracks of their electric train sets to watch it spark. Frank's version of this experiment was to take a bicycle and throw it on *real* train tracks.

By high school, Frank was using the electric current in the family's apartment to melt copper and weld things. His eyes itched and watered all the time from the exposure to ultraviolet light; sometimes he couldn't see clearly for days. One time a maid got her hand burned during these experiments. "[My parents] gave me terrible hell for all these things I was doing," he said.

Their scolding had little effect. When he ran his fingers over a circular meat knife in a butcher shop and cut himself badly, his mother said, "That will teach you not to touch everything." Of course, it didn't. "In fact," Frank later noted, "I am so addicted to touching that I started a science museum built around the permission to do so."

He took apart bicycles and alarm clocks. He took apart his father's player piano, and then had to stay up all night putting it back together before his parents came home. He spent a lot of time wandering around New York City. On Centre Street, he found a place that sold secondhand electric motors, and became a regular customer.

Certainly one of his most unusual activities was sneaking into New York's rooftop water tanks. "The call of the water tank would come to me on winter evenings when the moon was bright and the wind strong off the North River," he wrote. "It would lure me away from my homework and into the night." First he had to get past the doormen ("a suspicious lot"), then take the back elevator to the roof. Once there, "the world would be mine alone," he wrote. He'd climb up the ice-covered rungs of the ladder that rose up the side of the tank, feeling as if he were climbing aloft on a storm-tossed ship, struggling to hang on while the wind from the Hudson River howled, "trying to tear me off the ladder."

At the conical top was a little hatchway. He'd brush off the snow and unlatch the rusty hook at the bottom, then crawl inside onto a partial platform about six feet above the water. There he spent many hours, warm and content, listening to the creak of the valves, the gurgle of the water, the tugs and barges on the river with their "myriads of sparkles." "Only after I had returned home," he wrote, "would I realize what a sooty and grimy a place had enchanted me."

Despite the fact that Robert was a rather remote presence during Frank's early years, largely due to the age difference between them, they developed a "very special" relationship that started when Frank was about nine. "I don't think it ever stopped," Frank later said.

They climbed cliffs together to chop garnets out of granite for Robert's mineral collection. Once they climbed up a fire tower on Long Island, broke the lock with a rock, and had a fine time looking through the telescopes. When they were ready to go, Robert left behind fifty cents and a note: "This is to get a new lock."

They enjoyed endless sailing adventures aboard the twenty-seven-foot sloop *Trimethy* (short for trimethylamine), which Julius had purchased for Robert. On one occasion, they approached a pier with too much headway and rammed it so hard that a little girl standing on the other side fell into the water (it was shallow enough that she could stand). Most often, they'd sail to what is now Fire Island, and walk around naked all day, getting sunburned, seeing no one. Sometimes they'd run aground and have to stay out all night. Naturally, their parents worried, and Julius would ask one of the "revenue cutters" that chased bootleggers during Prohibition to go look for them. Frank remembered "weeping bitter tears" the time they took the *Trimethy* out in a violent storm and it got so beat up that it had to be towed back.

Robert was a father figure to Frank, showering him with advice and encouragement. When Frank was fourteen, Robert wrote (referring to his childhood Einstein imitation): "I don't think you will enjoy reading about relativity very much until you have studied a little geometry, a little mechanics, a little electrodynamics. But if you want to try, Eddington's book is the best to start on. I remember that five years ago you were dressed up to act like Albert Einstein; in a few years, it seems, they won't need to disguise you. And you'll be able to write your own speech."

Presaging Frank's later approach to teaching, Robert urged his younger brother to always try to understand, "thoroughly and honestly, the few things in which you are most interested; because it is only when you have learnt to do that, when you realize how hard and how very satisfying it is, that you will appreciate fully the more spectacular things like relativity."

Many of their letters discuss ethical questions; others went deeply into art, music, philosophy. Robert gave Frank guidance on everything from schoolwork to girls, and some of it was preachy or just plain strange. For example, Robert told Frank not to worry about girls, "and don't make love to girls, unless you have to: DON'T DO IT AS A DUTY." Somewhat less puzzling, and far more useful, Robert wrote Frank a year later: "one cannot aim to be pleasing to women any more than one can aim to have taste, or beauty of expression, or happiness; for these things are not specific aims which one may learn to attain; they are descriptions of the adequacy of one's living. To try to be happy is to try to build a machine with no other specifications than it shall run noiselessly."

Frank must have gotten annoyed at Robert's superior air on more than one occasion. During the family's 1924 European trip, Frank locked Robert in their hotel bathroom as the rest of the family was going off to visit a museum. Frank then told his parents that Robert had decided to stay behind. "I guess I must have been kind of jealous or deep-down angry," Frank said.

Still, Robert was proud of his younger brother's ability to grasp and simplify difficult concepts, getting to the core of a question — an ability we all benefited from in later years. "What pleased me particularly in the letters was your surprising gift for digging up cardinal and leading questions," Robert wrote in 1930, "for reducing a specific and rather complex situation to its central irreducible 'Fragestellung.'"

For the most part, their letters overflow with mutual admiration and affection. When Robert went to Europe for the first time as a young man on his own, the captain of the *Lorelei* took Frank out into the harbor to meet Robert's ship so that he could wave goodbye. "It was a very, very tearful time for me," Frank said. On Frank's twenty-second birthday, Robert wished him "peace and beauty and a good life, with the world pure in your eyes and unaltered."

The brothers missed each other when they weren't together, and said so without hesitation or embarrassment. When Frank was working in Italy as a young man, Robert wrote him: "Like you I feel an overwhelming frustration at the distance, such a desire to see you and talk to you as makes me play with the notion of going to Europe instead of the mountains for the five weeks I have . . . I would give my vacation . . . for one evening there with you."

Frank's last memory of Robert is poignantly familial, domestic, ordinary. Robert was lying in bed, in great pain from throat cancer, in many ways a ruined man. Frank lay down next to him, and together they watched a *Perry Mason* show on TV.

Hot Dog!

"My brother and I got to know each other and were always very sad when we departed."

The site of the brothers' most intimate outings—and also the international center of theoretical physics for many a summer to come—was a cabin in the Pecos Valley, New Mexico, known as Perro Caliente—hot dog in Spanish. It entered the family history in 1928, when Frank was sixteen, accompanying Robert on a trip to the Southwest. While visiting friends Robert had met on his previous excursions, the two happened upon a two-story hewn log cabin sitting at 9,000 feet, surrounded by 150 lush acres, close to the Sangre de Cristo Mountains. Upon seeing the place and learning that it was for rent, Robert said, "Hot Dog!" The Oppenheimers leased the property from 1929 to 1947, when Robert finally bought it.

The beautifully made cabin had a porch all around, where they generally slept. They had no electricity and brought water from a nearby spring, but seemed to have everything else they needed. Robert became known for his crêpes suzette, Frank for his cakes. Physics was also on the menu: frequent guests included a veritable Who's Who of physics notables, including Hans Bethe, George Gamow, Victor Weisskopf, Ernest Lawrence, Charles Lauristen, and Robert Serber. The main activity was long horseback rides in the mountains, fueled mainly by peanut butter and Vienna sausages and whiskey. There was always a gallon of corn liquor for drinks before dinner and a nightcap in the evening. Frank would play the flute in front of a roaring fire. "It was a wonderful time for all of us," Frank said. "We'd get sort of drunk when we were high up and we'd all act kind of silly, I guess."

But primarily, Perro Caliente was a retreat from the world, the place where Frank and Robert forged their friendship. They rode on horseback to Santa Fe and Taos, over 13,000-foot peaks and around the back side of the Jemez Mountains; they rode into Colorado; they once rode 60 miles in one day; Frank guessed they rode 1,000 miles a summer. They rode at night as well as during the day. Because Robert had poor night vision, it would be Frank's job to go on ahead and yell warnings of low-hanging branches. Once, he didn't yell in time, and Robert got knocked off his horse. "He was very thin anyway," Frank said. "And I came up behind and saw him and here was this guy, just a little bit of protoplasm on the ground, not moving. It was pretty scary, but he was all right."

During one particularly long and arduous ride, Frank wanted to veer off to see a wild canyon that looked as if it had never been explored. It turned out to be an old gold mine, abandoned by a Frenchman who was said to have killed his own brother, and a lot of people had been searching for it for a long time. Frank loved the feeling of finding something that everyone had been looking for, sliding down into the mine amid a great ledge of fallen timbers.

That first summer, the brothers also visited their parents, who were staying in Colorado Springs. Julius bought them a Chrysler Roadster to take to Pasadena, where Robert was by then teaching at Caltech, and the brothers spent a week learning to drive. Robert decided they should go through the Four Corners, down to the Colorado River, but the roads were rutted and altogether in bad shape. Because Frank was considered the more adept of the two at mechanical things, he took the wheel as they drove over an 11,000-foot pass.

No one who ever survived a drive with Frank would be surprised to hear what happened: he skidded on gravel, causing the car to go up on a bank and roll over into a ditch, wrecking the windshield and smashing the top. Also broken was Robert's arm. When Robert went into a store to get a sling, he bought a bright red one to cheer Frank up — because Frank was feeling so bad about having broken his brother's arm. The next night, they got stuck on a boulder. Frank had to jack up the car, and Robert suffered great pain, helped only by spirits of ammonia. Still they soldiered on. "All the rest of the way to California my brother was terribly impatient with me because I was being somewhat timid because of the accident, drove too slowly. He was in constant pain and it seemed to him that we never got to Pasadena."

Julius and Ella also went to Perro Caliente in those first years, but then Ella was miserable most of the time with leukemia; she would die in 1931, when Frank was nineteen. A strong sense of her — and also of her sons' feelings around this time — emerges from letters Robert wrote to Frank preceding what must have been a final visit: "I am afraid that you will find mother pretty weak and miserable," Robert warned. "The reports have not been very encouraging . . . Surely when you see mother and father so forlorn you will feel a little guilty to have had so marvelous a summer. But they will be happy in your happiness, in your strength and huskiness; not for the world would they have wanted anything else. I can trust you to do everything you can to cheer and comfort them."

In a letter to the physicist Ernest Lawrence, Robert describes their mother around this time as "tired and sad, but without desperation; she is unbelievably sweet." He describes their father as "brave and strong and gentle beyond all telling." Julius died in 1937, when Frank was twenty-four.

Eavesdropping on these letters and in general getting to know Frank's family, albeit vicariously, I can't help but consider it something of a blessing that Ella and Julius never lived to see the cruel hand history was later to deal their sons.

A Very Exciting Place to Be

"In Rutherford's lab, everybody met for tea, every day . . . You saw all these people you'd heard of, and listened to them . . . and really felt part of it."

For college, Frank chose Johns Hopkins, largely because his best friend, Roger Lewis, was going there, loved the school, and knew a lot of the faculty. The choice may also have been a way of separating himself a bit from Robert, who had gone to Harvard. Frank was at the top of his class his first year, and graduated Phi Beta Kappa three years later, in 1933.

Between 1930 and 1933, when Frank was getting his undergraduate education, Europe was the seat of a revolution that was turning physics inside out, and Robert was both a participant and a bearer of news. When Robert came to visit him at college, Frank would hear about all the new developments. It was an exciting time when all manner of old paradigms were crumbling, replaced by mysteries of the most profound and delicious sort. Suddenly there were more confusing things, Frank said, "than you can possibly imagine."

Until the late nineteenth century, atoms had been thought to be the fundamental, indivisible ingredients of matter — unbreakable, impenetrable, and perfectly homogeneous, rather like the smooth billiard-ball atoms that persist in the popular imagination. Then everything changed at once. The atom, in effect, unraveled. It had internal working parts; those parts had parts; and the newly discovered subatomic particles behaved in ways that were entirely alien not only to science but also to common sense. What had seemed sedately certain was now a churning, chaotic mix of mysteries, yielding, if at all, only to wildly imaginative (some would say crazy) solutions.

Robert Oppenheimer was one of the young physicists who was trying to make sense of this mess of mysteries — among them the fact that light appeared to be both a wave and a particle, and that atoms emitted and absorbed light in only specific wavelengths (which showed up as sharp spectral lines) rather than a continuous spectrum like a rainbow. Eventually, the emerging theory of quantum mechanics provided answers (although hardly final ones) but at the same time introduced a thicket of thorny complications, such as the innate uncertainty that seemed to pervade everything at the subatomic level, allowing particles to be, among other things, in two places at once.

It was during these same turbulent decades that Einstein published his Special Theory of Relativity, which revealed that space and time were elastic and that energy was equivalent to matter ($E=mc^2$), and then his General Theory of Relativity, in 1915, which revealed that gravity is the warping of four-dimensional space-time.

What little research Frank did during his undergraduate years was, by his own account, spectacularly unsuccessful. In his third year, he told one of his professors he'd like to try his hand at an experiment, so he was given a task that involved looking through a long black tube at the spectral lines of light emitted by an element and photographing what he saw. To test his technique, he started with mercury, because the spectrum was simple—just a yellow and a green and a blue-green line. But when he looked into the tube, he saw too many lines—an extra green line and also an extra blue-green one. At first he thought he might have discovered something significant. But when he photographed the spectrum, he got the right number of lines. He took more photographs, with longer exposures, but still the camera couldn't detect the second lines his eyes so clearly saw.

He did this for weeks before he realized that he was looking into that black tube with both eyes—and since his brain couldn't fuse the two images, he saw double lines where there was, in fact, only one. "I felt so dumb," he said. "I was ashamed of myself for being so dumb." Still, he found the process interesting because it was his introduction "to this terrible frustration of doing something, and you can't figure it out, and you try this, and you try that, and you just get gloomier and gloomier."

Before going on to graduate work, Frank spent from 1933 to

1935 working in European laboratories — initially in the famed Cavendish Laboratory at Cambridge, where the electron had been discovered. It was a particularly happy interlude. All the "big names" were there (Rutherford, Cockcroft, Walton, Gamow, and occasionally Bohr and Lawrence), everyone met for tea every day, and Frank loved being part of it all. Rutherford himself would stop by Frank's tiny lab to talk about his research.

In many ways, the atmosphere at Cavendish seemed much the same stew of freedom and individual attention that Frank later cooked up in his Exploratorium. The senior physicists turned him loose, and he wandered around and saw what was happening and picked something that looked as if it might be interesting to work on. All and all, Frank said, "it was a very exciting place to be."

Frank's research focused on nuclear physics — which is to say the study of the glob of protons and neutrons in the atom's core and how it disintegrates into smaller particles (the process called radioactivity), often in a series of complicated steps. The bits that come out — electrons or pieces of the parent atom or gamma rays (a highly energetic form of light) — can be measured and used as clues to determine what happened during the process of decay. In the early 1930s, physicists were also learning how to use the particles that flew out of these disintegrations as probes, although much had yet to be discovered. Frank worked for eighteen months on developing techniques for making accurate measurements of the results of these experiments.

The only downside for Frank was that Rutherford would shut the laboratory at night; the great physicist's feeling was that if someone hadn't done enough during the day to think about at night, then there was no reason to come back.

During his European years, Frank remained a bold adventurer. He got a pilot's license, flying a Gipsy Moth biplane, and went on excursions all over the continent. Once he decided on the spur of the moment to walk from France into Spain over the Pic du Midi, a forty-mile hike in the snow. He broke a ski in the Pyrenees and wound up having lunch with the family of the owner of a tiny local ski factory. He explored the mountains of Spanish Morocco, and later wrote of scrambling up steep cliffs and hopping from slippery rock to slippery rock over water. (Uncharacteristically, he declined a swim in an inviting channel that ran into a cave because he was wary, having seen some sharks.)

Following his work at Cavendish, Frank spent eight months at the Osservatorio di Arcetri in Florence, where he found the atmosphere even more exciting and intense than he had at Rutherford's lab — the scientists in Florence worked late into the night. Frank spent whole days at the Uffizi Gallery, memorizing the paintings, getting to know them "personally," and speculating about their meaning. "I tried to understand the changes that were taking place," he recalled. "I realized from observing the progressions in Italian art the kind of things that happen in an art or a craft as people use their increased skill to elaborate and embellish. Years later, I realized that a similar entrapment by technique and technology happens to physicists (as well as to three-star generals)."

For all his happy times, he couldn't escape the dark forces threatening Europe. Outside his lab in Florence, a brigade of soldiers in black uniforms sang and cheered. "No one thought it was that serious then; not like Germany." He'd been to Germany the year before, in 1934, and "felt the menace." Crowds of people were marching down the streets; his relatives told him of "terrible things"; they had to wear armbands and couldn't get work. What made it more horrible was that things looked so normal on the surface, Frank told me, while underneath, "the whole society seemed corrupt."

Alias Frank Folsom

"On all these fronts there was the sense that I could make a difference and my friends could make a difference."

In 1935, Frank followed his brother to Caltech to get his doctoral degree, and the four years he spent there were ones of great personal and political as well as scientific turmoil.

Most important on the scientific front, in 1938 Otto Hahn and Fritz Strassman had bombarded uranium atoms with neutrons and found lighter elements in the debris. In 1939 Otto Frisch and Lise Meitner realized what this meant: the atom had been split; they had discovered nuclear fission — the force behind the atom bomb.

Robert at the time had joint appointments at Caltech and Berkeley. He moved to Pasadena each spring, followed by carloads of devoted students, and became "a sort of patron saint" of both institutions. A student who was there then described Robert as "always at the center of any group — smooth, articulate, captivating." Frank, in contrast, "stood at the fringe, shoulders hunched over, clothes mussed and frayed, fingers still dirty from the laboratory."

Frank went to work with physicists Charles (Charlie) Lauritsen and William (Willie) Fowler; he already knew Lauritsen, who'd been a guest at Perro Caliente. Lauritsen had pioneered the use of x-rays in cancer treatment, and when Frank first walked into his lab, Lauristen was taking an x-ray tube apart. "Frank," he said, "do you want to smell a vacuum?" In those days, shellac was used to seal vacuum tubes, but the x-rays and electrons knocked apart the organic molecules in the shellac. "So Charlie had just opened his tube," Frank said. "And . . . there was the most foul smell from all these disintegration products of shellac."

Frank already knew what he wanted to do at Caltech: build a beta-ray* spectrograph, a device that could sort out electrons of various energies that emerged from radioactive sources. Caltech didn't have such a device, and Frank thought that was a "hole" in the institution's nuclear physics facility.

* The use of the term "ray" in physics is confusing, primarily because historical precedents established terminology before phenomena were fully understood. A beta ray is actually a particle — an electron. Cosmic rays are nuclei of elements impinging on Earth from space. Gamma rays and x-rays, on the other hand, are truly rays in the sense that they are highly energetic forms of light. However, since we now know that all waves have particle (quantum) aspects, and all particles have wave-like properties, the terms can still be somewhat muddy.

It took him a year and a half to do it. He had to build all the components himself. He even went to the foundry where the magnets were made and marveled at the little carts that moved along the rows of ovens that forged the magnets. "A man with a fork would go over and open the door and take out the forging as he would a cake or something, look at it, and see if it was ready," Frank recalled. "If not, he'd put it back. It was beautiful to see that."

The device he built was smaller than a desk, and the research he did "was not terribly good," he said, although it sufficed for his Ph.D. thesis.

What made Caltech "a wonderful place" for Frank were the gatherings every Friday night at Charlie Lauritsen's place, where five or six graduate students would sit around talking about physics and what they were doing. He got to see a great deal of his brother and became part of his circle of friends.

Of course, Frank also continued his tinkering—as well as playing his flute. He built a phonograph for the physicist Robert Serber and his wife, Charlotte. "He finished it around midnight one evening, and turned it up loud to test it out," Serber said. "He must have woken up the Caltech campus as far away as the Athenaeum [the Caltech faculty club]." He played a benefit concert to support the Loyalists in Spain with Ruth Tolman, the wife of physicist Richard Tolman.

The Spanish Civil War was much on people's minds in the early 1930s. General Franco's bloody attempt to oust a democratically elected government (ultimately successful) was all the more troubling in the face of rising fascism in Germany and Italy. At home, the Depression lingered, and many people were unemployed. As Robert wrote to Frank at one point, "I think that the world in which we shall live these next thirty years will be a pretty restless and tormented place; I do not think that there will be much of a compromise possible between being of it, and being not of it."

Frank chose to be of it. This was not a big departure, as he was raised to put morality first, and even in high school he'd been involved in leftist activities: once, he and a group of friends went to a concert at Carnegie Hall that had no conductor. "It was a kind of 'down with the bosses' movement," Frank recalled.

Many of Frank's friends were involved in political movements — the most important of them a young Berkeley economics major from Vancouver named Jacquenette Quann. Jackie had a job as a waitress to raise money for school, and was proud of her working-class origins. She eagerly joined the Young Communist League in Berkeley. In one letter (in which Frank calls her "my angel" and "Lady, who alone in this world exists"), Frank tries to explain that even though he agrees with many of the league's goals, he doesn't see a particular need to join.

Frank and Jackie were married in 1936, Frank's second year as a graduate student. She soon moved to Pasadena, where she enrolled in junior college, worked a little for Willie Fowler and also for the *People's World,* a radical newspaper. Frank got a scholarship to cover tuition, and for other expenses they relied on his family's money. At the end of the month, they'd always dine at the Athenaeum, "because you could eat and charge it," Frank said.

Jackie got involved in Frank's work from the outset. She accompanied him to watch the magnets for his spectrograph being forged, and went to Lauritsen's Friday-night gatherings. One time, Jackie was so bold as to ask a question. "It was something about nuclei," Frank recalled, "and everybody just stopped talking and stared at her for a moment, and then without answering, just went back to talking. They were just so astonished that some stranger would interrupt them."

Despite the rise of Hitler and Franco, and the Depression in the United States, Frank and Jackie were optimists who thought radical change was both necessary and possible. As Frank often pointed out when speaking of this period of his life, many people were trying out new kinds of social and economic arrangements: there was the New Deal, the rise of organizations of the unemployed, and, not least, the experiment with communism in the Soviet Union. Science itself offered the hope that advances in agriculture would mean that no one would have to go hungry. The arts were changing as well, and for the first time, reproductions were becoming readily available so that paintings confined to museums could now reach a much broader audience. The phonograph had done the same thing for music. "Imagine!" Frank said. "Records, music getting into everybody's home on a greater scale than ever before, and good music."

On all these fronts, he felt, "there was the sense that I could make a difference and my friends could make a difference."

So, soon after they were married, Frank and Jackie joined the Communist Party. At the time, it seemed like a sensible thing to do. Many patriotic Americans, driven by social conscience, joined the party, as the Communists were the only ones seriously fighting racism and unemployment on the home front. At the same time, the CP seemed to be seriously antifascist, and the rise of Hitler and Mussolini was frightening people.

Communism flourished to such a degree in the United States during the 1930s that Ted Morgan, in his book *Reds,* refers to this period as the Pink Decade. Going to Russia to observe the Soviet experiment became a popular fad, "akin to the Hula Hoop," Morgan writes. The news that travelers brought back was mostly positive. Even Thomas Edison allowed that "the Reds have done pretty well, but they are cruel." To many, it seemed like a perfect society. As the official Los Alamos historian David Hawkins put it, "You could most accurately describe the [U.S.] Communist Party then as the left wing of the Roosevelt New Deal."

Even though Stalin was already eliminating his political opponents through murderous purges, and even though the Soviets did, in fact, have spies embedded both in the U.S. party and in various front organizations, those who had doubts about the Soviet experiment tended to keep their suspicions to themselves.

Were people like Frank and Jackie fooled by clever Communist propaganda? Or, in their youthful idealism, did they fool themselves? The answer is probably a bit of both. On the one hand, the Soviets did use the American Communist Party as a recruitment tool for spies; they did manage to infiltrate the government, and certainly the labor unions; they even penetrated the Manhattan Project, in the person of Klaus Fuchs. On the other hand, most of the people who joined the party at this time were, like Frank and Jackie, simply civic-minded Americans who joined because the CP was fighting racism and poverty at home and fascism abroad. Many liberals were attracted to front groups with "innocent-sounding names."

The turning point for Frank was outrage over failed efforts to integrate the Pasadena public pool: blacks were allowed in the pool only on Wednesday; the pool was drained before the whites came back on Thursday. And the CP seemed to be the only political organization concerned.

And so in 1937, Frank and Jackie clipped an ad from a newspaper — probably the *People's World* — and mailed in their application for membership. "It was months before anybody came by," Frank said. "So it was that kind of casual thing. But then we became very active."

Forced to take a party name, he chose Folsom, after the California prison, because he thought it was absurd to take a fake name, and he never used it again. The FBI, nonetheless, took "Folsom" to be his alias.

Frank and Jackie became part of a street unit in a mostly black neighborhood in Pasadena. They had discussion groups and met with organizations that were trying to do something about the widespread unemployment among blacks. The party asked him to start a group at Caltech, and he got "six or eight or ten people" together. Most were secretive about their membership, afraid of losing their jobs. But the secrecy never sat well with Frank. "Jackie and I were quite open about it."

Robert didn't approve of Frank's decision to join the party,* and he didn't approve of Jackie either. At one point, he accused Frank of being "slow" because it took him so long to earn his

* While some authors have recently suggested that Robert did indeed join the party, the clear consensus seems to be that he did not — though he did contribute money to the Spanish Loyalists, and joined Communist front organizations such as the teachers' union.

Ph.D.—mostly due to the fact that politics took up so much of Frank's time. As for Jackie, Robert made no secret of his disappointment with "that waitress." He once called the marriage "infantile." The feelings became mutual. Jackie later regarded Robert and his wife, Kitty, as pretentious, phony, and parsimonious. When they went to Robert and Kitty's for dinner, Jackie complained, there was never enough to eat.

Frank was as alienated from what was left of his "little royal family" as he'd ever been. And while the war would bring the brothers together again, it seems clear that they never regained the intimacy and bonds of mutual understanding they'd shared as boys. In later years, Frank rarely talked about Robert, and when he did, it was mostly with a melancholy air.

Despite the brothers' closeness and their many commonalities—right down to the twinkly blue eyes and the chain-smoking—they were in deep ways opposites. Frank was boisterous, full of mischief. Robert was quiet and cerebral. Robert got through Harvard in three years, went on to Cambridge, and three years later was a theoretical physicist. Frank didn't get his Ph.D. until he was twenty-seven; he took more side trips, political activism among them. Robert was arrogant, picky about his company. Frank would talk with anyone, and did. He later made friends with his FBI tail. Robert's magnetism emanated from his brain, Frank's from his soul. In ways that would turn out to matter most, the two saw things very differently.

In any event, Frank soon realized that he wasn't cut out to be a Communist. From the start, he was a thorn in the party's side. He felt it was too authoritarian, and not as interested in social justice as in petty bickering. And while it was supposed to run on a policy called "democratic centralism," Frank said, "there was centralism, but no democratic. So we got fairly upset by that."

Besides, he'd never been able to think of himself as a Marxist. "The notion of being a Marxist," he said, "always struck me just as ridiculous as being a Newtonian. I do believe that Newton had some good ideas, but I don't call myself a Newtonian just because he made a big difference in physics." (Apparently Robert felt much the same way. A party member discussing Robert with a comrade while the FBI listened in on a wiretap said that although Robert was a brilliant scientist, he was "just not a Marxist.")

Frank and Jackie officially left the party in 1940. By then, Stalin had joined forces with Hitler, and much of the party's prestige in the United States evaporated. Franco had taken Spain.

The party was just one thing among many that wasn't turning out as Frank expected. While socialism "held promise," he said, it "didn't ever work out quite the way people had hoped it would." Having reproductions of paintings in people's homes didn't mean people suddenly revered the arts or aesthetics.

But Frank didn't let his frustration affect him permanently. He didn't believe that disappointments should discourage him, he later told me, because he wasn't brought up to believe they should. If they had, he said, "I wouldn't have started the Exploratorium."

3

THE UNCLE OF THE ATOM BOMB

"We're leaving you a world that runs like clockwork. And the clock it runs like is a cuckoo clock."

The first time I heard Frank talk about the atomic bomb test at Trinity, he left me with an image I was never able to shake: a desert turned to glass. Thirty years later, I went to look at the blast site myself, and was astonished by how ordinary it all seemed—scarcely a hint of the violence or import of what had occurred there. The green desert glass had long since been plowed under. There was a small memorial obelisk with a plaque, maybe a hundred tourists milling about, an old bomb casing, some brochures and photos, and not much else. I picked up a few leftover shards of dark green radioactive "Trinitite" and slipped them into my pocket when the guards weren't looking, and tried in vain to find a single reference to "Oppenheimer." The absence was unsettling, partly because of the pivotal roles both brothers played in the test, and also because of the consequences that terrifying "success" and its aftermath were to have on Frank and Robert—and especially on their relationship.

By the time Frank finished his Ph.D. work at Caltech, war was already raging in Europe. To Frank, as to so many others, it felt as if a wave of fascism were sweeping over the Continent and threatening to "walk over" the United States as well. Frank got more deeply involved in politics, and ever more distracted from physics.

He was not offered a postdoctoral position at Caltech, which he suspected was due at least in part to his political activities. Instead, Robert arranged for him to work with Felix Bloch, a physicist at Stanford, trying to get a broken-down accelerator running so that Bloch could study neutrons. Frank didn't succeed, and considered the time he spent there largely wasted. He wasn't interested in neutrons, and he didn't like Bloch. "Stanford was awful," Frank said.

The feeling was apparently mutual. His Stanford fellowship wasn't renewed, and Frank felt, in essence, that he'd been fired — partly because he wasn't doing much physics but mostly because Stanford was embarrassed by his political activity. In fact, Frank was so immersed in politics that at one point he was forbidden to hold any more meetings at the Palo Alto Civic Center.

Then, once again with Robert's help, Frank obtained a position at the University of California Radiation Laboratory under Ernest Lawrence, the inventor of the cyclotron — the innovation that made possible all the behemoth particle accelerators in use today. Lawrence knew about Frank's politics and didn't approve of them, but he was fond of Frank and very close to Robert — relationships that were to unravel painfully in the not so distant future.

Only a few months after Frank arrived at Berkeley in 1941, Pearl Harbor was attacked, and the United States entered the war.

What was happening in physics at the time was equally momentous: invisibly to most, but inexorably nonetheless, physicists were learning how to tap the awesome energy of stars. The energy of the sun and other stars comes directly from Einstein's famous $E=mc^2$. Matter, Einstein revealed, is a kind of frozen energy, and that energy can be released under the right circumstances. The letter "c" in the equation stands for the speed of light — 186,000 miles per second — so that even a tiny amount of mass equals a monstrous amount of energy. Before the atomic age, nothing on Earth had even approached such stupendous power.

The beginnings of the atomic age have been well chronicled in many wonderful works,* so what I provide here is only the barest minimum needed to understand the circumstances in which Frank and other physicists found themselves in the early 1940s.

Shortly after physicists Otto Frisch and Lise Meitner understood that the uranium atom had been split in 1938, it was clear that the energy released could be used to create weapons of enormous destructive power. And given that the scientific findings were public, there was reason to fear that Hitler (Germany being a formidable power in physics before the war) would be able to build an atomic bomb. In 1939, Albert Einstein — at the urging of the physicist Leo Szilard — wrote to President Franklin Roosevelt warning that the splitting of the uranium atom "may be turned into a new and important source of energy in the immediate future," and that "this new phenomenon would also lead to the construction of bombs . . . A single bomb of this type, carried by boat and exploded in a port, might very well destroy the whole port together with some of the surrounding territory."

But it was not until two years later, in mid-1941, that Roosevelt took the matter seriously enough to call for an all-out effort to build an atomic bomb — a project that became known as the Manhattan Engineering District (later the Manhattan Project), part of the Army Corps of Engineers.

One of the key challenges in making the bomb was producing enough of the right kind of uranium. While most uranium atoms (U-238) have 238 subatomic particles in the nucleus, one rare variation (or isotope) contains three fewer neutrons, and only this U-235 could produce a nuclear explosion. A nuclear bomb would require a substantial amount of U-235, which amounted to less

* Among these, I put Jon Else's documentary film *The Day After Trinity* near the top of the list.

than 1 percent of natural uranium and was extremely difficult to extract.

Frank's war work centered on figuring out ways to turn a "dirty" mix of uranium isotopes into "pure" U-235. The most promising method was something that might be described as a salad spinner for uranium atoms. A beam containing a mix of isotopes would be twirled around an oval-shaped track, steered by strong magnetic fields. The U-235 atoms, being easier to steer, made a tighter circle — and could therefore be separated and collected. Frank started working on these "racetracks" at Lawrence's Berkeley laboratory in 1941, before Pearl Harbor. Lawrence first dedicated an old 37-inch cyclotron to the task, and later committed his brand-new 184-inch cyclotron as well. Civilization was under siege, and like everything else, pure physics would have to wait. The machines were officially called Calutrons, after University of California.

The work was highly chaotic and frequently dangerous. Because they were under so much pressure to produce enough U-235 quickly, the physicists had to proceed by trial and error rather than taking the time to really understand the phenomena. "We tried everything we could think of and let nature do the calculating," Frank said. "It was extremely hectic. Very few people walked around the laboratory; they usually ran. I can remember one episode when Lawrence tried to get a little more current out of the apparatus than we had succeeded in getting. He sat at the control desk and the rest of us ran madly about the large 184-inch cyclotron lab with extinguishers to put out the fires in the various cages of overtaxed electrical equipment."

Everyone was tireless and frantic, but as Frank's physicist colleague Ed Lofgren later pointed out, "I think you can believe me when I say that probably the most frantic and tireless was Frank. He also gave a sense of purpose to many of the people at the lab."

Next to Lawrence himself, Robert Wilson later said, Frank was the person most active in developing the Calutrons, and then overseeing their operation. But even so, the process was difficult and slow — and couldn't produce enough enriched uranium (a few pounds) for a bomb.

By 1943, a sprawling complex dedicated to the separation of uranium isotopes using racetracks was being created at Oak Ridge, Tennessee, about twenty miles west of Knoxville.* Frank spent the better part of a year supervising the manufacture of Calutrons at a Westinghouse plant in Pittsburgh and then getting the racetracks at Oak Ridge into operation. His job, as he later described it, was training people to fix what broke, redesigning things when necessary, and ensuring that no one slacked off on the job.

It Just Seemed to Hang There Forever

"The only way we could lose the war was if we failed in our jobs."

While Frank was busy with his racetracks, most of the bomb research was going on at Los Alamos under Robert's inspired direction. Robert had even suggested the site for the secret lab; he knew the area from his rides in the mountains (it was a forty-mile horseback ride from Perro Caliente). He remembered a boys' school on a 7,200-foot-high mesa atop a steep canyon—well hidden from outsiders, a "magic place" that looked west to the 11,000-foot Jemez Mountains and east to the 13,000-foot Sangre de Cristos. The school had a lodge with a dining room and kitchen, a guesthouse, and a few other buildings. Robert took Ernest Lawrence to check out the school. They introduced themselves as Mr. Smith and Mr. Jones—fooling no one, as both were already well-known physicists.

Suddenly, top physicists began disappearing from posts at universities all over the country, only to reappear on a lonely mesa in New Mexico. A town of five thousand shot up virtually overnight. People were stuffed into trailers, dormitories, and clusters of identical houses painted a bilious green, and outfitted with army fur-

* Oak Ridge was one of several facilities created in connection with the Manhattan Project. In addition to Los Alamos, there was another complex in Hanford, Washington, where nuclear reactors produced plutonium. A new element, discovered only in 1941 at Berkeley, plutonium, like U-235, is high-grade bomb material.

nishings. The higher-ranking scientists got to live on what came to be called Bathtub Row, because houses located there had the only bathtubs in town. There were chronic water shortages, and the unpaved roads were either dusty or muddy. The stress was salved by almost constant parties, with dancing and plenty to drink, including homebrew. Almost everybody was young (Robert himself was thirty-eight). At one point, the army complained to Robert that too many babies were being born.

Of course, censorship and secrecy came with the territory. Drivers' licenses were anonymous (numbers replaced names). The address was "Box 1663." Mail was censored; families left behind were completely in the dark. Occasional excursions were permitted to Santa Fe, thirty-five miles to the south, but it was taboo to use the word "physicist" there. Still, it was obvious to the locals that something strange was going on, so the army tried spreading false rumors. Robert and Charlotte Serber trawled bars in Santa Fe in an effort to "let" people "overhear" them as they talked about the "electric rockets" they were building. But they made terrible spies, and no one was interested in their secrets.

Despite the obvious need for tight security, the commander at Los Alamos, General Leslie R. Groves, a career army officer and West Point graduate, was eventually persuaded by Robert that science could not flourish in an atmosphere of secrecy, so the physicists largely kept their scientific autonomy and enjoyed freedom of speech among themselves — though other government officials objected strongly to this policy. Frank liked Groves, and thought he "made very good judgments." Groves, in turn, liked Frank, and later defended him when he was attacked during the anti-Communist purges.

Having impressed Groves and others with his work on the racetracks, Frank was invited to Los Alamos to help Harvard physicist Kenneth Bainbridge prepare for the first atomic bomb test, and also to help his brother. Robert Wilson met Frank at the airport when he got there in 1945. Jackie followed with their daughter, Judy, and their son, "a very tiny Mike."

Frank and Wilson were already close friends. They'd met at Princeton in 1942 when Ernest Lawrence sent Frank to check up on Wilson's efforts to separate the isotopes of uranium. At the time, Frank was a leading expert in that subject, and their friendship "blossomed." Wilson had grown up in Wyoming, and he shared Frank's love of ranching. Like Frank, he was a pacifist. Yet it was wartime; their country needed them; so they worked day and night on an unprecedented means of destruction. "We both recognized the irony," Frank later said.

The arguments for building the bomb were certainly persuasive. Though many understood the devastation an atomic bomb would bring, the threat of Hitler getting it first was all too real. "We felt caught up in a just war," Wilson later said. "We both felt that our individual efforts could help reverse a long series of defeats of the free world by Hitler's forces. It is hard now to remember the idealism that pervaded this nation as we met that Nazi challenge." In answer to a questionnaire many years later, Wilson said that building the bomb was "inevitable," given that the Nazis seemed to be winning the war on every front. "I can't think of any alternative except rapidly to have won the war sooner – or to have lost it!"

And yet the work didn't stop when Hitler was defeated. In retrospect, Frank was amazed at this. "Nobody slowed up one little bit . . . And it wasn't because we understood the significance against Japan. It was because the machinery had caught us in its trap."

Frank later elaborated on the psychology at work: "Even before the Pearl Harbor attack, the group of us who were working on the project kept things going twenty-four hours a day . . . most of us . . . were working from twelve to sixteen hours a day. Yet when the Pearl Harbor attack occurred, the quality of the work, for me, and I imagine for all of us, changed. Before the attack, getting the job done had been a challenge; afterwards it was an obligation. As long as there was any chance of defeat for the Allies, every wasted hour made more possible the spread of fascism over the world. Later, when our victory seemed assured, delays seemed no less criminal, for we all felt that the time when the killing would stop might in large measure depend on when we completed our job."

And the lure of the science itself helped sustain momentum. "Nuclear explosives have a glitter more seductive than gold to those who play with them," the physicist Freeman Dyson wrote. "To command nature to release in a pint pot the energy that fuels the stars, to lift by pure thought a million tons of rock into the sky, these are exercises of the human will that produce an illusion of illimitable power."

Even Robert Oppenheimer, who was later to express such guilt and remorse over the bomb, echoed these thoughts during a speech he gave on his last day as director of the Manhattan Project. "When you come right down to it," he said, "the reason that we did this job is because it was an organic necessity. If you are a scientist, you cannot stop such a thing. If you are a scientist you believe that it is good to find out how the world works; that it is good to turn over to mankind at large the greatest possible power to control the world and to deal with it according to its lights and values."

The site chosen for the first atomic bomb test was a dry, flat, scrubby desert that was part of the Alamogordo Bombing Range. Appropriately, it was called Jornado del Muerto – the journey of death. Robert named both the site and the test itself Trinity. He later wrote to General Groves that he really didn't know why he chose the name, but thought he might have been inspired by a line in one of John Donne's Holy Sonnets: "Batter my heart, three person'd God . . ."

Frank liked to joke that if his brother was the "father of the atomic bomb," that made him the "uncle." But in truth, he was more of a midwife. He described his job as that of a "safety inspector," making sure that workers wore hard hats and mapping escape routes through the desert. He set off smoke bombs on the cliffs above to chart the winds. He was also responsible for a crucial calculation that showed the radioactive cloud would rise high in the atmosphere on a "very long stem." Just in case it didn't, he organized plans for evacuating his own shelter.

In the days leading up to the test, people worked feverishly, sleeping when they could in barracks, shaking scorpions out of their boots in the morning. They built wooden shelters, protected by concrete and earth, and a hundred-foot steel tower to hold the bomb. The bomb – referred to by almost everyone as "the gadget" – had a plutonium core surrounded by explosives and detonators. No one quite knew what would happen when it went off. For a while, physicists had seriously considered the idea that the blast might set the atmosphere on fire.

The whole enterprise had such an otherworldly feel that ordinary life came as something of a shock. One day Frank was driving to the site with Ken Bainbridge, and they stopped to help someone fix a tire. "I just felt suddenly something happened and we became human again," Frank said.

When the plutonium core of the bomb was delivered to Trinity from Los Alamos (in the back seat of an army sedan driven by Philip Morrison), the general who signed for it said it felt warm in his hands. "I got a sense of its hidden power," he said. "It wasn't a cold piece of metal, but it was really a piece of metal that seemed to be working inside. Then, maybe for the first time, I began to believe some of the fantastic tales the scientists had told about this nuclear power."

Everything seemed full of import, Frank later reflected. The night before the test, it sounded as if all the frogs in the area converged on a pond by the base camp. They copulated and squawked all night long, and he remembered there was "a kind of funny significance . . . the only living things around there, coming together."

The night of the test itself brought fierce thunderstorms; as Wilson climbed the hundred-foot tower to make a final adjustment, lightning flashed all around him. The sky didn't clear until just before dawn, when Groves gave the go-ahead. Most of the dignitaries watched from Compania Hill, twenty miles away. Groves watched from the base camp, ten miles away. Frank and Robert lay side by side in control bunker S-10000 at the nearest station, about five miles away.

"It was at the end of this long, rainy night and we'd gone to the tower and inspected it over and over again," Frank remembered. "[We] stood around waiting and wondering what to do, and finally saw that it was clearing, and not much time left before the light. There was this great worry that somebody might have been left behind, or was too close, all kinds of tensions started building up. Then I remember the countdown — this complete ignorance of what might happen . . . And then somehow despite all the advance training, somehow you just didn't expect the kind of explosion that actually happened."

The brothers lay face-down, eyes closed, with their arms covering their heads. "But the light of the first flash penetrated and came up from the ground through one's lids," Frank said. "When one first looked up, one saw the fireball, and then almost immediately afterwards, this unearthly hovering cloud. It was very bright and very purple and very awesome, and one still had the feeling, maybe it's going to drift over the area and engulf us. And all the time that this was happening, the thunder of the blast was bouncing bouncing back and forth on the cliffs and hills. The echoing went on and on."

The cloud, Frank said, "just seemed to hang there forever." Later, he couldn't remember what he and his brother said to each other. "I think we just said: 'It worked.' . . . I think we embraced each other, but that's all I remember."

The shock wave broke windows 120 miles away, shook people as far as 160 miles away. The army explained that a munitions dump had exploded.

Morrison was ten miles away, and even at that distance, he said, the heat felt like an oven door opening.

"It . . . bored its way right through you," I. I. Rabi said. "It was a vision which was seen with more than the eye . . . A new thing had just been born."

Ken Bainbridge called it "a foul and awesome display," and later said to Robert, "Now we all are sons of bitches."

The Harvard chemist George Kistiakowsky saw the doom of humankind. "I am sure that at the end of the world — in the last millisecond of the earth's existence — the last man — will see something very similar to what we have seen."

Robert himself famously quoted the ancient Hindu scripture the Bhagavad-Gita. "'Now I am become Death, the destroyer of worlds.' I suppose we all thought that, one way or another."

In some ways, the most touching remark — perhaps because it was so understated, and also ultimately so true — was made by one of the guards: "The long hairs have let it get away from them."

The horror they had seen, however, was tempered by the hope that this new thing under the sun would cause people to see the world in a new light; it would compel them to understand that everything had changed and nations would have to behave differently toward each other. As Robert later put it, "The atomic bomb . . . has made the prospect of future war unendurable." Like Alfred Nobel* before them, the physicists hoped their horrific new weapon would one day bring about world peace.

"Those were the days when we all drank one toast only," Robert said. "No more wars."

Woe Is Me

"If you want to describe it as something you are familiar with, a pot of boiling black oil . . ."

Well before Trinity, people were already agonizing over how and whether the bomb should be used. By some estimates, the cost of *not* using it to end the war in Japan could have resulted in half a million more Allied casualties (perhaps twice that number for the Japanese).† Many others, however — among them General Dwight

* According to some accounts, Alfred Nobel hoped his invention of dynamite would also mean the end of war; his recompense to society, was, of course, his prizes in peace, literature, and science.

† Many historians now believe that this number was a politically expedient fiction. Despite President Truman's subsequent claim that dropping the bomb on Japan would save half a million American lives, apparently no one in power at the time believed U.S. casualties would exceed forty thousand, and other estimates were even lower. But saving American troops was not really the point in any event; having spent two billion dollars to build the bomb, it would be used, if nothing else, to signal U.S. military supremacy in the world, and especially to the Soviet Union.

Eisenhower and General Curtis LeMay — argued that since Japan was essentially already defeated, dropping the bomb would serve no legitimate military purpose and could cost the United States dearly in terms of prestige and bargaining power. "The reputation of the United States for fair play and humanitarianism is the world's biggest asset for peace in the coming decades," said Secretary of War Henry Stimson. "I believe the same rule of sparing the civilian population should be applied, as far as possible, to the use of any new weapon."

Robert Wilson wanted to hold meetings at Los Alamos as early as late 1944 to consider the moral and political implications of the bomb. Robert Oppenheimer tried to talk him out of it, but attended a meeting anyway, and eventually persuaded most people that the physicists should stay out of politics. Frank remembered meetings at Oak Ridge where people talked about what should be done with the bomb. Most agreed it should not be dropped on a population center.

The world's most eminent physicist at the time, Albert Einstein, was almost entirely barred from the atomic bomb effort because of his well-known socialist and pacifist views. Still, he was enlisted by Leo Szilard and others to write a letter to President Roosevelt introducing a memorandum that he, Szilard, had written arguing passionately against the use of the bomb and warning against a nuclear arms race. Roosevelt died before he had a chance to see the memo. According to some accounts, the letter was sitting on the president's desk at the time of his death; in any event, it was never passed on to President Harry Truman.

Szilard also circulated a petition raising objections to a surprise atomic attack; if the bomb were to be used, the petition said, the Japanese ought to be warned and allowed to surrender. The petition was signed by more than 150 Manhattan Project scientists. Another group of physicists, at the University of Chicago Metallurgical Laboratory, proposed staging a demonstration, setting off a bomb in some deserted area to which Japanese officials could be invited. This would be far preferable, they argued, than setting a dangerous precedent by using the bomb on civilians. Truman apparently never saw these documents either. (Not that it would have made much difference; those making such critical decisions had little use for the views of most physicists, and the bombs were seen as one more weapon in the arsenal — ready to use as necessary until Japan surrendered.)

Even some physicists worried that such a staged demonstration could fail; what if the bomb were a dud? Philip Morrison, among others, argued that at the very least leaflets should be dropped over doomed cities to give civilians a chance to evacuate. Others argued that this would only alert the Japanese to watch out for the plane carrying the bomb, making it easy to find and destroy.

In the end, of course, an atomic bomb nicknamed Little Boy was dropped on August 6, 1945, on the pristine city of Hiroshima — kept blissfully and deliberately untouched by conventional U.S. bombs, the better to assess the damage. "By and large," Szilard later said, "governments are guided by considerations of expediency rather than by moral considerations. Prior to the war I had the illusion that the American government was different. This illusion was gone after Hiroshima."

In an instant the city was flattened, its people reduced to charred cinders, survivors hobbling around with their skin peeled off and hanging from their bodies like rags; many were so badly burned that their faces didn't look human, and their neighbors couldn't tell if they were looking at them from front or back. One of the American airmen riding in the plane that dropped the bomb said, "If you want to describe it as something you are familiar with, a pot of boiling black oil." Another airman said: "I don't believe anyone ever expected to see a sight quite like that. Where we had seen a clear city two minutes before, we could now no longer see the city."

A *New York Times* reporter who was riding in a B-29 directly behind the plane carrying the bomb wrote: "It was a living thing, a new species of being, born right before our eyes. The mushroom top was even more alive than the pillar, seething and boiling in a white fury of creamy foam, sizzling upward and then descending earthward, a thousand geysers rolled into one. It kept struggling in an elemental fury, like a creature in the act of breaking the bonds that held it down. In a few seconds, it had freed itself from its gigantic stem and floated upward with tremendous speed, its momentum carrying it into the stratosphere to a height of about sixty thousand feet."

That bomb killed 140,000 people instantly, and left as many as 200,000 dead within five years.

Frank vaguely remembered standing outside his brother's office when the news of the bombing came over the loudspeaker at Los Alamos. His first reaction, like that of many other physicists, was simple relief that it worked. "But before the whole sentence of the broadcast was finished one suddenly got this horror of all the people that had been killed," he said. "Up to then I don't think I'd really thought of all those flattened people."

Little Boy made use of the very enriched U-235 that Frank had worked so hard to produce. Three days later, on August 9, Fat Man, a plutonium bomb, was dropped on Nagasaki, immediately killing 70,000; in five years 140,000 would be dead. In both cities, the death rate was well over 50 percent of the population.

It is hard to fathom what all this must have felt like for those largely peace-loving scientists who had worked so tirelessly and feverishly to produce these devices — devices that, in the end, worked exactly as they were supposed to. The physicists had succeeded brilliantly at their task. And yet for many the achievement was a horrific moral failure; they had succeeded only to betray their highest principles. No wonder it tortured so many of them. There was no making sense of it.

Frank's friend Robert Wilson became physically ill; he felt betrayed that the physicists had built the bomb only to have it dropped on civilians without warning and without discussion. Einstein — whose equations first unveiled the energy inherent in matter — said only "*Vey iz mir,*" a Yiddish phrase meaning roughly "Woe is me."

As the diplomat George Kennan said at Robert Oppenheimer's funeral, "On no one did there ever rest with greater cruelty the dilemmas evolved by the recent conquest by human beings of a power over nature out of all proportion to their moral strength."

"The physicists have known sin," Robert later said, "and this is a knowledge they cannot lose."

Certainly the Manhattan Project physicists (as well as many government officials) were acutely aware of the enormity of the turning point: for the first time, the human race had the means to annihilate itself. Truman, however, was matter-of-fact. He said in a statement broadcast over the radio after the bombing of Nagasaki: "Having found the bomb, we have used it."

When Robert famously confessed to Truman, in the fall of 1945, "I feel I have blood on my hands," Truman told Undersecretary of State Dean Acheson, "I don't want to see that son-of-a-bitch in this office ever again." A year later, he called Robert a "cry baby."

A Blueprint for Peace

"If it ever was so that American security depended on her military might alone, it will not be so in the future."

Given that there was (and is) no possible physical defense against nuclear weapons, the only effective strategy for self-protection seemed to lie in political change — that is, in the creation of social inventions as revolutionary and powerful as the bomb itself. As makers of the weapon, physicists felt a special responsibility to take on this new role. "The whole experience became a kind of existential revelation of the horror of what we had in this country done," Wilson said at Frank's memorial. "And it convinced us both to dedicate the next years to try to explain the danger of nuclear bombs and the necessity of internationalizing nuclear energy. If we and our colleagues were not successful, it was not for lack for trying."

Frank saw a close parallel between the postwar struggles to contain the atom bomb and the challenges he'd faced while working on his "racetracks." Success had depended on the efforts of all kinds of people. "We had hoped we could say to the manufacturer, 'Here is a model; build it,' and to the operators, 'Here it is; run it.' But it did not turn out that way at all," Frank said. Rather, the physicists found they had to work closely with the manufacturers at every turn — including training thousands of people "fresh from farms and woods to operate and repair the weirdest and most complicated equipment . . . Our job, in general, turned out to be that of showing people how to adapt their existing techniques to the solution of our new and outlandish problems."

In the same way, he said, it would never work to foist the atom bomb on the world and simply say, "Here is a fearful bomb, control it." Rather, everyone would be called upon to do their part in securing the peace.

The physicists had a hard task ahead of them. For one thing, they would have to change the way people felt about the very idea of war, convince them that they must find alternative ways of influencing each other. This went against the whole of human history, in which nothing has been more celebrated than the will to fight for a cause. "And now the physicists were saying: 'Well, there's no way you can fight for what you believe in anymore.' It wasn't enough," Frank said.

Still, there was a sense of optimism among the physicists; many believed that they could turn the tragedy of Hiroshima and Nagasaki into "a creative re-ordering of international relations," according to the historian Jessica Wang. "For a brief, exciting, tumultuous period, the atomic scientists moved to the center of American politics as the leaders of a fledgling political movement."

Within weeks of the Nagasaki bombing, Frank, along with Hans Bethe, Edward Teller, and others, had drafted a statement warning against an arms race and pushing for international control of atomic energy; they were backed by five hundred Manhattan Project scientists in the newly formed Association of Los Alamos Scientists (ALAS). They hoped that Robert would use his influence in Washington to circulate the statement widely, but instead he told them, in essence, to sit tight.

Robert thought there wasn't time to bring the public in on the debate; he preferred to use his considerable fame and power to influence policy, betting on his ability to maneuver in Washington. Frank, ever the populist, thought the only hope was to take the battle to the streets; he firmly believed that the greatest wisdom lies in ordinary people. The brothers argued often and intensely, "in some instances a real sense of outrage with each other," Frank remembered. "Nobody was explaining anything."

Frank expressed his disgust at what he considered his brother's futile and elitist approach in a long, unpublished essay that begins with a metaphor that seems just as apt today as it did in 1945:

> Certain current trends in this country are very alarming. It is as though we were all in a great bus, traveling hell-bent for leather at 85 miles per hour down Hollywood boulevard. Everybody is scared stiff but a few odd characters among the passengers, motivated by somewhat obscure private considerations, proceed to pull down the shades and to tell us that our bus is equipped with new blow-out proof tires and that therefore we are really quite safe. They further assure us that it does not matter whether or not we run into anybody because our bus is so substantial. Obviously no one advocates having any sort of a crash. Nevertheless those who suggest slowing down and those who encourage the passengers to figure out some saner method of getting to their destination frequently are . . . bodily thrown from the bus.

The problem of keeping the country out of war was extraordinarily difficult, Frank acknowledged, and the only hope lay in telling as many people as possible all the relevant facts and then encouraging them to think and talk and write about them and then take a vote. "When one is confronted with a really important and difficult political problem one cannot trust a group of experts to make the decisions," he said. "They are too likely to be wrong."

Very few people agreed with him. On the contrary, they distrusted the ability of the general populace to contribute meaningfully to such a critical discussion. The source of this mistrust of the hoi polloi, Frank thought, stemmed largely from the tendency of people to credit their own success to a single personal characteristic, which they then "idolize" and use as a yardstick to measure everyone else. "There are some people of my acquaintance who judge all people by whether or not they are smart," he wrote. They conclude that most people are too stupid to make sensible decisions. At best, the public could recognize smart people when they see them, and elect them.

But Frank argued that the "smart" and the "stupid" people really weren't that different: the ability to solve a differential equation was only an infinitesimal advance over the ability to speak and write. The ability to read *Hamlet* was unimpressive compared to the ability to recognize one's mother. The ability to win a game of tennis was only slightly more wondrous than the ability to eat with a knife and fork. "The ability to dance gracefully," he wrote, "is rare and a delight to find, yet it is only a slight improvement on the ability to walk."

As for the objection that most people were too uneducated to make policy decisions, Frank argued that when people believe they are being listened to, they educate themselves. "All of us have seen especially during the war," he wrote, "the enormous increase in the competence of people that results from a sense of responsibility, from a sense that what they do or say may have some decisive effect."

Of course, the American public would be "flabbergasted" if Congress actually started heeding their advice, Frank wrote. But if they believed their answers mattered, then he thought that every local meeting in the country and every physics department would respond: "And if, by some chance, some Rotary Club had forgotten to send their answer, then you should have sent them a wire telling them that you were waiting."

In the end, trusting the hoi polloi was the only way to get good answers to such impossibly difficult questions as how to keep peace in a nuclear age. "If you don't speak out . . . tell them all you know and trust them," Frank wrote, "you are going to get yourself, and me, into a nice long, deluxe and king-sized war."

Determined to speak out himself, Frank returned to work at Lawrence's lab in Berkeley, and promptly began giving speeches "all over the map" — to bankers and teachers and doctors and naval officers and labor unions. Despite his participation in the Communist Party during his graduate school years, Frank didn't consider himself a particularly political person ("I have never devoted more than the smallest fraction of my time to politics") and found making these speeches "a great effort." He wanted to get back to doing physics. But he felt obligated to make the effort, and so he did.

One of his first talks was at the Berkeley Democratic Club, and even though he'd cleared the content with the lab's security staff, he still managed to get in trouble with Lawrence over it — although for unexpected reasons. The club had arranged for him to speak in a very small hall; people complained, resulting in some public fuss, which wound up being reported in a local newspaper. As it turned out, the club hadn't used a larger hall because that hall had separate seating for blacks. "And Ernest Lawrence saw that in the paper and said, 'See what you've done, Frank? You've brought race relations into the radiation lab!'"

Above all, Frank tried hard to convey in his speeches that the atomic bomb was not just a bigger bomb, but a huge leap in destructive power that could be controlled only by making equally huge changes in human behavior. Even the smallest, earliest atomic bombs were a thousand times more powerful than previous bombs. Human brains can scarcely grasp what that means. Imagine, Frank said, that your salary was $20,000 and it was reduced to a thousandth of that. You would make $20 a year. Or that instead of preparing dinner for 4 people, you had to feed 4,000 — in the same kitchen, using the same equipment.

His other constant and related theme was that the old methods of peacekeeping wouldn't work in a nuclear world, and that radically new approaches were needed. The plan Frank promoted tirelessly came out of a "board of consultants" commissioned by Undersecretary of State Acheson and chaired by David Lilienthal. Robert had a hand in it; in fact, he probably wrote most of the board's report himself. It proposed that an international agency take over all aspects of atomic energy, and that all atomic activity in all nations be entirely open and transparent to everyone.

From today's perspective, it may look unrealistically optimistic to think that nuclear nations would cooperate over the control of atomic energy. However, atomic weapons were so new that Frank (and many others) thought they presented a golden opportunity. Unlike other ideas being promoted at the time – world government, for example – atomic power came with no vested interests. Therefore, atomic energy was the easiest domain in which to make real changes in the way nations dealt with each other. "It is the place where we must begin," Frank said.

The Acheson-Lilienthal plan was not a panacea, but it seemed a reasonable, workable start. International control of atomic energy could at least give people the time to inaugurate sane policies and create an atmosphere in which the United States and other countries "can straighten ourselves out," Frank said. It could establish a pattern of cooperation among nations. "If, on the other hand, we do not have this control then we are surely sunk."

Secrets

"It is said with some justice that people who have lived around high circles in Washington develop a curious crook in the neck from the quantity of whispering they are forced to practice."

Perhaps the most important and far-reaching aspect of the Acheson-Lilienthal plan was its call for total transparency of all atomic activity. This was a radical idea at a time when both politicians and the general public seemed convinced that strict secrecy was required to protect the United States from our enemies. "A widespread presumption emerged that 'atomic secrets' existed whose possession would allow other countries immediately to build their own bombs," noted David Kaiser, a historian of science at MIT.

But this impression was false. Kaiser quoted Henry DeWolf Smyth, a Princeton physicist and Manhattan Project consultant, as stating flatly: "There is no 'secret' of the atomic bomb in the sense of a mysterious formula that can be written on a slip of paper and carried in the sole of a shoe or the handle of a hunting knife." Robert Wilson emphatically agreed: "There is *no secret* of the atomic bomb," and therefore no reason why it should not be internationalized.

Even if the "atomic secrets" hadn't been widely known, the nature of science is such that it's impossible to keep fundamental knowledge a secret for long. For Frank, secrecy and dishonesty were two sides of a coin, and he had an abiding disdain for both. "We, as a nation, are becoming inured to misrepresentation, withholding of information, in almost every phase of public life and endeavor," he said in one postwar political speech, pointing out that the worst thing about secrecy is that you don't get the best advice. If people don't have all the information, they can't help you make the best decision. (Of course, without secrecy there is far less need for military control, so the idea of openness was anathema to the armed services.)

Much of what Frank said is unnervingly prescient of the political situation we've faced in recent years, where public discourse is dominated by messages stronger on ideology than on truth. "The lack of public honesty," he said, "tends toward disunity in our own country because it allows different and prejudiced groups to trust only those sources of information which appeal to them." What eventually defeated the Nazis, he argued, was their "utter disregard for the truth."

Frank feared that the same disregard for truth and openness threatened the scientific enterprise he loved so well. "Today . . . science and scientists seem to be infected with the same disease that infects so much of our public life. We too are acting under the false principle that it is possible to withhold and distort information and at the same time make progress."

The more atomic physics came under the control of the army rather than of scientists and civilians, the more secret the entire field became. Frank and others worried that the secrecy would spill into research and teaching. Military control would mean not only "extreme secrecy" but also increasing "compartmentalization." "Both of these," he said, "are the mortal enemies of scientific progress."

In the end, the Acheson-Lilienthal plan went nowhere. The press barely reported it, so it did not become a basis for public discussion as Frank and others had hoped. Truman gave the task of presenting it to the United Nations to the seventy-seven-year-old investment banker and political adviser Bernard Baruch, who altered key aspects of the plan in ways sure to make it unacceptable to the Russians—including taking away their veto power in the UN. "It was presented in a very haughty sort of uncompromising way," Frank told me, "with some variations . . . that made it clearly unacceptable to any other country that didn't already have that bomb, and it fell flat."

Frank believed that the main reason it failed, however, was lack of openness. He reminded people in his speeches that the report had been written by men of very different backgrounds and starting points. Yet because they had all had the same information in front of them, they had come to the same conclusions. "The board had unrestricted access to information on atomic energy and the Manhattan District Projects," he said. "It is highly significant that a group of men who met without any basis for initial agreement came to a unanimous conclusion and to one which is supported by the scientists who worked on the project and by those who are familiar with its nature."

Despite the failure of the Acheson-Lilienthal plan, Frank's faith in the possibility of progress never seemed to wane. As he wrote in a letter to the *New York Times* at the time, just because this particular effort had failed didn't mean we should give up trying to come up with the "social inventions" required to deal with a nuclear world. Surely other approaches could and should be tried. "By insisting that [the plan is] the only possible system of control, we merely endanger our prestige and our security . . . No matter how personally discouraged our delegation may be, they should make use of every opportunity, however remote."

It's probably fair to say that the Manhattan Project physicists in general believed they had an important and legitimate role to play in containing the monster they had created. Their special knowledge of the subject, they felt, made their contributions especially valuable. As the Chicago Metallurgical Lab scientists had written to Secretary of War Henry Stimson, "We feel that our acquaintance with the scientific elements of the situation and prolonged preoccupation with its world-wide implications impress on us the obligation to offer to the Committee some suggestions as to the possible solution of these grave problems."

Nevertheless, many people in high military and government circles, and even a few physicists, thought that scientists should stick to science — a point of controversy that was to become central in the debate over the vastly more powerful hydrogen bomb. "Their science gives them no special insight into public affairs," Edward Teller said sarcastically during the H-bomb debate in 1954. "There is a time for scientists and movie stars and people who have flown the Atlantic to restrain their opinions lest they be taken more seriously than they should be."

Echoing Teller's sentiments, the *Minneapolis Morning Tribune* published an editorial warning readers that while Einstein might be brilliant at physics, he was not qualified to speak on political questions. Frank responded with a letter to the editor saying that by the same reasoning, the editor's ability to write and edit newspaper copy did not qualify him to have political opinions either. In Frank's view, both Einstein and the newspaper editor had a right and an obligation to express their opinions: "The very essence of our political democracy lies in the fact that we consider the political judgments of all sane people equally authoritative."

But Frank's brand of patriotism did not fit the times, and it was soon to bring ruin on himself and his family. A clear picture of his "dangerous" ideology emerges in a note Frank wrote, apparently to himself, after reading in the *Denver Post* about a gambler who infuriated his interrogators by pleading the Fifth Amendment at a congressional hearing: "The frustrated committee members apparently became angrier and angrier and finally Senator Mundt shouted to the man: 'Do you think that you are bigger than the United States government?' The answer is, of course, yes. Sure, he is bigger than the United States government; everybody is. That is precisely the point. It is the essence of the great social invention we made in the 18th century."

4

UN-AMERICAN

Dear Lawrence,

What is going on? Thirty months ago you put your arms around me and wished me well . . . told me to come back and work whenever I wanted to. Now you say I am no longer welcome.

Who has changed, you or I? Have I betrayed my country or your lab? Of course not. I have done nothing. Has anybody said I did a bad job during the war or since? I doubt it. Does anybody think that I ever let any classified information leak out, intentionally or unintentionally? No one has ever suggested such a thought to me, no one would have any reason to think so. Have I been charged or accused of any crime or misconduct whatsoever? No. You do not agree with my politics but you never have. There are no new rumors about my distant past floating around . . . you told me of such rumors yourself during the war . . .

I am really amazed and sore because of your action.

Sincerely,
Frank

Particles from Heaven

"Our equipment is probably still out there, the shiny aluminum sphere floating on the bright water off the shore of the Grand Cayman Island."

Frank was, first and foremost, a physicist, and after the war he was eager to get back to the business of exploring and discovering — preferably the kind of seemingly useless "sightseeing" that so often produces spectacular rewards. He liked to point out that the atomic bomb itself was developed out of "obscure and apparently useless research on the tiny dense nuclei of atoms."

When he wasn't making speeches, he was working with Lawrence, for the first time putting the new 184-inch accelerator to work in the service of science instead of war. It was the largest accelerator in the world, and Frank did some of the first experiments on it. He also worked with Wolfgang Panofsky on constructing the first proton accelerator, even building scale models of the subunits. "He was well versed in the witchcraft of accelerators," said Panofsky, who later became director of the Stanford Linear Accelerator Laboratory. "He did a lot of really nice work."

For all the wear of the war years, Frank was still, at heart, a tinkerer, an explorer — an only slightly grown-up version of the boy who took apart player pianos and bicycle brakes and climbed trees to watch lightning storms. Unlike the ever-cerebral Robert, who never learned to split wood or drive a car, Frank loved getting down and dirty, squishing his hands into whatever the world had to offer. So the challenge of building machines that would pump tiny bits of matter up to near light speed served both his talents and his passion.

But the Berkeley job was only a research position, so when Frank was offered a regular faculty appointment at the University of Minnesota in 1947, he accepted. The physics department there was in the midst of assembling a group to study high-altitude cosmic rays — mysterious particles of unknown origin that rain continuously on Earth from space. The Minnesota physicists were motivated in part by the fact that the General Mills Corporation had just developed huge helium balloons capable of carrying eighty-pound payloads to an altitude of twenty miles — plenty high enough to study the particles before they were absorbed or transformed on their journey through Earth's atmosphere.

Frank was intrigued. Cosmic rays, he thought, were "truly fascinating." No one knew exactly what they were or where they came from, but clearly these fast-moving cosmic messengers had interesting tales to tell about the sources that produced them and accelerated them to speeds approaching that of light. Just as important to Frank, there was "no apparent connection between [these] investigations and anything practical." He saw this work as an opportunity to escape politics and return to what he considered to be his true self.

He arrived in Minnesota with his family in the spring of 1947 and immediately began building cloud chambers for the balloons to carry aloft. The detectors were encased in thirty-inch-diameter spheres packed with sensors that could pick up the wakes of particles passing through. Before one launch, Frank decided at the last minute to add a stack of photographic plates to the instrument package.

Until these experiments, people had thought cosmic rays were primarily protons — the particle that is the nucleus of the simplest atom in nature: hydrogen. But when Frank studied the plates, he found one particle track that was much too heavy to be a single proton. "Just seeing this suggested that maybe all kinds of nuclei were coming in," he said. "We had made a discovery!"

And a very significant discovery it turned out to be. If nuclei of all the elements, from airy hydrogen to weighty gold, were truly streaming toward Earth from space, then perhaps they were coming from the very stellar furnaces that cooked up these more complex elements in the first place — exploding stars. The mysterious cosmic rays, in other words, might be well-traveled sneezes from violent events in our galaxy or even distant parts of the universe.

The discovery, Frank said, immediately changed the focus of his group's research. They began to study the origin of cosmic rays and their behavior as they moved about the Milky Way. They become more involved with cosmology than nuclear physics.

The group also discovered that while Earth's magnetic field allowed only some of the elements through its protective shield at the equator, all of the elements got through at the poles. Frank published many scientific papers during this period that were considered to be "landmark research."

Even the great physicist and Nobel laureate Hans Bethe later praised Frank for having invented new techniques that quickly spread to other universities. "I have been impressed by his great ability, by his experimental ingenuity, and by his enormous capacity for hard work," Bethe wrote.

While Frank still made the occasional speech to a local citizens' group, his heart was in his quest and he was doing what he loved best: poking his nose in nature, wondering how stuff worked, figuring out ingenious ways to ask interesting questions about the basic whys and wherefores of matter itself. In most photographs from this time, he is fiddling with some kind of apparatus, wearing a dark suit and white shirt, his thin tie hanging down in front — asking for trouble. As is typical in physics experiments, his detectors were handmade from leftover junk (much of it sophisticated versions of the "junk" his brother gave him).

Also typically, the work could be tedious and frustrating. Frank's son, Michael, often accompanied him to the lab and watched time and again as his father tried to ensure that a tight vacuum was maintained inside a detector chamber so that stray air molecules didn't confuse the results. The two halves of the spherical chamber were held together with "a thousand bolts around it," Michael remembered. Frank would tighten all the bolts, then put the sphere in a water bath, and if he saw bubbles he knew that the seal wasn't tight, and he had to remove all the bolts and start over again. "There was a lot of cursing and cussing," Mike told me.

Tropical Adventures

"One of the young men offered his machete after he took one look at the big kitchen knives that I had swiped from the Officers' Galley . . . I guess we looked pretty inexperienced."

The cosmic ray experiments plunged Frank into the kinds of adventures he'd always relished – and he spoke and wrote about one in particular with special fondness. He and his group had talked the U.S. Navy into letting them use an aircraft carrier for their experiments – the USS *Saipan,* which was on its way to Guantánamo Bay to carry out maneuvers. It had a crew of a thousand men, two squadrons of corsairs, and a helicopter nicknamed Garuda. Frank's group boarded the carrier with a ton of gear and 150 tanks of helium for their balloons.

After launching the balloon from the ship, the trick was to track it and retrieve it so the physicists could study whatever tracks the detectors might have picked up. Often the balloons came down in hard-to-reach places (or disappeared altogether). During one such balloon chase, Frank exulted in the ride along a Cuban coast he later described as more spectacular than Big Sur. He made it into something of a joy ride, zooming in close enough to the surf to follow schools of fish, near enough to cliffs to sense their texture, so close to houses he could see the flowers and the small, hilly fields planted with a hodgepodge of sugar cane, coffee, tomatoes, beans, bananas, and pineapples.

He and a handful of colleagues sighted their lost balloon and parachute tangled in a tree at the bottom of a ridge, but the nearest landing site was twelve miles down a canyon. So they studied the landscape as best they could in order to retrace their steps. "The fact that we were able to walk back up through twelve miles and locate our equipment in the midst of a dense woods still amazes me," Frank wrote.

He never would have succeeded if it hadn't been for a remarkable entourage of helpers Frank's crew had gathered along the way. The first addition to the party was a small boy who warily watched from a hillside with his elderly father as an enormous pile of miscellaneous junk was unloaded from Garuda: C-rations, walkie-talkies, pole climbers, kitchen knives, a coil of nylon rope, various small tools. Realizing the physicists would never make it on their own, the old man offered his ten-year-old son as a guide. Soon another young man arrived, and seeing the woefully inadequate knives Frank's group had brought, he immediately offered his machete.

They set off, and every mile or so stopped at a thatched house surrounded by bright red and purple flowers. At each house they gestured toward the sky and toward the mountain that marked the site of the balloon. And at each house they picked up another boy or young man. Within hours, the group had swelled to thirty-three.

Knowing how difficult the trek would be, the locals set a terrific pace through the rough terrain so they could make it back before dark. Huge smooth boulders, waterfalls, and cliff-walled canyons made walking close to impossible. "Our heads swam with the exertion and the heat," Frank recalled. "Our mouths, our throats, even our bellies cried out. The trail seemed to be covered with a layer of ball bearings and we constantly lost our footing climbing up and down the sides of the steep draws. We had no hats and fought off the sun with our arms." Occasionally the boys would go off in search of sugar cane, which they chopped and peeled and passed around. "We chewed it eagerly and felt the warm sweet sap give energy to our legs."

The physicists drew maps for the Cubans, and the Cubans drew maps for them. After a while, Frank realized the maps didn't matter. He was part of a precession and could not have influenced the course in any event.

At the last house before the ridge, they picked up yet two more fellow travelers for the final push. There was no trail, but thirty-odd machetes slashed at the vines. The ridge was so steep they could not walk up it, but had to pull themselves up with both hands.

At long last, the ensemble arrived at the sphere and parachute. Frank tried getting up the slippery tree trunk using the climbing spikes the navy had given him, but he kept sliding to the ground. Then one of the Cubans asked if the sphere would explode. Assured that it wouldn't, he took off his shoes and "just walked up the tree," Frank said. The sphere was eventually retrieved, and the party returned to their helicopter, led by Cubans carrying torches made from flammable vines.

Alas, few of his adventures turned out so well. After one recovery effort in the Caribbean, Frank returned with his "spirit completely broken" after losing one of his spheres in a futile chase over water. "I cannot recognize myself," he wrote. "I walk slowly, dragging my legs. I eat enormous quantities and lounge at the dinner table nursing lukewarm cups of coffee. At the moment, the prospect of going downstairs in order to fill my fountain pen requires a most intense and disagreeable effort."

One of his spheres caught fire at high altitude. Another was lost after its supporting cords broke. One burned up before leaving the ground. Another collided with a lamppost and smashed. Yet another made it up safely, but was lost in the northern woods of Wisconsin. It snowed heavily the next day, so there was no chance of finding it. The physicists would work day and night for weeks preparing the equipment, only to return empty-handed. At one point, a despairing Jackie asked, "Why don't you guys give up?"

As Frank wrote to Bob Wilson, then at Cornell, "We didn't *always* lose the equipment." Then he described his latest attempt as a "heartbreaking fizzle." Fifteen minutes after takeoff, the balloon had encountered violent wind shear and "convulsed so frightfully that it broke the cords holding our sphere, which then dropped free fall from fifteen thousand feet. We were unable to locate that splatter . . . it is sickening hard luck."

In his own mind, Frank struggled to reconcile the enormous effort and cost (some $90,000 a year) in the face of what he called the "doubtful benefit of the nation and humanity that would result from the completion of our experiments." As he wrote to Wilson, "[I] have the continuing sense that it is extraordinarily foolish to continue what I am doing just because it is fun when every time I read the newspapers such a terrible fate stares me in the face."

Still, it was a very good time for Frank. Perhaps most of all, he enjoyed the opportunity Minnesota gave him to teach seriously for the first time — and not surprisingly, his tenure left lasting impressions. One memorable letter to Frank from an otherwise unidentified "John" — a former graduate student who hadn't seen Frank for fifteen years — recounted an oral exam he had taken in Frank's office which he called "unforgettable" because of what Frank had made of the occasion. The student was at first dismayed when Frank started asking him about quantum statistics, because he hadn't expected the question and wasn't prepared.

"Then a marvelous thing began to happen," John wrote. "You managed, by your remarkable kindliness, to get me to say a few words relevant to your questions. Building on that — regarding it, by your sympathetic interpretation, as partly correct — you persuaded me to say more. You showed me how this, also, was partly correct . . .

"I went out of your office glowing. I think I hadn't ever *felt so good* about physics before."

Dozens of similarly moving testaments surfaced among Frank's letters and interviews with former students.

But while Frank was happy in his research and teaching — and especially to be through with politics — it turned out politics wasn't through with him. All the time he was chasing balloons and mentoring students, the FBI was tracking his every move, bugging his phones, even wiring hotel rooms as Frank and Jackie and the kids moved from place to place.

Alias Frank Folsom

"He was a very interesting window into how this nation devoured its own children. The nation chewed up and spit out Robert in one way, and it chewed up and spit out Frank in another way."

As it turned out, the FBI had been keeping tabs on Frank as far back as October 1941, when special agents from the San Francisco office reported seeing him "enter the home of [crossed out] and join a group of persons already assembled in [crossed out] living room . . . [crossed out] stated that Communist songs were sung." The bureau's reports often make note of Frank's "alias," Frank Folsom, the name he chose mainly as a joke at the expense of the Communist Party and the silliness of having to acquire an assumed name.

FBI surveillance was put on hold during the war years (military intelligence took over), but picked up promptly afterward. FBI informants either knew about or attended virtually every postwar talk Frank gave.

Although Frank hadn't been involved in the Communist Party since 1940, the ideas he was promulgating were unpopular with a great many powerful people. The very idea of giving away our atomic "secrets" — much less handing control of atomic power to civilians — must have seemed like treason in some quarters, even though the proposals Frank supported originated at the highest levels of government.

And so federal agents followed at Frank's and Jackie's heels like so many faithful puppies, trying to sniff out whatever evidence they could of subversive activity. When Frank and Jackie took a trip to Mexico in January 1946 ("allegedly for vacation," according to the FBI), their every move was monitored, every detail carefully noted, including the fact that Frank's pants were "a little short" and neither he nor Jackie spoke much with other passengers on the plane. A thorough baggage check secretly arranged through U.S. Customs failed to uncover anything suspicious, but did turn up among their possessions "two original paintings" and a half-dozen books, including *Life and Works of Abraham Lincoln* and *A Practical Spanish Grammar.*

By March 1946, when Frank was still working with Ernest Lawrence, FBI headquarters in Washington advised the San Francisco field office to submit reports "promptly and regularly" on Frank's activities — "particularly the results of the technical surveillance" (i.e., phone taps). Among Frank's apparently suspicious activities was his membership in the Northern California Association of Scientists (NCAS), somewhat ominously described as "a group opposed to the present military control of the atomic bomb project."* Apparently damning as well were Frank's attempts to "organize, support and advocate such causes as . . . the Consumers Union" and Jackie's active involvement in the Fair Employment Practices Commission.

Some of the information contained in the FBI files was public knowledge, but much came from "reliable" and "confidential" sources and unnamed informants, one of whom baldly stated that Frank was in contact with Soviet agents. Such "confidential informants" were among the primary sources of information the FBI used during these years.

To be sure, the "red menace" was a palpable threat to most Americans at the time – after all, a tiny political party in Russia had overthrown the whole government and brought in a reign of terror – so the presence of "pinkos" in the population was perceived as a real and present danger. In fact, a growing consensus in the late 1940s and 1950s perceived the American Communist Party as perhaps the gravest threat to national security. Under increasing pressure to protect the country from Communist infiltrators, the FBI became, in effect, an "ideological police force."

Shortly before Frank left Berkeley, the FBI's San Francisco office contacted agents in St. Paul requesting that they initiate "a discreet, comprehensive investigation of Oppenheimer upon his arrival in Minneapolis." On March 6, 1947, a coded teletype to FBI director J. Edgar Hoover stated that the family would reside temporarily at the Hotel Curtis in Minneapolis; "loose surveillance" was requested. On March 28, another coded teletype to the St. Paul office said: "Technical surveillance installation completed at 8:00 A.M. this date on subject's telephone, suite 122E, Curtis Hotel."

* The NCAS was closely associated with the Federation of American Scientists (also suspect), of which Robert Wilson was president.

When Frank arrived at the university, the chairman of the physics department told him that federal agents had already paid a visit. They wanted to know if Frank had arrived yet. "I don't even know if he's left," the chairman had responded. "Oh, yes, he's left," the agents had replied.

Even Frank began to see the increasingly ominous signs. When a Professor Woxen in Stockholm invited Frank to spend six weeks in Sweden to give a series of twelve lectures at the Royal Institute of Technology, the State Department refused to renew his passport. Frank received a letter stating, "It is the considered opinion of the Department that a passport should not be issued to you at this time."

What Frank didn't know was that his passport application had prompted a flurry of memos between Hoover and various agents who tried to ascertain just how big a threat it would be to allow Frank to travel out of the country. Basically, the FBI was covering its butt. As one memo candidly put it: "It is felt that in the event Frank Oppenheimer did go to Sweden and was then reported by CIG to have gone to Soviet Russia . . . this bureau would be subject to considerable criticism."

The worst blow came after Frank received an invitation to return to Berkeley to collaborate with former colleagues. When Frank left Berkeley for Minnesota, Lawrence had put his arm around him and said, "Come back any time you want to." But overnight, it seemed, everything changed. Frank remembered seeing a wire from Lawrence that read, "Frank Oppenheimer is no longer welcome in this laboratory."

Frank was crushed. Lawrence was an old friend. What caused him to turn on Frank so abruptly? Frank thought it might have been connected with widely spread rumors that he had taught at the California Labor School in Berkeley, "a pretty red hot outfit," Frank admitted. In truth, Frank hadn't taught there. He had talked with school officials, but didn't agree with their views on the bomb (they didn't seem at all concerned that the Russians were building one) and so declined the offer to teach. But he was listed in the catalogue, so "all the FBI and everybody else assumed I had taught there."

Frank also found out that Luis Alvarez, a physicist who worked closely with Lawrence, had been sitting on a train next to somebody who identified himself as a "security person"; the man told Alvarez that Frank had been a spy. While "McCarthyism" itself was still several years in the future, such widespread suspicion was a precursor to the persecution that would accelerate in the 1950s.

Whatever the reasons for Lawrence's turnabout, he never responded to the letter Frank wrote him (reproduced at the beginning of this chapter). Frank never had the chance to see Lawrence before he died, in 1958, which left him feeling sad and frustrated. "I really was fond of Ernest," he told me, "and he was fond of me."

A Terrible Mistake

"It was a dumb thing to do. I was angry . . . I was a little scared. I didn't know [how] saying I had been a Communist would affect my brother. I just should have said 'No comment.'"

Frank never mentioned his previous membership in the Communist Party when he accepted the position at Minnesota. A few months into his new job, he drove to meet Jackie and the kids at a lake where the family was spending the summer with some friends. As soon as he arrived, his friends told him there had been a call from Washington. "I sort of anticipated something good happening," Frank told me. "I don't know why."

It wasn't to be. When he returned the call around midnight, a reporter from the *Washington Times-Herald* read him part of a story that was to run the next day, accusing him of having been a Communist, of having gone to Mexico to meet Communists, of having given away atomic secrets, of "all these terrible things," Frank said. Asked to confirm the report, Frank said in a fury that it was all lies. Of course, it was not all lies. He *had* been a member of the Communist Party.

On July 12, 1947, the *Times-Herald* ran a front-page story by a James Walter under the headline "U.S. Atom Scientist's Brother Exposed as Communist Who Worked on A-Bomb." It stated, among other things: "This newspaper today can reveal that Dr. Frank Oppenheimer, brother of the American scientist who directed development of the atomic bomb at Los Alamos, was a card-carrying member of the Communist Party who worked on the Manhattan Project and was aware of the many secrets of the bomb from the start." It was obvious that the reporter had had access to Frank's FBI file, since he revealed (among other privileged information) Frank's party membership numbers.

The following day, Frank talked to his dean, who told Frank he'd have to make a public statement denying the charges. So Frank wrote a rather vague statement saying he supported social justice "and stuff like that." But after continued pressure from the university administration, he found a lawyer and asked him to draw up a statement. The statement contained a bald-faced lie: "I am not now and I never have been a member of the communist party." It was a mistake that put an end to Frank's career.

Why did he lie? To some extent, the lie must have been the fruit of youthful naiveté, a failure to accurately sense the tenor of the times, and a combination of willfully blind optimism and poor judgment. What he later told me was "I don't know. It was a dumb thing to do. I was angry, and I didn't know what they thought they might have on me. I was a little scared. I didn't know [how] saying I had been a Communist would affect my brother. I just should have said 'No comment.'" Frank was only beginning to grasp how fear of the atomic bomb and fear of communism reinforced each other, creating a volatile mix that put people like himself in a very dangerous situation. His lie didn't help.

An article some years later in the *Minneapolis Morning Tribune* suggested Frank might have lied to protect Robert, and in fact Frank told one of his colleagues at the university that he feared the *Times-Herald* piece may have been part of a campaign to attack his brother. Bob Wilson also concluded that Frank probably had lied to protect Robert, who was at the time playing such a pivotal role in framing the Acheson-Lilienthal proposal—the only hope, Frank had thought, of avoiding a nuclear arms race.

Certainly, the lie was influenced by the hysteria gripping the country. In the past, Frank had always been honest about his party membership; now things were different. Communism was no longer just another political party but a large and looming threat. Scientists in particular—celebrated as heroes right after the war—were becoming increasingly suspect. Physicists knew bomb secrets, and they were strange characters to boot. As the MIT physicist and historian David Kaiser concluded, "American scientists bore the brunt of the loyalty-security investigations of the Cold War era."

The fact that Frank's brother Robert enjoyed such power in Washington made Frank a tempting target; the fact that he'd been a Communist made him an easy one. So if he was afraid, he had reason to be. The ground beneath him was shifting so fast that what had seemed solid only a few years before was suddenly and frighteningly uncertain. Frank confided to one Minnesota colleague that he feared a campaign of intimidation was afoot, aimed at stopping the efforts of atomic scientists "to make people understand what is in store for them in an atomic war." An article in the *Tribune* around that time referred to "what seems to be a studied campaign to discredit the most highly-placed civilian officials involved in atomic energy."

About a week after the damning *Times-Herald* story appeared, a reporter called Frank to ask whether any policemen had been bothering him. When Frank said no, the reporter responded, "Well, we heard you'd been arrested." Frank locked the doors of his house, turned out the lights, and went to bed. "It was a fairly scary sort of situation," he told me. "I didn't know all kinds of things."

To the FBI's credit, the agency tried hard to find out who had leaked Frank's files to the *Times-Herald* reporter. The bureau also rebuffed aggressive efforts by the University of Minnesota's vice president of academic administration, Malcolm Willey, to conduct an independent investigation of Frank's political entanglements. First Willey got in touch with FBI agents in St. Paul, trying to get information about Frank; turned down there, he wrote directly to Hoover, pleading for "advice in this matter that you are in a position to give us." Hoover also turned him down, explaining that "it has long been the practice of this Bureau to hold its files confidential."

Willey's colleagues were disconcerted by his activities, to say the least, and strongly protested his actions. Dean T. R. McConnell wrote Willey with evident sarcasm, "I am not sure we should ask Mr. Hoover for 'advice.'"

Undeterred, Willey suggested contacting the *Times-Herald* to dig up the paper's sources. He wrote to several of Frank's colleagues, including his supervisor at Los Alamos, Kenneth Bainbridge, who was offended by the letter. "The questions you ask me are strange indeed," Bainbridge wrote Willey. "There is absolutely no question in my mind of his loyalty."

Like the FBI, Willey seemed to be interested primarily in public relations. In a letter addressed "Dear Lew" — most likely the university president, James Lewis Morrill — Willey (who described Frank as "a very strange person"; this was a common theme) wrote that he didn't "want to find myself subject to outside queries with nothing from Oppenheimer in our hands." In another letter five days later, he explained to Morrill that even if the university's attempts to get information from the *Times-Herald* and the FBI were turned down, "at least you are in a position to tell the Regents what you did . . . . It would be complete protection against the charge that the University had done nothing."

There were no such "charges" against the university, of course. Willey himself admitted that the press "have dropped the matter." But he wanted to be prepared. He even told Hoover he needed the information about Frank because "we are in the position of being subject to *incipient* criticism" (emphasis added).

Ultimately, it appears that Willey's behind-the-scenes crusade* forced Frank's hand. It was Willey who had insisted that Frank's initial statement wasn't good enough. And Willey had gone much further: he argued that Frank should sign what some faculty members thought smelled of a loyalty oath: "It would be helpful administratively," Willey said, "if Dr. Oppenheimer indicated forthrightly his support of the American proposal for the handling of the atomic bomb." But Willey's actions, like Frank's, can also be seen as a window into the poisoned atmosphere that pervaded universities in the postwar period. Everyone, it seemed, was running scared; blameless people were afraid of getting caught in the crosshairs and willing to do almost anything to protect themselves "just in case."

This early McCarthyism, according to the historian Ellen Schrecker, "produced one of the most severe episodes of political repression the United States ever experienced." And as she points out, it was a very American style of repression, "nonviolent and consensual."

Most of Frank's colleagues at Minnesota never questioned his honesty. A dozen of his associates, including the chairman of the physics department, stated that "our confidence in the personal integrity of Dr. Oppenheimer is so great that we do not question his denial."

Not surprisingly, this unquestioning support only caused Frank further regret about his lie, but he was still too frightened and confused to do anything about it. For once, the idealized world he'd made up in his mind out of pure conviction was no match for the real world that lay outside his powers and probably outside his understanding as well. Until this time, he'd lived a relatively sheltered, even charmed, life. He'd gotten away with a good amount of mischief and firmly believed in the basic goodness of people.

* It is ironic that among Willey's "most notable contributions to the University of Minnesota," noted in the university archives, was his authorship of the faculty tenure code in 1938, which stated that faculty "had the freedom to write and speak about any issue outside the University."

But the unstoppable optimism that would later serve him so well was now his own worst enemy.

The irony of the lie was never lost on Frank. "Just before that incident where I lied to everybody, I had written this great paper about how lying was the worst thing anybody could do, that withholding information was just as bad as falsifying it," he told me. "I got myself into a real pickle."

Happily, the whole thing blew over, and Frank went back to his research and teaching. Even the FBI seemed to have lost interest—though not entirely. One memo reports "it is known that he is engaged in highly secret experiments involving nuclear energy" (untrue); the bureau continued to monitor his travels and phone calls and make use of "reliable confidential sources" in the university to look for "possible violations of the Atomic energy act." But in general, the agency had apparently decided that Frank didn't appear to be misbehaving. "Contacts close to Oppenheimer have advised that the subject has thrown himself wholeheartedly into his new work," an agent in the St. Paul office wrote to Hoover, explaining why he was submitting no report at that time.

Resigned

"At many times in many places, the irrational fear of subversion has gripped a people. In 1949 in the United States what came to be known as McCarthyism counted Frank Oppenheimer among its victims. As so often happens, the victim of the hysterical fear is the least to be feared."

On June 3, 1949, Frank and Jackie received a summons to appear before the House Un-American Activities Committee. Knowing that he'd have to tell the truth about his past, Frank stopped in to see J. W. Buchta, the head of the Minnesota physics department, and admitted he'd lied. Yes, he *had* been a member of the Communist Party for several years, but had quit before the war. He gave Buchta a letter of resignation just in case, stating that his continued employment might "confuse and endanger the strong and fine stand [the university] has persistently taken on all matters relating to academic freedom." But it was understood that the letter was merely a courtesy; Frank was assured that it wouldn't be acted upon.

An FBI informant must have been close at hand during the conversation: a memo from the St. Paul office to Hoover on June 14, 1949, said that "informant at University of Minnesota advised this date that subject a week ago voluntarily appeared before his dean and stated that a letter of denial which Oppenheimer had written was untrue."

Once Frank testified at the House committee hearings, the word about his Communist past, and about his lie, spread quickly. While he was in Washington, Frank received a phone call from the University of Minnesota: it had accepted his resignation. He returned to the university a few days later and tried to talk President Morrill into letting him keep his position, but to no avail. Buchta later said he felt they had no choice but to let Frank go. Frank left the office in tears.

Colleagues quickly rallied in support, as did students, staff, Minnesota alumni, and physicists around the country. "Nothing could possibly be more asinine than your acceptance of Dr. Oppenheimer's resignation for rejecting an untruth and accepting a truth," one telegram read.

A former student wrote that the incident made him "and many like me lower our heads with a feeling of shame." Another said it indicated a "lack of courage on the part of the university." A third said Frank's lie would be "the instinctive protective reaction of most of us in his place . . . His family, job, everything that he had achieved was in jeopardy."

More than fifty physicists — including Hans Bethe — at the Cosmic Ray Conference in Idaho Springs, Colorado, asked Morrill to reconsider accepting Frank's resignation. Edward Teller added a letter of his own: "[Frank's] discovery of the heavy particles among the primary rays has changed our outlook about the nature and origin of Cosmic Rays," he wrote. "I never agreed with Frank Oppenheimer on politics. I always liked him. I strongly believe in . . . the freedom to make mistakes."

Everyone from staff secretaries to former classmates from the Fieldston School joined the protest. Students were the most vocal of all; most wrote personal letters praising Frank's "unusual ability for helping people think independently and critically," in the words of one. Twenty graduate students sent a petition. One of the most articulate letters came from a graduate student who pointed out that Frank was "not alone" in his guilt. "It is a guilt which everyone as individuals or collectively as institutions must share," she wrote. "We are responsible for the temper of the times and unfortunately it has been such that one cannot help feel afraid of any mistake made in the past."

Naturally, some applauded the university's action. "Congratulations Oppenheimer matter," one telegram crowed. Another Morrill supporter wrote to say he thought the university handled the case well, and that while Frank was "an excellent scientist," he was also "queer as a witch." A third said Frank's lie was clear proof that "a Communist's word of honor isn't worth a thing."

But mostly Morrill found himself facing a storm of criticism, which he responded to by denying that Frank was either "fired" or "dismissed" (he had resigned) and arguing that lying was the issue, not academic freedom. It didn't do much good. For months to come, the faculty, staff, and students of the Minnesota physics department continued to plead with Morrill to find some way to bring Frank back.

As a state university, Minnesota was particularly sensitive to public opinion, so any hope that Frank might ultimately get away with the lie was no doubt unrealistic. But the university was also somewhat disingenuous about the reasons for his firing. While the public stance was that the firing was entirely a response to Frank's lie, in fact it was as much about "his continuing association with leftists," according to Schrecker — in particular, his attendance, along with Paul Robeson and several Harvard professors, at a rally in support of Henry Wallace's bid for the presidency.

Whatever the real reason for Frank's departure, it was certainly "scary" for a young student like Clark Johnson. "That was one of the things that drove me into joining the ACLU," he told me.

The son of a woman Frank played the flute with, Ed Emerson, was also greatly upset. Even though he was only in the eighth grade then, he remembered Frank playing the Bach B-minor sonata with his mother – and the dark FBI sedan parked outside the house. "I think Frank made a joke about it." After finding out that Frank had gotten fired, Emerson said, "I couldn't believe anything bad about him. He was the smartest, gentlest man I had ever met."

Un-American

"This individual, this citizen, has been sacrificed to the Congressional headline hunger as cheerfully as though he were a rabbit . . . His life has been needlessly, pointlessly, broken."

An article about Frank and Jackie by Joseph Alsop in the *Minneapolis Star* captures something of how the two were perceived as they arrived in Washington for the House Un-American Activities Committee hearing on June 14, 1949.

Alsop described Frank as a bohemian with "excessively ostentatious sideburns," "as silly about politics as he is clever about physics," the kind of person who believes that "the good will triumph . . . that men will grow virtuous because one wants them to." Frank, the columnist concluded, was precisely the kind of fellow "Congressmen like the least. This thin, dark youngish man, with his nervous, intense manner, obviously suffers from idealism almost in the way that so many of the Congressmen's favorite lobbyists suffer from gas on the stomach."

On Frank's behalf, Robert asked the eminent New Dealer and political activist lawyer Clifford Durr to represent his brother before HUAC – and even Durr remarked at the time that Frank and

Jackie impressed him as "very decent but very disturbed young people." Durr was one of the few lawyers who would take on those accused of Communist affiliations — even fewer would take on those who actually had been party members. Durr's clients could rarely pay him, and he himself was hounded relentlessly by the FBI. Years later, he represented Rosa Parks when she refused to give up her seat to a white passenger on a Montgomery, Alabama, bus, thereby playing a pivotal role in setting off the civil rights movement.

"Daddy loved Frank, and it went both ways," Durr's daughter Virginia (Tilla) told me many years later. The two men had much in common: both were gentle, and stood up for what they thought was right; both were always searching and questioning; both believed that all people should have the power of speech, and that civilization depended on keeping the dialogue active and open.

Relations between Frank and Robert were beginning to fall apart again, however. After Robert helped Frank secure Durr's help, he removed himself from Frank's case almost entirely. "He was not being terribly supportive of Frank," Tilla Durr said. "He was worried about what would happen to him. Jackie was absolutely furious, and that was causing a lot of pain in that family . . . I don't think anybody came away without terrible scars, because it went to the heart of the family, who was supporting who and how."

Tilla's mother, Virginia Durr — herself a prominent political activist and the sister-in-law of Supreme Court justice Hugo Black — sided with Jackie, while Clifford Durr tried hard to understand Robert's position. "My mother felt Robert was getting off the hook," Tilla said, "but Daddy was aware of how much Frank loved his brother."

Tilla was only seven or eight at the time, but the memories remained vivid when we spoke almost fifty years later, because Frank was so kind to her and also because it was the first time she'd ever seen someone in so much obvious distress. She remembered Frank pacing up and down in their garden with an intense, worried look on his face; she remembered her father's concern that Frank was thinking of suicide.* "I gather that the support of his brother . . . was less than what could and should have been at the time and this caused much family trauma that seemed to go on forever." Frank was also Tilla's introduction to what "the whole Communist hysteria was doing to actual individuals and their families."

After consulting with Durr, Frank decided not to plead the Fifth Amendment, as so many others had done, but rather to speak plainly and openly about himself — but not say anything about anyone else. There was no legal basis for this plan, and Durr warned that it could be risky: he and Jackie could be cited for contempt. Once a witness talked about himself, he automatically waived his protection against self-incrimination. And while Frank was never cited, many witnesses who came after him and took his position were.

The hearing itself ("Hearings Regarding Communist Infiltration of Radiation Laboratory and Atomic Bomb Project at the University of California, Berkeley") began at 11:30 A.M. in room 226 of the Old House Office Building. Frank was accompanied by Durr. Richard Nixon was one of the committee members present. Jackie was not allowed in the room, even though it was purportedly an open hearing (later, Frank was allowed to remain in the room during Jackie's testimony).

Frank was, of course, asked about his membership in the Communist Party — not that the committee members learned anything they didn't already know. Unlike Frank himself, they knew the identification numbers the party had issued Frank: #56385 in 1937, #60493 in 1938, and #1001 in 1939.

One after another, the committee members reeled off the names of people Frank knew or might have known and asked what he knew about their political affiliations. "I do not wish to talk about the political ideas or affiliations of any of my friends," Frank re-

* It seems unlikely that Frank considered suicide, although later there was talk among his rancher neighbors in Colorado that he had had a nervous breakdown.

sponded. The more they badgered him, the more he tried to explain himself. "I believe they have to be questioned on those themselves . . . In those cases where there might be any indication of illegal action or inimical action, I would certainly report those to you. I know of no such cases." It was not enough in these hearings to admit one's own sins; naming names was required, and those with "derogatory information" in their files had the responsibility to clear themselves.

There were implied threats as well—suggestions that Frank could get in a lot of trouble if he didn't answer their questions. ("I think the witness should be warned of the penalty of his refusal to answer without sufficient grounds.") But he continued to refuse. "The people whom I have known throughout my life have been decent-thinking and well-meaning people," Frank told the committee. "I know of no instance where they have thought, discussed, or said anything which was inimical to the purposes of the Constitution or the laws of the United States."

In Frank's support, a letter from Major General Leslie Groves, dated September 28, 1945, was read into the record. It stated that Frank's work on uranium isotope separation during the development of the atomic bomb was "an essential factor in our success. Your further work in Tennessee and in New Mexico were also of great value. Your skill and judgment in the field of science are beyond praise . . . In behalf of the War Department as the agent of our fellow-Americans, I wish to express to you grateful thanks for your indispensable part in our success."

All the while, Jackie was sitting in an outside office, stewing as she looked over the reading material provided to the public: "100 Ways to Tell a Communist in Education"; "100 Ways to Tell a Communist in Church"; "100 Ways to Tell a Communist in Labor." She later reflected:

> The House committee is in a wonderful building with marble Walls. It is surrounded by beautiful parks, carefully kept grounds and is quite lovely. But as I sat there in the office I looked out the window

across the street and saw the rows of tumble houses. The little kids, most of them negroes, were running around in the street barefoot. They all looked rachitic and most seemed undernourished.

All they had to play with was junk they found in the street. As I sat there reading and listening and looking out the window, I found myself alternately worrying about what the Committee was going to try to do to me and getting madder and madder at the fact that I had been called down here so that some fellow could question ME about being Un American.

When the committee was finished with Frank, Jackie took her turn, and the ritual was repeated. Did she know Frank Malina? Paul Crouch? Sylvia Crouch? Kenneth May? On and on. "I wouldn't like to say," she responded. "I just don't like to testify on such things." Some questions were so clearly designed to elicit traitorous testimony that Durr moved to object. For example: "Do you believe that the Un-American Activities Committee is a greater menace to our free institutions than the Communists are?" Jackie answered anyway: "I think [such committees] are a good idea. I think this is the way you find out what kinds of laws should be made."

At 4:45, the two were dismissed and the committee went into executive session.

"Frank wanted to do the right thing," a colleague later said. "He really wanted to be honest, and he believed that if he was honest and talked about himself, but refused to talk about anyone else, that would be okay; that he would be vindicated; that right would win out over insanity. And he was wrong, and he lost everything."

Frank was worried about far more than just saving his own professional skin. He feared that the HUAC hearings had stifled conversation and therefore thoughtfulness at a time when new ideas and new ways of thinking about things were crucially needed. No one was going to come up with desperately needed social inventions if they couldn't talk to each other and toss ideas around openly.

Remarkably, Frank continued to teach and guide his graduate students throughout this period. In fact, exams were scheduled for the day he was to appear before HUAC. "All of the pain he must have been going through in the months preceding his appearance was never shared with me and his other students," Phylis Freier said. "The things he taught me were many—but probably the most impressive was—that a love of doing science—of learning and teaching about it—can survive throughout adversity."

Blackballed

"You have to cooperate with us."

Despite being fired from Minnesota, Frank didn't give up on physics, and he tried hard to find a job. Over and over again, he'd hear that this or that physics department wanted him, only to get a letter a month or so later saying that it was wasn't going to work out. Frank was pretty sure that he was being blackballed, but he didn't connect it with the FBI until an agent told him flat out, "'Well, don't you want an academic job? You have to cooperate.' . . . Then I realized what the wall was."

Unable to work in the United States, Frank turned abroad. In December 1949, he received an offer from Dr. H. J. Bhabha, director of the Tata Institute for Fundamental Research in Bombay, to continue his work on investigating cosmic rays in India.

Frank admired Bhabha, and was inclined to accept. But once again, Frank was unable to get a passport. This time he protested directly to Secretary of State Dean Acheson, asking for a hearing on whatever charges might be behind the refusal. "Over six years of continued investigation and constant surveillance by the Federal security authorities cannot have revealed any transgression of acts of disloyalty as I have committed none," Frank wrote Acheson, explaining that since he was unable to get a university appointment in the United States, working abroad was the only chance he had to continue his career. "If it is actually believed that my going to India might place our National policy in jeopardy, then I would like very much to have a hearing on this question."

For unknown reasons, these arguments were entirely mischaracterized in a memo to J. Edgar Hoover from the chief of the Passport Division, a Mrs. R. B. Shipley. Shipley wrote that Frank's letter "in substance was a plea from Oppenheimer in which he asked forgiveness of his wrong doings as being politically wrong in the past but that he had mended his ways."

Despite Frank's lack of involvement in weapons research (and, for the most part, politics), his banishment from physics continued through the late 1950s. A December 16, 1958, letter from Oak Ridge National Laboratory stated that getting clearance for a visit "might now present a few difficulties." A letter from Wolfgang Panofsky in January 1959 said, "Dear Frank: I am sorry to report that the job we discussed . . . has evaporated."

The blacklisting of academics who refused to cooperate with HUAC was pervasive. Had there been any meaningful opposition from the universities, Schrecker writes, "people like . . . Frank Oppenheimer could have found teaching jobs in the U.S. They didn't."

Toward the end of the Oppenheimers' stay in Minnesota, Frank and Jackie had a goodbye party at their house. At one point, Frank found himself standing outside on the sidewalk talking to a former colleague, Charles Critchfield. Critchfield started attacking him, saying that he'd been a Communist "and everybody knew it, that it really wasn't a secret." And then: "You're still a Communist!"

Frank's rage and frustration boiled over. "I punched him in the belly," Frank told me. "Really hard."

5

EXILE

"I think that if you've been beat up by your nation for your political beliefs and you've built an atomic bomb and seen it used on 200,000 civilians, and after a career of cosmic ray research you find yourself shoveling horseshit up in the mountains of Colorado — it sounds corny, but it really is a pretty good way to figure out what's important in life — to sort out the real values from the artificial values."

In the summer of 1948, Frank shared a house in Nambe, New Mexico, with his friends Bob Wilson and Phil Morrison and their families. At one point, Wilson and Frank took a six-day horseback ride together and chanced upon a two-room cabin in a lovely valley known as Blanco Basin, about twenty miles southeast of Pagosa Springs, Colorado. It turned out to be for sale, and so Frank bought it as a summer place (his FBI file notes that he bought the cabin on September 8, for $24,000). The property had 830 acres, much more than he and Jackie wanted, but she thought it might be nice to have a place to live "someday."

Someday came within a year. Having no place else to go, Frank and Jackie decided to weather the rest of the McCarthy storm there.

They left Minnesota at the end of June 1949 and soon were rejoicing in their new surroundings. "We have been in the Basin less than two months and have not in the least become used to its beauty," Frank wrote to Wilson. "Every time I walk out of the house or go to the can or look up from a post hole the sight of the meadow and the mountains seems to tingle and refresh the deepest parts of the nervous system."

Frank's emotions were a jumble. He took enormous pleasure in his new life on the ranch. But at the same time, he wanted to get back to physics, and felt he should be out in the world doing what he could to prevent the nuclear arms race he so clearly saw coming.

"Oh Bob," he wrote Wilson, "I should feel even worse than I do for shirking responsibilities about myself and the world and letting other people sweat it out for me." He found Blanco Basin so beautiful, he wrote, "that leaving seems almost like dying." And yet he also feared he might "go nuts without a job and without physics. But even if I didn't, I don't think it would be right to stay. Somehow I should be in there pitching and working."

Frank was also still smarting at the treatment he'd received from Minnesota: he wrote Wilson about punching Charles Critchfield in the stomach with obvious pride.

Some months later, when Frank returned to Minnesota to collect the family belongings, he was so depressed that he couldn't bear to talk with his former colleagues about their balloon flights and experiments because "it hurt so not to be doing it with them."

Of course, he was pleased to learn about the flood of letters and petitions of support. But he was also angry. The situation was all the more frustrating because it was clear that his former department chairman was trying his best to help, and felt genuine regret that he couldn't do more. But there seemed to be no hope of getting Frank's job back.

"I am sorry because, despite the nincompoops in the department, I would like to have continued flying a cloud chamber and associated gear," he wrote Wilson. "The fact that the faculty who I have been most closely associated with simply sat down and take

. . . my being kicked out burns me . . . At the moment it seems that all organizations that men create are either impotent or monsters."

As usual, Frank saw his own situation as merely a symptom of a much broader and more serious problem: universities were becoming "tongue-tied." The Minnesota physics department seemed to want to have him around, he wrote to Wilson, but didn't know how to explain their thinking to others.

That universities, the "cream of the literate world," should be unwilling or unable to "make anything they believe clear in simple terms" was deeply upsetting to Frank. After allowing that he himself had "acted badly" at times because he didn't want to explain things, he wrote: "It is a much worse error for a university. I think if one does not try to explain what one believes in and still pretends to be an intellectual, then soon one ceases to believe in anything."

Job offers continued to flow in, but they either quickly evaporated or turned out to be temporary, and in any event, Frank was already beginning to settle in as a rancher. He wrote to a colleague at the University of Chicago that September would be a good time to begin what seemed to be a promising appointment because "I have made some commitments regarding crops and cattle that should have my attention during the summer." (It wasn't a problem, because the appointment never materialized.) To another colleague he admitted he didn't want to uproot his family for a temporary position and that "we have made something of a go of this ranch during the past season."

The Wilson letters reveal Frank's internal conflict most clearly. "I have wanted for so many years to own and fix up and feel that I lived on a real place that I am chronically and completely delighted — so it will make me sad to leave," he writes. ". . . Nevertheless, I am becoming reconciled to leaving the aspen and the snow and the spring if I can really get back into the world the way I want to be in it."

Robert's rather dismissive attitude only added to Frank's general angst and confusion. Robert had always been a bit condescending toward his brother, and now he made it clear that he thought the idea of becoming a rancher a little silly, as well as beneath him. He wanted Frank to return to academia, although there's little evidence that he helped his brother as he had in the past. And he let Frank know that he didn't think he could make it as a rancher in any event. All this hurt Frank a great deal. For one thing, no matter how much he missed physics, there didn't seem to be anything he could do to reenter that world—at least not for now. But in addition, Frank did take ranching seriously. "I really felt like a rancher," he insisted. "[I] *was* a rancher."

It was a time when the brothers were particularly estranged. By now the director of the prestigious Institute for Advanced Study at Princeton, Robert was enjoying his star status, and Frank felt he could no longer reach him. "I saw my bro in Chicago," Frank wrote to Wilson. "We did not have too much time to talk, not nearly enough to convince each other of anything. I fear that I merely amused him slightly when, in brotherly love, I told him that I was still confident that some day he would do something that I was proud of . . . I cannot live down the conviction that the appropriate way for us contemporaries to achieve immortality is by establishing pre-conditions for a warless world. It seems to me that both by his actions and his words he has deserted this project which was once so major a preoccupation of his . . . Of course, so have I, damn it."

Aliens

"What hay?"

Thus did the gentle Jewish intellectual from Manhattan wind up settling in the land of pioneers and homesteaders in a primitive cabin where the nearest neighbor was a mile away. The town of Pagosa Springs had a population of about 1,500; the entire county only about 3,000. Just five other families lived in the entire Basin.

For the locals, it was as if aliens had landed. "The normal folks were wearing tight jeans and cowboy hats, and here was a rancher who didn't wear a hat," said Pete Richards, who lived on one of the neighboring ranches at the time. "He was skinnier than a rail, he was really hyper. Both he and Jackie swore like sailors. And they were atheists!" Frank didn't wear a tie for haying, but he came close: "He wore the kind of shirt you could wear with a tie," his daughter, Judy, told me.

The cabin sat at eight thousand feet on land that comprised steep hillsides covered with pine, fir, aspen, and scrub oak, interspersed with "little grassy parks." "It was a magical place," Pete Richards said. "One of the most beautiful valleys anywhere in the world. It was like the Swiss Alps." The Oppenheimer house had wood shingles, two small rooms, and no electricity. There was no foundation, and it rested on blocks. The washtub and scrub board where Jackie did the laundry still sit outside.

The first summer on the ranch was an adventure—a relaxed time for the family with lots of reading and long rides. Friends came to visit: the Wilsons and the Morrisons and the Durrs and Linus Pauling.

But as the job offers evaporated, Frank decided to make a real commitment to raising cattle. He shipped the family possessions to Denver, then loaded up a trailer and headed for the Basin. As he went over a rather precarious section of Wolf Creek Pass, the trailer started fishtailing. Frank was afraid it might pull him over the edge. Ever resourceful, he took off his ever-present necktie and used it to secure the trailer.

In many ways, Frank was well suited for his new occupation. He was good with his hands and a master at coming up with ingenious solutions on the fly. The only problem was that neither Frank nor Jackie knew the first thing about ranching. One day a few months after they arrived, they were enjoying the waves the wind made in the tall meadow grasses when a neighbor asked, "What are you going to do about your hay?" "What hay?" Frank responded. Frank thought hay was something special you had to plant, "not just any old grass."

Realizing the extent of their ignorance, Frank and Jackie got busy learning how to irrigate their land and fix fences. They built a new section onto the house, hauling the lumber and the gravel for cement themselves. A neighbor helped Frank buy some cattle, "because I knew nothing about a cow," he said.

They bought an additional 640 acres and were soon raising a couple of hundred head of cattle – which they continued to do for nearly ten years. Some of the bills were paid by selling off one of Frank's van Goghs, and Frank's letters to Bob Wilson suggest the income wasn't enough to comfortably support them. "Our bank account has been chronically sick so fixing the place up has been slow," he writes. "The $1200 we spent on lawyers and travel to the Un Am Committee would come in mighty handy."

(Frank's relationship with his family's wealth was always something of a muddle. During their years on the ranch, Michael would ask his parents for some particular thing he wanted, only to be told that they couldn't afford it. "Then we'd go into Albuquerque and Frank would pay cash for a car," he said. "I grew up very confused about money." No doubt, Frank often felt poorer than he was. While his income was certainly well below expenditures, he did have some valuable works of art – including a Picasso that was valued at roughly a million dollars in the 1980s. Yet it must have felt awful to part with paintings that he'd grown up with and referred to as old friends.)

Relying on neighbors and a great deal of reading, Frank and Jackie taught themselves how to raise cattle. Frank studied books on animal husbandry. He described to me how he learned to deliver newborns in the snow – pulling on the legs of the emerging calf while Jackie read instructions out loud from a veterinary manual. Sometimes he tied a rope around the calf's feet and used a block and tackle to pull it out.

At calving time, Jackie would sit by the window, watching with binoculars from inside the cabin as the cows outside prepared to give birth. When the time came, they'd both run out, not only to help along the delivery, but also to keep the newborns from freezing. Sometimes the calves would have to be propped up "so that they can get a good drink of milk and survive the cold spring," Frank wrote to his friend Phil Morrison. Frank was proud that he and Jackie were among the first ranchers to freeze some of the mother cow's vital colostrum after a calf was born, and to use antibiotics on baby calves "with the scours."

"We became good obstetricians," he later reflected. His long arms and fingers allowed him to "get in there and rotate the calf," according to his daughter Judy, who went on to become a pediatrician.

Life on the ranch was both difficult and rewarding, and starkly different from anything Frank had experienced before. As he told Morrison, "I find it hard to write because there are many unsatisfactory parts to this life and the very satisfactory parts are still so new to me that I feel they must be almost unintelligible to anyone else."

Frank and Jackie worked seven days a week, most of the time with only one hired hand for help. They had to replace the rotting posts along twenty miles of fence, build dams of rock and burlap and bales of hay. There were gates to make, plumbing to install, roofs to shingle, bogs to drain, irrigation ditches to clean. Judy and Mike — nine and six when they first arrived — both chipped in with haying and caring for the animals.

"In order to raise cattle," Frank later explained, "you also have to doctor them, and you have to raise hay, and you have to irrigate the hay, and you have to fight the river, and you gotta ride the forest to find the cattle that are out on the national forest, and you've got to fix fences, and you've got to drain the meadows, and reseed some of the fields . . . You've got to undo the beaver dams that dam up the drain ditches, and you've got to trap the muskrats that are eating holes in your dams. So it's just one continual job of trying to undo the property of nature to increase entropy and disorder."

Frank shared much of his frustration in long letters to Bob Wilson. "I wish you were around so that we could organize the place together," he wrote. The relentless hard work reminded him of Ernest Rutherford's dictum that if you hadn't done enough during the day to think about at night, then you'd wasted your day. "Almost all physicists work night and day," he said, "but I think [Rutherford] was right, it's worth taking the time to reflect on what you're doing and that's probably why they invented the Sabbath a long time ago." Whether the directive came from Rutherford or some god, Frank thought that people seemed to need someone or something to make sure they took time off from "doing" to think and reflect. But it just didn't happen.

The best times were the summers, when old friends came to visit; Frank got caught up on physics, and Mike and Judy had other children to play with. But summers were short. Sometimes the snow lasted from October through April — six feet deep in places. Sometimes they had to use a team of horses and a sled just to feed the cattle.

The long winters "were very empty of people," Frank said. Left largely alone with his rage and disappointment — over both his personal situation and that of the world at large — Frank blew up frequently and indiscriminately. One day he got furious because he couldn't find a match to light a cigarette, Pete Richards remembered. "So he swore and stuck 'em [the matches] in his pocket and then he went into one of his pace routines. He was pacing back and forth in the kitchen, bitchin' about something, and then all of a sudden the matches went off. And he says, 'Fuck!' They weren't safety matches, and so there they were all in there kind of grinding away in his pocket."

His temper could be aimed at almost anything, including his children and even livestock: he once set a fire under a stubborn mule on a pack trip to get it up and going. On another occasion, Frank was having trouble getting a cow into a pen, and Michael and Pete Richards heard a cascade of cursing from a hundred yards away. "You could hear Frank yelling 'Fuck!' — and that 'Fuck!' would go across the valley, hit a mountain, come back, and reverberate, and you'd be getting these multiple fucks every two or three minutes," Pete told me. "And so we weren't looking forward to when he finally came back to the house because by then we knew he would be in a really nasty mood."

The premium Frank now put on truth, no matter what the consequences (or the context), made for a painful kind of honesty in the household. White lies were neither respected nor permitted. It was a kind of "rigid honesty" that allowed no exceptions for sparing people's feelings. Michael remembers his father coming into his room and saying, "Don't perjure yourself."

But Frank's often literal directness also had a sweet side. Many years later, just after Jackie died, Frank was talking with Michael on the phone and, as was his habit, Michael ended with the rote sendoff "See ya later." And Frank replied, "Oh, really? Great. When?"

Despite the loneliness, hardships, and his often uncontrolled anger, Frank remembered the ranch years primarily as "a wonderful time." The life was much harder, in many ways, on Jackie. She was angry at what she considered betrayal by friends, emotionally undone by the spying, and not at all convinced that Frank was going to be able to make a living as a rancher. Raising small children alone while tending a cattle ranch wasn't easy, and neither was putting up with Frank, although he did appreciate her enormously and often said so. "[She] really worked like hell," Frank said. "She would go out every day with [the hired hand] and feed the cattle. It was lonesome and hard for her."

"I had two very depressed parents," Judy explained to me. "What saved Mike and me was that my father had the good sense to marry my mother."

Jackie was tough, solid, straightforward, funny, and intolerant of pretense. She was far more available, physically and emotionally, to the children than Frank, and she knew how to connect with them. Pete Richards remembered how much Jackie had respected the children, and how he'd enjoyed his visits to their place: "I loved Mike and Judy, and they had great toys and did interesting things."

Frank loved Jackie too, though he cruelly hurt her by having love affairs — something I resisted believing for a long time, in part because it didn't fit the Frank I knew and wasn't something I learned about until just before he died. This reflects, to some extent, my own naiveté. To me, Frank was a kind of Yoda. He wasn't an obvious womanizer, and he respected women in ways I'd rarely seen in men of his age and position. Certainly, he was self-indulgent. Perhaps he felt that the rules didn't apply to him—a lifelong theme that mostly served him well. Perhaps he simply didn't wish to see how much his actions hurt Jackie. But even in his infidelity there was an incongruous innocence. A dear friend of Frank's from Los Alamos days, the philosopher David Hawkins, told me that during a blowup with Jackie over a girlfriend, Frank pleaded with her, "Why can't I have *two* nice girls."

"Frank had his own kind of wickedness . . . with women. But we all forgave him," Hawkins said. Jackie, unmoved, threw Frank's suitcases down the stairs after learning about one of these escapades. (Before Jackie died, she told me one thing she really regretted was that she never had an affair herself.)

I caught up with Michael Oppenheimer at his home on Lummi Island, near Bellingham, Washington, in the fall of 2004. Mike has Frank's eyes as well as his warm and gentle manner, and also his father's stubborn sense of honesty and serious respect for play. Mike is a wonderful artist: his expansive hay-covered lawn sprouts sculptures that seem to grow out of the landscape as if someone had seeded the place with magic beans. There were fields of brightly colored dichroic flags, long sinuous arms that waved like welcoming neighbors; other pieces sang and chattered and hummed while weaving about in the wind. Appropriately, the name for his budding enterprise is Windy Hill Art. His childhood years on the ranch were still very much with him.

"There are certain events in your life that seem to crystallize something very important," Mike told me. One of those was the day Frank and Mike dissected a pig's head. Earlier in the day, Frank had butchered a pig, and that evening he brought the head into the house. He asked Mike if he'd like to dissect it with him. "We put newspaper down on the kitchen floor, and we got knives and saws and all sorts of things to start taking apart this pig's head," Mike remembered. "We took the eyes apart and we looked through the lens. We took the brain out and looked at the brain. And it was the most glorious time. It was messy and gooey, but it didn't bother me at all; it just seemed very natural. And we spent until midnight or so, just me and my dad, on the floor in the kitchen with this head of a pig.

"It only takes a few of those things to take you through a lifetime. As a family, we hardly spent time together. But that one evening of being together was worth many years."

Frank and Mike similarly dissected a big John Deere tractor together (it needed a new gearbox, but they decided to take the whole thing apart anyway). "I don't know if Frank had ever taken apart a pig's head before," Mike said, "but I know he'd never taken apart a tractor. It was all discovery for him, too. And I can remember everything about it — the smell of the oil and the feel of it."

The two of them made a carbon microphone, "and damned if it didn't work wonderfully on the family hi-fi amplifier," Mike said. Frank taught Mike enough about electronics to wire Judy's dollhouse with lights when he was only six years old.

Mike also remembered that his father was always asking him interesting questions. Once it was Would you rather be the person who invents a machine to talk to extraterrestrial beings? Or would you rather be the person who does the talking? "I thought it was a great question," Mike said, "though I don't know what I answered."

Both Mike and Judy went to a one-room schoolhouse with a big potbellied stove in the corner and a half-dozen students; one year, they were the only students. Frank and Jackie helped build the "teacherage," as Frank called it — a two-room cabin behind the school. He helped to "corral" a twenty-three-year-old teacher. The kids rode to school on a horse named Old Snorty.

Tilla Durr accompanied Judy and Mike to school during a period when her parents were in Colorado. It was a difficult time for Clifford Durr. Unable to support himself defending clients who mostly couldn't pay, he got a job with the National Farmers Union in Denver. But he soon lost the job because his wife, Virginia, signed a peace petition, along with Linus Pauling, protesting rumored plans by the United States to bomb China during the Korean War. The union wanted her to say she'd been duped into signing by Pauling; she didn't, and her husband was fired.

Frank taught Tilla to ride, and after school he'd take the kids horseback riding on trails up into the mountains; they'd pack a picnic lunch and ride to beautiful places, explore waterfalls and woods and find spectacular views. He was always giving the children hands-on science lessons and made nature come alive for Tilla, explaining "everything that was going on in the physical world." At night, Tilla would drift off to sleep to the sound of Frank's flute.

Hard-Boiled

"The people in Archuleta county really told the FBI to go to hell."

Mike and Judy's childhood was unusual in other ways: FBI agents visited them regularly. "I remember that pretty vividly," Mike said. "I remember them driving up in their car to the ranch and I never understood my parents' disdain, because they seemed nice to me. They gave me money, I guessed because we were dressed in rags. They thought we were dirt poor. I remember getting fifty cents from one of them. I was thrilled!"

Almost as soon as Frank arrived on the ranch, he received another subpoena to appear before HUAC, this time on September 5, 1949. He wrote a long letter (which he most likely never sent) informing the committee that nothing had changed, and that he didn't want to be uncooperative, but that any attempt on his part to testify about others would do them an injustice without furnishing the committee the information it was seeking. "It appeared to me virtually impossible for one person to give a reliable account of the political activities and motives of any other person," he wrote.

The second hearing never happened, but each time the FBI paid a visit, they asked about the very same people HUAC had grilled Frank about in Washington. And each time, they told Frank that if he wanted a job in physics, he'd better cooperate.

The family got to know the FBI agents quite well. Usually they'd call and ask about the road conditions to the ranch; it was a hard dirt road, and if it had recently rained, they wouldn't come. Once when Jackie saw them approaching, she said to Frank, "I'll be damned if I'll feed them." But they stayed and stayed, and Frank got very uncomfortable. He saw several hard-boiled eggs sitting on the counter and weakened, asking them if they'd like some. They said no, but after that, whenever Mike saw the FBI coming up the road, he'd say, "Here come the hard-boiled eggs!"

The agents also pestered the neighbors, asking whether Frank was doing experiments or sending radio signals to Communists in Mexico. They'd flash their FBI badges and ask if the neighbors were absolutely sure that nothing underhanded was going on. Had they been in *all* the buildings on the Oppenheimer ranch? Every chicken coop and toolshed and pump house?

Some nearby ranchers simply told the FBI that Frank was a good neighbor, that he had gotten them a schoolteacher and fixed up the school. One said it was "a sad commentary that Oppenheimer was a brilliant man who could be doing valuable work for the government." Several noted that he was "eccentric" or "odd in his manner."

Frank eventually concluded that the FBI wasn't really trying to investigate him; the agents already had answers to most of the questions they were asking. Rather, the bureau was trying to poison the atmosphere in which he lived. It was trying to punish him for his left-wing politics by making his friends and neighbors and colleagues suspicious of him. "They try to do by stealth and suggestion what they cannot do openly," Frank wrote in a 1949 essay he titled "The Tail That Wags the Dog." "Because of what they believe to be my politics, they try to make me an outcast."

The FBI didn't get much cooperation from Frank's neighbors,

however. The people in Blanco Basin were fiercely independent and friendly to exiles. At least one draft dodger holed up there during World War II, and the place had its share of eccentrics and alcoholics. The motto of the *Pagosa Springs Sun* was "Independent in everything; neutral in nothing." So when the FBI tried to suggest that a mad scientist had moved in next door and was doing dangerous experiments and sending secret radio messages to Communist allies abroad, the locals just shrugged. "The McCarthy era really didn't touch the people there in that sense," Frank said.

And in truth Frank *was* a very good neighbor. He became president of the Blanco Basin Telephone Company, chairman of the Archuleta County Soil Conservation Board, and was an elected delegate of the Archuleta County Cattlemen's Association who appeared before the state legislature's agricultural subcommittee.

As the cattlemen's representative, he described his job as going around the county finding out "what people's troubles were." A big problem for the ranchers was that raising cattle required expensive equipment that went unused for most of the year. Frank suggested that the equipment be placed in a common pool that everyone could draw from when needed. One of the ranchers objected, "That's just what they do in Russia!" Frank said to himself, "My God, I can't get away from it."

(This same rancher apparently saw fit to share his views with the FBI. One report from the Denver field office relates a conversation with a neighbor who was of the opinion that "since the arrival of Oppenheimer, 'there has been a Communist atmosphere in this area.'" Oppenheimer, the report says, had apparently been trying to get the neighbor involved in community activities, and lectured the neighbor "concerning the necessity for all farmers to stick together." This neighbor, the FBI report states, then "threatened to push [Frank's] face in unless he stopped." When Frank later told me what I assume was this same story, his said the neighbor had threatened to "strangle my skinny neck.")

Frank also got involved in issues such as access to public lands, suggesting ways that more land could be made available for recreation without intruding on the rights of ranchers — "so that the rural/urban fight doesn't have to continue," he wrote. As a city kid, he appreciated the importance of recreational facilities, but he also understood the problems faced by ranchers. He believed conflict was unnecessary, and existed largely because national forests under Republican administrations were no longer seen as a public service; rather, they were run as a business. The government was heavily involved in timber and mining and revenue-producing activities, but it wasn't building trails and access roads or generally protecting its resources. If the government took care of the public lands, the problem of people intruding on private lands would disappear, he argued. "The government really, rarely does enough to think of the needs of people . . . It was much more interested in making money off the national forest than making it a place that people could get to."

A Nice Little School

"He might have come from Mars."

The ranchers liked Frank. He was hardworking, good company, interested in music and the arts as well as physics. He volunteered at the 4-H Club, teaching the kids how to make electric motors.

So when there was an unexpected vacancy at Pagosa Springs High School, Frank seemed a natural to fill it. At first he couldn't get hired for what he suspected were political reasons. His suspicions were confirmed when he lost a bid for a seat on the school board, and one of the ranchers told him, "It's a fine thing when they have to have a Communist on the school board!" When Frank saw the rancher later, he told him he wasn't a Communist. The rancher said, "Oh, I'm sorry, I thought you were."

Then Frank couldn't get teaching credentials because he lacked the requisite college courses in education. Stanley Fowler, a former student of Frank's, recalled that the community was up in arms over the issue; they couldn't believe that Colorado wouldn't give Frank Oppenheimer a license to teach.

Eventually, he was allowed to teach with temporary credentials, and he began taking correspondence courses in education to try to fulfill the requirement. The research paper he wrote to get his permanent credentials is classic Frank. In the introduction, he compared the development of mathematical tools to the development of tools such as a carpenter's level, which helps a builder know whether the foundation of the house he's constructing is flat (he said it was similar to the development of calculus as a tool for understanding the motions of planets). But sometimes, Frank noted, mathematics is more like play. "I am certain that mathematicians must frequently run into some object that they want to play with or investigate much as one is always tempted to play with magnets or gyroscopes or Silly Putty," he wrote. He went on to examine a problem involving the base of natural logarithms, *e*—demonstrating how one can discover the properties of a number simply by fooling around with it mathematically.

Properly credentialed in 1957, Frank Oppenheimer began to teach general science, biology, chemistry, and physics to just over a hundred students; he taught the first physics course the school ever had. The community of 850 people worked mainly on ranches, raising cows and cutting trees for lumber. Kids went to school only because their parents made them. "The only thing they knew was basketball and football," said Fowler, who today is a professor of pathology and an associate dean at the University of South Carolina School of Medicine. "We could never get a science teacher. Then in walked this super high-class physicist who was working on the atomic bomb." It was, Fowler said, as if Frank had dropped in from Mars.

He must have been a strange sight. Other ranchers wore overalls. Frank wore a tie, which he used to clean off the blackboard. Covered head to toe in chalk, he'd twitch and pace in front of the class, then suddenly disappear into the storage room for a smoke.

Yet Frank understood that his students neither knew nor cared much about science, so he started coming up with ingenious experiments to grab their interest. (Twenty years later, Fowler stopped into the Exploratorium and was amazed to see some of the same setups. "I thought: Oh, I know that one!" he recalled.)

In fact, it seems Frank tried anything and everything to reach his students, including killing and dissecting kittens in hopes of showing them the cochlea of the inner ear (which he never found). He felt so guilty that when he discovered that one of the kittens had escaped, he made a pet out of it. He took the kids to the junkyard at the edge of town with a box of tools and spent the afternoon with them taking things apart. "We were girls and guys and we were taking apart the carburetors and distributors, and everybody was totally into it," Pete Richards remembered.

A friend of Frank's later recalled seeing him at a blackboard explaining basic math concepts to a very young girl. "It was the kind of thing you ordinarily don't get until college algebra," the friend said. "The way he told it to this girl was really delightful."

Pagosa Springs was a small, isolated community with an inferiority complex, and before Frank appeared, school sports were the town's primary source of pride. He argued that this was a pernicious—even malignant—state of affairs. Few students actually participated in sports, and the wins and losses seemed to be mainly about the egos of the spectators, he said. The system took the fun out of sports, making them competitive and vicarious. "Can't we start playing ball just for the fun and gaiety of it?" he wrote in a long letter to school officials.

With Frank's arrival, things began to change. His students started winning science fairs in Durango, and even at the state level. The parents of a student who won at the Denver science fair told Frank, "Now we don't need a football team!"

A physicist at the University of Colorado, Albert Bartlett, was a judge in the combined Colorado-Wyoming science fair during those years. At one awards banquet, Bartlett told me, most of the prizes went to the kids from Pagosa Springs—many of them bused in by Frank, many of them Hispanic, many of "ordinary origins." As the evening progressed, the faces of the kids from the Denver-area high schools got longer and longer. "You could see the jaws drop," Bartlett said. "Pagosa Springs? Who?"

As Frank's students graduated and went on to college, word of their ability spread. A Caltech physicist, Hal Zirin, then at the University of Colorado, remembers that a stream of stellar students suddenly started arriving at the university from Pagosa Springs.

And then there were students like Stanley Fowler, who went on to work for Nobel Prize winners but never met anyone who could make biology come alive like Frank Oppenheimer. "His grasp of biology made me soar," Fowler said. "He made every chapter fascinating. It was as good as any course I had in college . . . To this day, because of him, I read articles about physics."

When he was in Frank's class, Fowler made a seminal discovery when he noticed "some green stuff" growing around the rocks of the Pagosa hot springs, and he wondered what it was. Along with Frank, he studied the literature; they couldn't find any mention of algae that could grow in such hot water. Fowler's father took him to pawn shops in Denver, where they found a secondhand microscope, and Fowler tried to identify and draw the algae. "It was a real discovery. No one knew there was life!" he recalled. He took his project to the Colorado state fair. "I learned how you go about research." By the time Fowler was ready for graduate school, he got a full stipend from Rockefeller University, something he squarely attributed to Frank. "I turned down Harvard and everything!"

In the summer of 1958, Frank moved the family to Boulder, in part so that he could scout out teaching opportunities that might lead to a position at the University of Colorado. He took part in developing and teaching the National Science Foundation's new science curriculum, known as the Physical Science Study Committee Physics Course (PSSC), part of the country's post-*Sputnik* response to the fear that the United States would fall behind the Soviets in science and technology. Frank made "inspiring contributions" to the program, according to one colleague.

He also became an instructor in the Summer Institute for High School Physics Teachers, also in Boulder. David Hawkins, who was at the University of Colorado at the time, said that the high school physics teachers in this region "were completely turned around by meeting Frank and seeing the experiments that he made, because he had a way of making them come alive in the hands of students."

During the same period, Frank had started teaching special physics classes to students from all over Jefferson County, selected on the basis of an exam for aptitude in science. The superintendent of schools hired him despite severe criticism for bringing in a "Communist" to teach young people. But his students were devoted.

One of these was James Heckman, now an economist at the University of Chicago and a Nobel laureate. He still has a collection of books Frank had read, music he'd listened to. What is astonishing about this is that Heckman last saw Frank when he, Heckman, was a seventeen-year-old high school student.

"We knew he'd been a great physicist, that he'd worked on the Manhattan Project. That he was raising cattle now. We knew his assumed name, Frank Folsom. We knew all about him," Heckman told me over dinner in Chicago as if it had all happened yesterday. The students could scarcely believe that Frank Oppenheimer was dedicating himself to getting a bunch of rural high school kids fired up about science.

Early every weekday morning, they met with Frank for two and a half hours. He introduced them to his friends, including George Gamow, one of the first physicists to realize that our expanding universe must at one time have been compressed into an infinitely dense point in space-time before exploding in what we now call the Big Bang. Frank put on music they'd never heard before — the Beethoven 131, Heckman remembered — a very late string quartet. "He played it over and over again. This guy was from another planet.

"We devoured the physics and we devoured the man. I've thought a lot about him over the years, because he was the first glimmer that I knew that any of this existed. I remember him as kind of stooped, slight, almost fragile, always hunched over a lab table. I've never known anyone quite as gentle, as understanding.

He was so encouraging, you were never intimidated. And he had such a presence! There was an intensity I'd never seen before. He was so curious! . . .

"We all felt we were experiencing something we'd never experienced before. It kind of took over our lives. Everything in science we could get our hands on we would read, because of him. Any book that he carried I would immediately go out and buy."

Forty-five years later, Heckman told me, "I still treasure my lab book. He gave me an A-plus."

In the fall of 1958, Frank moved the family back to the ranch so that Mike could complete his sophomore year at Pagosa Springs High School, and by this time, Frank was teaching teachers as well as students. He worked for the county, doing in-service training, meeting with a dozen or so teachers in the evenings throughout the year.

As a teacher, especially, Frank revealed just how different he was from his famously brilliant older brother. Robert needed to lecture at a big university, Frances Hawkins (David's wife) told me. But Frank was happy to set up experiments for a handful of high school teachers. "That's the quality that his brother didn't have," she said.

Frank developed much of his educational philosophy during these years — the philosophy that became the foundation for the Exploratorium. For example, he noticed that at first the teachers came to him in hopes of acquiring "jewels" that they could take back to their students. "[They] acted as though they were just transmitters from me — through them — to their students, without any involvement in the material themselves. And it took almost a year before the kinds of questions they asked showed that they were genuinely interested in the subject and became more interested in learning for themselves than just as robots for their students," Frank said.

He realized that the teachers themselves had to be excited about the material and engaged in discovery or they'd never be able to inspire, or even adequately teach, their students.

Frank persuaded the school district to pay teachers for summer work so they could develop their own curricula. And that made all the difference. "Telling teachers how to use materials that were prepared by somebody else does not make good teachers," Frank said. "That's one reason the nationwide effort to improve science teaching after *Sputnik* did not accomplish as much as it might have."

Most of all, Frank learned firsthand what extraordinarily hard work teaching was. He was in the prime of health. He'd been ranching and bucking heavy bales of hay and tending cattle for nearly ten years. "And yet that job wore me out," he said. "I never have had to work so hard, so intensively, as to get ready for those high school kids." And Frank's 130 students didn't come close to the load of an average California public school teacher. "I learned what I think is wrong with schooling," he said. It was impossible under such circumstances to pay attention to each of the students, to grade their papers, to find out what their individual needs were.

"I never felt even reasonably on top of the job," he said. And despite all the evidence to the contrary, he wrote Bob Wilson that he feared "only a fraction of my students are learning anything."

Frank concluded that the only way to fix education was to double the number of teachers. He acknowledged the expense, but argued that education must grow faster than the gross national product because much of the growth of GNP is due to automation, and education can't be automated. It gets relatively more, not less, expensive over time.

Robert

"I think to a certain extent it actually almost killed him."

All the while Frank was mastering cattle ranching and then re-entering the world of teaching (if not research), his brother Robert was being systematically destroyed. It wasn't something Frank liked to talk about.

In fact, when Jon Else first approached Frank about doing an interview for his film about Robert, *The Day After Trinity,* he was pretty sure that Frank wouldn't consent to be part of it. "He was really resistant and grumpy and difficult," Else said to me. Frank did eventually agree, and the interview, Else said, was "one of the most moving, breathtaking interviews I've ever sat through." Else developed such a respect for Frank that he showed him a rough cut of the film. It was the only time he'd ever shown a person in one of his documentaries the film before it was finished. "You never do that," he told me.

What brought Robert down was something of a perfect storm of converging factors: his own power and celebrity, the fear engendered by the Russian atomic bomb, the rise of Senator Joseph McCarthy, the revenge of enemies Robert had made by his own arrogance and naiveté, and, not least, his own rather pointless lies. It is a story well told in countless books, as well as Else's extraordinary film, so here I give only a brief summary.

After World War II, Robert Oppenheimer (or Oppie, as he was called by everyone who knew him) was America's darling. His porkpie hat was almost as famous as Mickey Mouse's ears. As the director of the Institute for Advanced Study at Princeton, he was Einstein's boss. He served on dozens of advisory committees and was on a first-name basis with the secretary of state. He was also adamantly and eloquently opposed to a nuclear arms race.

In August 1949, an American bomber detected evidence, in the form of radioactive fallout, that the Russians had exploded a Hiroshima-scale atomic bomb. Edward Teller, whom Robert had already alienated by passing him over for the job of head of the Theory Division at Los Alamos (he chose Hans Bethe instead), was hell-bent on a crash program to make a vastly more deadly hydrogen bomb. The Russian success gave Teller just the ammunition he needed. He didn't take kindly to Robert's response to his pleas to help get the ball rolling — which was essentially to "keep his shirt on."

If the destructive power of a Hiroshima-sized bomb is hard to fathom, a hydrogen bomb — which is a thousand times more powerful — is truly unimaginable. Instead of liberating energy by splitting large atoms (uranium, plutonium), a hydrogen bomb fuses small atoms, the same process that fuels the sun. The first American test of an H-bomb, in 1952, turned millions of gallons of water into steam and vaporized an island in the Pacific, leaving nothing but a hole in the ocean a mile in diameter.

Robert chaired the advisory committee of the newly created Atomic Energy Commission, which took up the matter of whether to recommend a crash program to build an H-bomb, or "Super." The committee met and voted unanimously against such a program, for technical reasons but also partly for moral ones. "The extreme dangers to mankind inherent in this proposal wholly outweigh any military advantage that could come from this weapon," the committee concluded. It stressed that this was a weapon of unprecedented and unlimited destructive power, and as such was a "weapon of genocide" and a "threat to the future of the human race which is intolerable."

Teller and many other H-bomb enthusiasts were apoplectic.

It was worse than that. Not only did Robert and his cohorts (though none had the power to persuade and the charisma to charm that Robert did) oppose the H-bomb effort, they also promoted international control of atomic energy — positions extremely unpopular with the right wing.

President Truman went ahead and ordered the fast-track program to develop the Super despite his advisers' firm opposition. But Robert was still out there, vocal and persuasive as ever.

At the same time, the McCarthy red scare was in full force. In his biography of Robert, Jeremy Bernstein wrote that "if one did not live through the McCarthy period of the 1950s it is difficult to convey what it was like." Countless lives were ruined (including those of some of the country's best scientists) based on nothing more than unsubstantiated charges, innuendo, and rumor.

Robert's leftist associations were well known before World War II (including the fact that his brother, and his wife, had been Communists). The FBI had been tapping his phones, opening his mail, and following him around for almost a dozen years. By the early 1950s, there was a four-and-a-half-foot stack of evidence detailing his activities.

So when an executive at the Westinghouse Corporation* decided in 1953 to write FBI director J. Edgar Hoover accusing Robert of being a Soviet agent, he had essentially no new information. Robert had been given a clean bill of political health in 1947. But times had changed — and now the letter was enough to start a stampede.

In December 1953, Robert was notified that his security clearance had been suspended. The letter he received read in part: "As result of additional investigation as to your character, associations, and loyalty . . . there has developed considerable question whether your continued employment on Atomic Energy Commission work will endanger the common defense and security."

The father of the atomic bomb was now considered a security risk. Robert requested a hearing, which was held in March 1954.

The hearings went on for four weeks. There were forty witnesses, and three thousand pages of testimony. But no reporters were allowed, and even Robert and his lawyers were barred from seeing much of the evidence. Since Robert's work for the Atomic Energy Commission was about to end anyway, it was pretty clear that the whole exercise was meant to humiliate him and strip him of power — and in that it succeeded brilliantly.

Robert wasn't accused of doing anything illegal. He was accused of being a sponsor of the Friends of the Chinese People, of having his name on the letterhead of the American Committee for Democracy and Intellectual Freedom, of being a member of the Consumers Union — all designated Communist front organizations by HUAC. He'd subscribed to the *Daily People's World* in the early 1940s. He'd given money to organizations supporting the Loyalists in the Spanish Civil War. Most important, he'd been insufficiently enthusiastic about the H-bomb.

* This was William Borden, who had recently left his position as executive director of the Joint Atomic Energy Committee of Congress and had long been suspicious of Robert's opposition to the H-bomb.

Robert didn't help himself either. For one thing, during the war years he told a series of complicated and ever-changing lies concerning the identity of a person who had passed along information about a third party who wanted to know if he, Robert, would be open to sharing atomic secrets with the Soviets. Robert reported the approach to the authorities, but seemed to change the details every time he told the story—at times implicating innocent friends. Robert himself didn't seem to know why. At the hearing, the best explanation he could offer was that he was "an idiot." "I wish I could explain to you better why I falsified and fabricated." Some people concluded that somehow Robert was trying to protect Frank.

Robert didn't even really try to defend himself, which puzzled many people. "I don't think he was especially a skillful politician," Frank said. "I don't think he tried to be. He trusted his ability to talk to people and convince them. He was pretty good at that . . . But he was up against people that weren't used to being convinced by conversation, but other kinds of more coercive forces."

His most poignant testimony involved Frank. Asked if his brother had ever been a Communist, Robert answered: "Mr. Chairman . . . I ask you not to press these questions about my brother. If they are important to you, you can ask him. I will answer, if asked, but I beg of you not to ask me these questions."

The final blow came from Edward Teller. Asked outright if he believed Robert was a security risk, he gave a roundabout but nevertheless decisive answer: "In a great number of cases I have seen Dr. Oppenheimer act . . . in a way which for me was exceedingly hard to understand. I thoroughly disagreed with him in numerous issues, and his actions frankly appeared to me confused and complicated. To this extent, I feel that I would like to see the vital interests of this country in hands which I understand better and therefore trust more."

On June 28, two days before the expiration of Robert's contract with the AEC, the commissioners released their findings: Robert Oppenheimer was "not entitled to the continued confidence of the government" due to "fundamental defects in his 'character.'"

At that moment, Robert Oppenheimer was effectively obliterated. As Robert Coughlan wrote in *Life* magazine: "After the security hearings of 1954, the public character ceased to exist . . . He had been one of the most famous men in the world, one of the most admired, quoted, photographed, consulted, glorified, well-nigh deified . . . Then suddenly, all the glory was gone and he was gone too."

And because Robert did not put up clear walls between his public and private self, "Oppie" was destroyed as well.

When Frank did talk about his brother, he often expressed sadness that Robert had allowed himself to be so easily seduced by power. Robert did this, Frank thought, not in order to promote himself, but rather to help steer the country away from a disastrous nuclear arms race. In doing so, however, he bought into all that comes with power and position. And in the end he failed to accomplish what was truly important to him — stopping the nuclear arms race before it began. As Freeman Dyson later put it, "He wanted to be on good terms with the Washington generals and to be a savior of humanity at the same time." This may have been, by definition, an impossible task.

Frank seemed to feel that his brother had let power and celebrity get the best of him, and that, in the end, destroyed him. Robert, Frank once said, had a "youthful cockiness, and some of it stayed with my brother a little longer than it should have." Years later, Frank told a friend who'd just enjoyed a major public success to "beware the applause factor" — the danger of letting yourself be seduced by the thrill of public acclaim.

As Dennis Flanagan, the editor of *Scientific American,* wrote to me after Frank's death, "Robert was always the fancy Dan, eaten alive by ambition . . . He betrayed his friends to the police. Frank could never have done that; that's why he had to spend 10 years in limbo . . . Frank wound up far the happier and more fulfilled of the two."

Robert's attraction to the high and mighty was something Frank said he never entirely understood. "He felt really injured by not being part of, by not being respected in, government and official circles," Frank said. "He wanted to get back into that. I don't know why, but I think it's one of these things where once you get the taste of it, it's hard to not want it."

Watching from afar and immersed in his own troubles, Frank felt helpless, sad, paralyzed, and a little distant. Beneath their differences, they always had enormous affection for each other. At the end of a long interview for the documentary *The Day After Trinity,* Jon Else asked Frank if there was anything he wanted to add to what he'd already said about his brother. Frank answered, "I just want to say that I loved him very much."

Of course, Frank and Robert were well aware that they were far luckier than most of McCarthy's victims. Not everyone could sell a van Gogh and go live on a ranch. Some people nearly starved.

The broader tragedy, for Frank as well as for Robert, was that Robert never had the kind of impact that might have made a difference. He'd hoped his influence in high circles could help prevent incipient nuclear insanity. He'd hoped that the introduction of a weapon that could not be used for war would bring about fundamental changes. But it didn't. The H-bomb, like the A-bomb, came to be viewed as just another weapon.

"What undid him," Frank said, "was not just his fall from official grace, but the fact that this fall represented a defeat for the kind of civilized behavior that he had hoped nations would adopt."

And it wasn't only Robert who failed. Grassroots efforts by Frank and others also seemed to have little effect in preventing an all-out nuclear arms race. "We were all defeated equally in our attempt to create a less dangerous world," Frank said.

By the end of the 1950s, each brother was exiled in his own way. Robert's ordeal all but drained him of life. Frank, on the other hand, somehow emerged with renewed confidence and conviction — tougher, wiser, and beginning to see how he might, in fact, create a world in which he could make a difference.

6

AN INTELLECTUAL DESERT — AND A LIBRARY OF EXPERIMENTS

"I found it an intellectual desert because there'd been fifteen years in which people had been scared to talk to each other."

Frank thrived on his teaching, but he also had an ulterior motive for these activities; he hoped this work would gain him entry into the University of Colorado, where many of them were based. More than anything, he wanted to get back into physics — designing experiments and tinkering with equipment and maybe even making discoveries. He wrote asking Bob Wilson to fill him in on "what new insights you have of the beauty and order of nature."

Frank's strategy worked, and he was offered a position as a research associate in 1959.

Accepting it, however, would mean moving to Boulder permanently, leaving behind the ranch — and his students. "There are twenty or so students that I hate to abandon to whatever monster of a science teacher they get," Frank wrote Wilson. On the other hand, he wanted Mike to go to a better school. He'd already "rented my grass," he said, and was seriously thinking of selling the ranch, though "up to now I haven't been able to get myself to list it as for sale."

Complicating Frank's choices further, Wilson had finally arranged a firm offer for him to work at Cornell. It was, alas, once again a temporary appointment. Still, Frank may well have taken it, if only to work again with his best friend, had not Cornell's president insisted that he sign what Frank thought amounted to a loyalty oath, a statement saying that he wasn't a member of the Communist Party.

Frank felt he could do a lot of "nice physics" in Wilson's lab, and would love to work in the same university, live in the same town. But the idea of signing a statement greatly upset him. "I'd hate to be responsible for starting the precedent of an oath at Cornell," he wrote his friend. In another letter, he confessed to Wilson that every time he sat down to craft a reply to the president's offer, "it turns out slightly nasty."

Wilson himself was furious. He wrote to President Deane W. Malott of Cornell that while Frank had made a "tremendous mistake" in lying about his past, he was an honest and principled man: "Even Klaus Fuchs, a real criminal, is out of jail. I will be ashamed of my nation and, in particular, of my school and myself if his condition of enforced unemployment continues because no place has the courage to make the first step at getting this gifted man back into the profession."

But the truth was, Frank had grown accustomed to his "comfortable isolation" in Colorado, and people had gotten used to him. The idea of a steady, presumably permanent income from the University of Colorado was hard to refuse. It made Frank "dream that I can start buying people presents once more," he told Wilson. Turning down a chance to work side by side with his old friend was obviously painful as well. "I would like to keep on writing to you," he said at the end of a long, rambling letter, "but it is so late that all that comes to mind are various unduly complex ways of expressing my affection."

Finally, Frank did gradually close up the ranch and move the family to Boulder, a university town. By 1961, he'd sold those parts of the ranch that needed constant attention.

Ever the alien, Frank seemed nearly as out of place in his new academic setting as he had been in Blanco Basin; newly acquired eccentricities only enhanced the old ones. "He had this great big silver belt buckle," physicist Al Bartlett remembered, "with a streak up through it. He would never use paper matches to light his cigarettes; he'd use the big wooden matches, and he'd strike them on his belt buckle. He just wore a groove in that silver belt buckle."

At the same time, Frank managed to be so quiet and unassuming that few people connected him with the famous brother who had recently fallen so publicly from grace. "He seemed an ordinary, friendly guy," one colleague remembered.

Nevertheless, Frank's Communist past continued to tail him, as did the FBI. When he went up for tenure, virulently anti-Communist members of the Board of Regents tried to block the appointment. During the meeting where Frank's fate was decided, Bartlett and others sat outside the room, waiting with their fingers crossed; they knew the outcome was anything but certain.

It helped Frank's cause a great deal that the chairman of the University of Colorado physics department, Wesley Brittin, went to bat for him. He got letters of support from esteemed physicists all over the country: Hans Bethe wrote a glowing appraisal; George Gamow used his considerable influence; Victor Weisskopf, another physics eminence, praised Frank as "a driving element and a source of new initiative in research." He went on to mention Frank's "unusual" qualities as a teacher, concluding that any university would be "very fortunate" to have Frank as a member of the permanent staff.

The most interesting letter came from J. W. Buchta – the physicist who had brought Frank to the University of Minnesota a dozen years earlier and who'd had a hand in firing him. "I believed that he was one of the outstanding young physicists among the group at California, or elsewhere, at that time," Buchta wrote. "His exclusion from the field of physics was a loss to the science and I have never been happy over the events that took place here . . . I think the department that acquires his services will find that he is a dedicated scientist and teacher and that he is loyal to the United States."

In the end, Frank was hired as an associate professor in 1961, and became a full professor in 1964.

As soon as he could, Frank plunged back into research — mostly on the vast array of new elementary particles being discovered, especially those involving very rare (and therefore especially mysterious) events. Since the University of Colorado didn't have its own high-energy accelerator, Frank put together a "users' group" that studied the tracks of particles produced at the Lawrence Berkeley Laboratory, where Frank had worked on "racetracks" before and during the war. The physics he was doing attracted close to a million dollars in grants from the National Science Foundation and the Atomic Energy Commission.

But Frank found physics sadly changed, and the work he'd once loved so dearly no longer held much excitement for him. Part of the problem was that physics had undergone a radical transformation since the war. When Frank went into physics, it was a haven for eccentric, adventurous intellectuals who mixed art and philosophy (as well as politics) with their science and generally looked down on material gain as irrelevant, if not somehow immoral. Now physics was suddenly a high-status occupation. A new generation of physicists began to look at it as just another job; they moved from bustling cities to placid suburbs and became increasingly conformist. In a nationwide survey of college graduates in 1961, future physicists chose "making a lot of money" as their prime motivator for going into the field — in fact, according to the survey, money was more important to them than it was to graduates in any other field.

"Physicists who had come of age during the earlier period wondered what had happened to their beloved pursuit," wrote David Kaiser of MIT, who had studied the issue in depth. "Gone was all talk of a lifelong pursuit of knowledge with little care, or even disdain, for material rewards."

The lively and open discussions Frank had so enjoyed with his community of colleagues — long explorations of the relationships among physics, art, philosophy, and politics; the passionate search for truth — seemed sadly absent from the brew. "Gone was all talk of 'philosophy' for these physicists," Kaiser wrote. "The famous puzzles and paradoxes that had so exercised the interwar generation of European physicists, including the architects of quantum mechanics, received nary a passing glance from the postwar Americans."

In addition, physics had gotten intensely competitive, and it seemed to Frank as if getting credit for a discovery had become more important than knowledge itself. So if someone wasn't the first to discover something — and therefore got the credit — the work was considered a waste of time.

He was both disoriented and distressed by these changes, despairing of the "cutthroat competition," pleading in an unpublished essay for physicists to "do a little soul-searching . . . I don't want to disparage fame and the high living that often comes with it. I just don't think that scientific discovery is the place to try and find it." Knowledge is knowledge no matter who discovers it, he said, and should bring satisfaction to everyone. "The pleasure comes from . . . the understanding of new phenomena and the insights into interconnections."

Academia had also changed by the time Frank returned to it in 1959. Ten years of McCarthy-driven paranoia (and swift reprisal for those who didn't play along) had taught people to keep their heads down and mouths shut. Academic life itself, Frank thought, had become "an intellectual desert" because faculty members had gotten out of the habit of talking openly with each other. Even students seemed affected. Many of the young engineers in Frank's classes didn't appear to want to know about how things worked, he thought. Just as the physicists seemed to have lost their taste for pure discovery, the students seemed to have lost their curiosity. "They just wanted to know how to pass their tests," he observed. "Even many of the students who hoped to do research in physics did not ask questions beyond those needed to understand their textbooks."

So while Frank immensely enjoyed his Colorado students and colleagues, he also felt somewhat lost and certainly bereft. Having successfully reentered the world he so missed during his ten-year exile, he found that world unrecognizable.

Into the Fire

"To escape reality is insanity, and leads to irrational and futile activity and childlike displays of aggressiveness."

Much else had changed during Frank's exile in the Basin. The country was now enmeshed in the Vietnam War, and as soon as Frank had come down from the mountains, he started speaking out. "As long as I was around people, I never stopped talking," he told me years later.

He regularly gave speeches, wrote essays (many unpublished), and sent letters to the *New York Times*, the *Washington Post*, and the *New York Herald Tribune*. He wrote to both houses of Congress, urging lawmakers to consider alternatives to war. He thanked Senator William J. Fulbright for his "courageous and wise" stand against the war. "It has given us the heart and the drive to express our views," he said, articulating the well-placed and pervasive fear of speaking out against the government. He repeatedly referred to the "fear-inspired" panic that led the nation to "essentially insane actions in Vietnam and elsewhere."

While the Vietnam War was not a nuclear war, it was still, Frank argued, very much a war "in the shadow of the H-bomb." The U.S. nuclear arsenal was something the country could always fall back on if large numbers of American troops found themselves in grave danger. There was a certain parallel between this situation, Frank wrote, and a kid who gets into trouble, knowing all the while he can send home for money. The Vietnamese, he said, were "fighting for keeps," while as a nuclear power, the United States would always have a sense that it "could find a way out of any really serious jam . . . To me this aspect of the war serves to accentuate its horror and immorality."

Yet most public debates about the Vietnam War ignored the existence of nuclear weapons. People acted as if nothing had changed since 1945 — either because they weren't aware of the threat or because they didn't want to be. "To escape reality is insanity," Frank wrote, "and leads to irrational and futile activity and child-like displays of aggressiveness."

He tried once again in his writings and talks to impress upon people the unheard-of destruction that nuclear weapons could release. The Hiroshima bomb was one thousand times more powerful than existing bombs at the time; the H-bomb was a thousand times more powerful than the one that vaporized much of Hiroshima. "Imagine the attendance at a University multiplied by one thousand, from 12,000 students to twelve million students, without increases in the size of the buildings or the number of professors," he wrote.

Many years later, I asked Frank why he thought that more than twenty years of pointing out these raw facts by dozens of physicists involved in the atomic bomb project seemed not to have made a difference. Why weren't people scared enough to do something about nuclear proliferation?

"We tried making people scared for twenty-five years, and it didn't work," Frank said. "Perhaps the answer was to make them angry instead."

The ideas stirred up during Frank's years on the ranch come through in all these writings. One that pops up frequently, for example, is the notion that coercion inevitably recoils against the coercer. The war in Vietnam, he wrote Representative Roy McVicker, was corrupting the United States. "We cannot let ourselves be callous to Vietnamese lives without being callous to all lives . . . Everything we are now doing in Vietnam has bad effects." In a letter to the *New York Times* he wrote, "When America bombs Vietnam, America changes."

Besides, Frank argued, coercion is useless as a means of effecting real change. Our total reliance on coercion in Vietnam had precluded attempts at persuasion. "Our napalm and our bombs can induce nothing but hate. There is no indication that these bombs have or will induce either submissiveness to our might or admiration for our right." Instead of dropping bombs on Vietnam, he wrote, it would be far more effective to airlift supplies and food. He ended one letter like a scolding parent: "We are a decent nation. Let's try to act like one."

Everywhere he looked, Frank found inspiration for what he called "social inventions," especially those that might serve as models for alternatives to war. While visiting Cambridge and Watertown, Massachusetts, one summer to work on elementary school science curricula, he became entranced with the notion that traffic control systems provided an excellent model for solving conflicts nonviolently. He noted that five lanes of traffic converged on Watertown Square, carrying cars, delivery vans, huge semitrailers, buses, and towering earth-moving trucks — and yet drivers were aware of one another's movements, yielding or slowing down in what he called a "symphony" of give and take.

It occurred to him that the world situation was not so different. "There are so many people with so many high-powered gadgets that there is always someone in the way," he wrote. Yet there were people in this country who "would like to act as though America were the only car on the road"; sometimes these people didn't seem to realize that other nations were also on the move — industrially, politically, and militarily. Such blindness inevitably led to crashes, and didn't get us where we wanted to go.

Traffic proceeds smoothly only because drivers make compromises, Frank noted. They stop at red lights, for example. But political conservatives, he wrote, complained bitterly about the "inevitably annoying traffic lights of government controls." They seemed to want to drive everyone off the road but the Cadillacs — making crashes all but certain. What the world needed was "some social and political analogue of super highways, underpasses and cloverleafs" so that people could get where they want to go safely, without always getting in each other's way.

A Library of Experiments

"If anything was missing from your lab, any piece of equipment, the first place you go look is Frank's lab. Chances are, he'd stolen it. But the chances are that he was making better use of it than [you] were, and nobody could get angry."

At the University of Colorado, Frank turned his attention increasingly to teaching—and most of the ideas he toyed with at this time were later embodied in the philosophy behind his Exploratorium. He proposed an alternative course structure that would allow students to concentrate on one subject at a time, for example, rather than dividing their attention among four or five. His reasoning was that the present system allowed no time for students to aimlessly explore, no chance to stumble upon those serendipitous "fairy slippers" that made real discovery so rewarding. A more focused approach, he said, would encourage students to become "intensely and personally preoccupied" with a subject.

Frank also proposed, in a letter to the university newspaper, that the grade of F be abolished. "If a course has left a negligible record on the mind of a student," he argued, "why is it necessary to provide a record of this fact on a piece of paper?" An F was essentially a punishment—"the epitome of the coercive use of grading," a kind of "brand" that says to anyone who looks at a student's record: "You are not to be trusted and are incapable of behaving the way you are supposed to." The fear of being so branded prevented students from taking courses they were afraid to fail but that might enrich their lives. (It's no wonder Frank was so taken by the fact that "no one flunks a museum.")

He wrote letters and essays defending students who were punished for speaking their minds, and protested the multiplicity of enforced rules and predetermined ways of doing things that were keeping blacks and Hispanics away from the university despite the administration's efforts to lure them. "Why should they trust our rules?" he asked in an essay entitled "The Stacked Deck." Those rules have led to a great deal of "unfairness" and "atrocities" directed at these groups.

Instead, he argued, the university should let Mexican Americans and blacks themselves decide how to use the resources that the university has to offer. There was no reason at all, he thought, that Mexicans and blacks should have to accept the "world we made up" as the only one possible. It was they who should decide "who comes and who graduates, what courses are needed for an education, what should be the class size, what to do about students who feel lost or discouraged. If, when it is all done and running, it looks very different from what we have been doing, it's still OK with us. In fact, some of the students who are here now may decide to do it your way."

Perhaps the academic world *they* made up would have features the university at large might want to borrow. But in any event, playing by established rules is always an insidious trap for outsiders. Not only does it feel alien to live in a world defined by other people's rules, it all but guarantees failure. And if outsiders protest, "Well, we don't like your rules," they are told: "See, we told you you couldn't hack it."

Frank wasn't just developing philosophies during his tenure in Boulder. He was also beginning to create concrete prototypes of the "trees" that would later fill his "woods of natural phenomena." Since the world in which he found himself didn't offer what he needed, he took over a mostly abandoned corner no one else seemed to care about — the introductory physics lab — and threw himself into making it come alive again. In doing so, in a sense, he brought himself back to life as well.

When Frank first arrived at the university, the physics lab, like most college teaching labs at the time, looked like something out of a scientific catalogue from the 1880s. Frank's faculty colleagues typically viewed teaching labs as "the pits"; the demonstrations were hard to put together, and no one got much credit for doing them. For students, labs were pure drudgery. So Frank got to work trying to remedy the situation. Part of the problem, he thought, was that experiments were left out for a week or two at most, then returned to a storeroom until the next year. Students had no choice about which experiments they'd like to do or in what order. Frank thought it would be much better to have six or eight really good experiments that were always available.

He found some space in the attic of the chemistry building and starting working. Colleagues would see him in the physics shop at all hours, weekends and nights. "He just commandeered stuff wherever he needed it," Al Bartlett said, "and he managed to jimmy up, with his own two hands, maybe twenty experiments, covering mechanics, electricity and magnetism, and optics. And these were incredibly ingenious things. He was one of the most ingenious, creative people that you can hope to encounter."

Along with a colleague, Frank applied to the National Science Foundation for funds, and they received enough (as much as $100,000 in total) to make multiple copies of the experiments. Before long, Frank had created more than eighty experiments.

He used the most mundane objects imaginable: wooden carts on roller-skate wheels, springs and pulleys and pendulums, Slinkies, sacks containing iron shot, waxed paper, bicycle wheels, toilet floats, piano wire, Polaroid cameras, crochet needles, aluminum foil, and light bulbs. Each experiment was designed to promote familiarity with physical principles, and especially to illuminate concepts that students found difficult to grasp.

"He liked doing things simply," said Jerry Leigh, who worked with Frank for a short time and then became the de facto caretaker of the lab. "He wanted to build little trinkets and toys to show people they could understand these things."

The measuring apparatus, in contrast, was often quite sophisticated, which was crucial for obtaining useful and accurate results. So Frank incorporated photocell timers, oscilloscopes, and stroboscopic cameras into his otherwise jury-rigged creations.

Leigh remembers Frank as a quiet presence, something of a "mystery man" who'd lean against the wall reading a newspaper and "fuss about" the day's events, dropping cigarette ashes everywhere. "He was extremely modest," Leigh said, and had the kind of devotion to teaching that was less missionary zeal than "daily determination, unfailing labor."

Leigh also remembered Frank's FBI tail, who hung around constantly. One day Frank decided to introduce himself and invite the agent out for coffee. "They became best of friends," Leigh said. (One of the last entries in Frank's FBI file is a memo from the San Francisco office, dated July 7, 1960, stating that Frank had been at Lawrence Berkeley Laboratory between June 15 and 19, doing unclassified work. However, the feds hadn't completely lost interest. The very first purchase order for Else's film *The Day After Trinity* came from the CIA. "The film hadn't even been released yet," Else told me. "I gave the purchase order to Frank. He just laughed.")

Around this time, the University of Colorado physics department obtained funds to build grand new laboratories, and Frank spelled out what he wanted: large open spaces so that students could see the experiments the other students were fiddling with, piquing their curiosity; inexpensive movable tables; electricity available from an overhead trolley duct. The laboratory space would wrap around the lecture halls in a big open horseshoe, making it both convenient and flexible. "We built the entire laboratory the way Frank wanted," Bartlett said.

When it was done, Frank's "Library of Experiments," as he called it, resembled a small museum. It was open from 8:30 A.M. to 9 P.M. on most days, and the demonstrations were "nearly as accessible as are the books in the stacks of a university library."

His "library" of sophisticated science toys operated in typical Frank style – which is to say with a large measure of anarchy. He insisted on not having a lab manual, for instance, because he thought it would be too confining and inhibit free exploration. He liked to work with students one on one, encouraging them to ask questions and always making suggestions: "Why don't you try this?"

"If you took that lab with Frank as your instructor, it would be an incomparable educational experience, even for a professional," Bartlett told me. "He was personally setting an example of how fascinating it was to be engrossed in the excitement of learning physics. His enthusiasm was contagious. He seemed to take so much pleasure getting other people to see the interesting things in what he was doing."

Frank couldn't teach all the sections, of course. And none of the other faculty or graduate students could match his mastery of the experiments. Eventually, he agreed to prepare some 8-millimeter film cassettes of instructions, which were later converted to videotapes. "It was an enormous help to us," Bartlett said.

Almost forty years after Frank left Boulder, I went to see what was left of his "library." I took an elevator to a subbasement of the Duane Physics Building, walked past the vending machine area and through a large set of double doors. The first thing I saw as I entered was a small, sad, faded photocopy of a photograph of Frank that someone had tacked to the wall. The big U-shaped complex was lined with metal equipment cabinets labeled *Experiment 1*, *Experiment 2*, and so on. Metal tables, stools, and lab benches were scattered about, but there were almost no people. I asked Jerry Leigh about the sad little photo of Frank. "Frank has been deleted," he told me.

Leigh showed me a few relics of Frank's legacy. There was a small brown box — just the right size for Schrödinger's cat, I thought — that was supposed to demonstrate the photoelectric effect. Inside the box was a little plastic fan, motors, and odd bits of equipment, all strung together in what looked vaguely like an overgrown mousetrap. It was all Frank's doing, including the box. There were also experiments involving radioactive decay, line spectra, and the Doppler effect; cars that slid along on air tracks; little carts on wheels that abruptly gained mass when a bag of sand was dropped from above; a Millikan oil-drop experiment.

During its heyday, Frank's "library" won national recognition, and calls came in regularly from other universities wanting to follow his lead. Yet after Frank left, it didn't survive for long. No one cared about it the way he did, Leigh said. "People complained that it was too much work." Development on new experiments came to a halt, and the project stagnated. Soon, instead of getting ten calls a week from other universities who were interested in finding out about the lab, Leigh was getting none.

One after another, the experiments began to deteriorate, and before long most of them were junked. "It really made me quite ill when I saw all his equipment out in the hall," Bartlett told me. He immediately phoned a friend who was a high school physics teacher and told him to bring his pickup truck. "I told him, 'They're throwing out all of Frank's equipment!'"

Leigh surveyed the sprinkling of remaining experiments. "When this goes, it will be the end of Frank," he said. "I think it's tragic."

Of course, it wasn't the end of Frank. It was just the beginning. He was already thinking about how to bring such a "library" to much larger audiences, but he didn't believe it could be done in an academic setting. "He wanted to create his own environment," David Hawkins told me. "He wanted a way to reach ordinary people."

In 1965, he received the first of two Guggenheim fellowships, and he used the grant to visit science museums in Europe (as well as doing work in bubble-chamber physics at the University of London). Frank and Jackie explored the South Kensington Science Museum in London, the Deutsches Museum in Munich, and the Palais de la Découverte in Paris. Frank became convinced that science museums could become exactly the kinds of adjuncts school-based learning needed.

He also picked up some novel approaches to teaching that intrigued him. At the Palais de la Découverte, for example, he saw young people demonstrating phenomena, answering visitors' questions, and helping to keep exhibits in working order. He realized that they were "using teaching as part of their learning."

Thirty years later, when I visited the Palais in order to retrace

Frank's steps, the signs on the exhibits were eerily similar to those on Exploratorium exhibits. *Que faire?* they asked. (What is going on?) *Que voir?* (What do you see?) *Comment interpreter?* (How do you interpret it?) The Exploratorium's labels typically read *To do and notice, What's going on?,* and *So what?*

At the science museum in South Kensington, Frank saw displays of everyday technologies and demonstrations of how they worked—from refrigerators to alarm bells to toilets. Some invited visitors to make something happen by pushing a button. When I was there, high school students in blue polo shirts were on hand to answer questions.

At the Deutsches Museum, Frank saw that teachers regularly came in for training, using the exhibits as props. He saw a technician working by himself on a "very fine milling machine that was supposedly only on display," Frank wrote. "He had plugged it in just for his own purposes." Frank watched him work, fascinated, and he realized that technology and tools could be an important part of a science museum experience.

In August 1966, Frank was asked to present a paper at a two-day conference in Vermont sponsored by the Smithsonian Institution and the U.S. Office of Education.* He talked about the connections between art and science, the idea that a museum must be as enticing and useful to casual visitors as it is to serious students, to adults as well as to children. He was already saying that a museum should be a participatory environment where visitors could change what was happening, experience "what does not work," and learn by making mistakes.

Frank noted that many scientists had been inspired to pursue their profession because of exposure to science museums during their school years; that a lot of scientific concepts are difficult to explain with words alone; that such science museums as existed tended to be well attended, and yet there wasn't a single physical science museum, for example, in the state of Colorado.

* It was in Burlington that Frank became enchanted with a museum devoted entirely to decoy ducks; the idea of having multiple examples became an important part of his Exploratorium.

It was a scandal, Frank said, that there were so few science museums in the United States, and he spelled out in elegant detail how science museums might become rich and exciting resources for schools and scholars and curious passersby. "This realization was to alter the rest of my life significantly," he later said.

The organizer of the conference, Charles Blitzer of the Smithsonian, said that Frank "stole the show" from the seasoned museum professionals.

While Robert was succumbing to throat cancer after a long decline (he died in February 1967), Frank was, in many ways, just coming into himself. He was, in fact, on the brink of making his own kind of world—a "museum of human awareness" that reflected the sum total of his ideals and his history: his love of tinkering, physics, and art; his hatred of secrecy, fakes, and elitism. It was a magical place that ran on undiluted idealism and unfettered exploration, belief in people and anything-but-blind optimism. The world he made up was his way of dealing with loss and betrayal and his fears for the future; it was his determination to reach out to the best in people while figuring out how to protect humanity from the worst.

Being inside the Exploratorium, many of his friends and colleagues noted, was like being inside Frank's brain.

Frank and Robert as children.

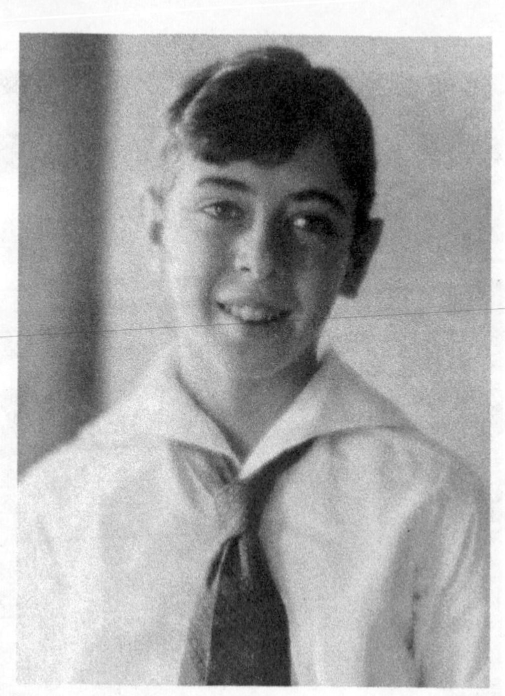

Frank as a student at the Ethical Culture School in New York.

Frank as a young man, probably around the time of the HUAC hearings.

Frank preparing one of his aluminum spheres to fly above the atmosphere in order to catch tracks of cosmic rays.

Frank and Jackie on their Colorado ranch.

Frank with a gyro during his time as a teacher at Pagosa Springs High School, late 1950s. (*Stanley Fowler*)

Frank taught at the University of Colorado, where he founded the Library of Experiments.

Frank talks with an unidentified person in the Palace of Fine Arts before the Exploratorium was created—his "blank slate."

Frank with a child at the Exploratorium.

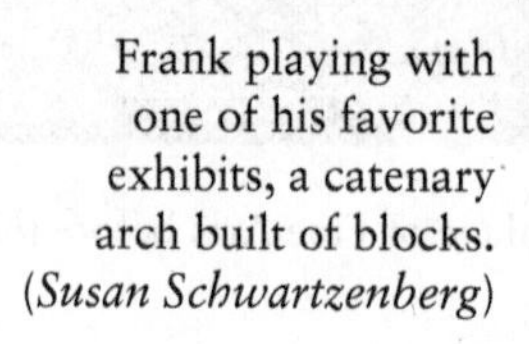

Frank playing with one of his favorite exhibits, a catenary arch built of blocks. (*Susan Schwartzenberg*)

Frank at play with a table full of pendulums, 1980s.

Frank in his office with a Picasso in the background, 1985. (*Nancy Rodger*)

Frank with his cane, not long before he died.

PART II

THE WORLD HE MADE UP

7

A MUSEUM DEDICATED TO AWARENESS

"I was surprised when we opened the doors and people came in and we actually had a museum. It seemed much easier to do than I had thought."

Almost immediately after the Vermont conference, Frank began to think about putting his ideas into action, and in the summer of 1967, he and Jackie headed to San Francisco to explore possibilities. Both liked the Bay Area and had friends there. Surrounded as it was by major universities and high-tech industry, the city also seemed to have the ingredients essential for the project's success. It looked promising, so they returned a year later.

Frank and Jackie sold their Colorado house and moved west, vowing to make a go of it one way or the other. (The University of Colorado granted Frank another in a series of leaves that were extended until Frank became professor emeritus in 1979.)

One of the first people Frank sought out was the *San Francisco Chronicle* columnist Herb Caen. Caen was intrigued by Frank's ideas and told him, "Get thee to a typewriter!" Thus encouraged, Frank wrote up what became his "Rationale for a Science Museum," which was later published in *Curator* magazine.

The "Rationale" is an impressive, well-thought-out document, detailed and inspirational, and a remarkably good blueprint for the institution Frank eventually created. Explaining science and technology without props, he wrote, was like trying to explain what it's like to swim without ever letting a person near the water.

Frank's teaching experiences at Colorado convinced him that young people had become dangerously removed from the real stuff of their own surroundings and desperately needed a "woods of natural phenomena" in which to reconnect. The students in his classes, he'd noticed, seemed strangely incurious. Only after they got involved with experiments did they begin to bubble over with questions. Most of them had had little experience with the natural world and few opportunities to build intuition about it. They hadn't climbed trees and sailed boats and collected minerals and taken apart bicycles and fiddled endlessly with electricity and motors. The result was that they showed up at their courses not knowing why they were there.

He thought this was "a scandal." "Their experience was so meager, their whole contact with the natural world so restricted, that I thought a place was needed where they could walk through a kind of woods of natural phenomena."

Like a lot of us of a certain age, I can remember as a child being sent outside to play—perhaps not even allowed to come indoors until it was too dark to see. We explored neighborhoods and built forts or dollhouses out of mud and sticks, climbed trees and watched clouds, played hide-and-seek and threw stones. We made model cars or planes, dressed dolls, played house or tag, spied on neighbors, chased dogs, stole flowers from the woman across the street, drew on sidewalks and lampposts (hopscotch or expletives). No one told us what to notice.

Our eyes, needless to say, were a lot more open before we became shackled to our schedules, tuned in to sights and sounds and experiences programmed by others (in turn programmed themselves). Whether kids were happier or more productive in those days, I can't say. But unmistakably, both children and adults were engaged in one-on-one contact with the world in a way that is almost impossible to achieve today, even for those who spend substantial time and money to try to make it happen.

In many senses, the tangible reality that used to be our playground no longer exists: that university of experience has been lost to television and computers and lessons and packaged entertainment and education, and also to the increased fears (often real) associated with wandering freely, poking your nose into things just for the heck of it.

And that has enormous consequences for science—as well as for the development of intuition and critical thinking skills in general. There was something about dealing with the real physical world that left you not only better informed but more grounded, more centered—less likely to be swayed by insubstantial claims or fluffy nonsense.

So with swimming holes and woods and cars that you could tinker with increasingly becoming the stuff of history books, Frank's woods of natural phenomena would serve as a substitute—not actually teaching so much as providing a means to rewire the brain's neural circuitry in a way that would come in handy when similar phenomena were encountered in books or on exams or in the everyday world. It would be a place where people could build their own personal repertoires of images and experiences.

He liked to call that place a woods of natural phenomena rather than a science museum because he envisioned it as containing "all manner of wonders" that people didn't normally associate with science but that "surpass flights of fancy in beauty and novelty": the inside of a star, the inside of a metal or a crystal, the inside of the brain or the eye. All, he said, were "miracles of life and nature" that few people ever have the chance to see. Yet he believed (and proved) they could be made accessible.

In creating his "woods," Frank was also clearly driven by the passionate commitment to public service he'd absorbed from his family, the Ethical Culture School, his experiences building the atom bomb (and its aftermath), and his years of exile. He believed that museums, like public education, were institutions that should serve the people, and that creating such an "Exploratorium" could play an important role in helping to build "a decent society." He imagined it as a place where both art and science could be used as vehicles for understanding. Most of all, perhaps, it would be a museum of hope — founded on Frank's optimistic conviction that an understanding of nature could eventually help solve human problems.

By 1968, Frank and Jackie had settled into a house high on San Francisco's windy Lombard Street, not far from the Palace of Fine Arts — Bernard Maybeck's pseudo–Roman ruin, which itself was built in a context of enormous optimism: the 1915 Panama-Pacific Exposition. Cars and airplanes were just then coming into general use, the Panama Canal had recently been completed, and science seemed the wave of the future. During the fair, the Palace housed fine art from all over the world. It then fell into disrepair, just as Maybeck had intended, and became a warehouse for telephone directories, an army garage, a home for tennis and badminton courts, and a venue for Christmas tree sales.

As luck would have it, Frank appeared on the scene just as the Palace completed a major restoration, and the cavernous interior beckoned irresistibly. Contemplating the immense structure, Frank remarked to Jackie that it would take a lot of nerve to take over such a place; of course, it was just the sort of challenge he liked.

Not surprisingly, many other groups already had designs on the space. Some wanted to install tennis courts; others, an art center. All were local, and all had well-cultivated contacts in the city government. Frank was a stranger. Nonetheless, he quickly charmed his way into the right circles.

Among the first people Frank won over was Scott Newhall, the *Chronicle*'s editor, and his wife, Ruth, who had recently written a book about the Palace's history. Ruth had studied science and math, and was immediately taken by Frank's ideas. The Newhalls arranged for Frank to meet with Mayor Joseph Alioto and the city's supervisors. Caen wrote a column mentioning that Frank was in town, wanted to create a science museum, and was interested in the building.

Newhall also introduced Frank to the *Chronicle* science writer David Perlman, a meeting Perlman remembered as if it happened yesterday. "Scott called me and I went into his office and there was this guy I'd never seen," he said. Newhall told Perlman about Frank's project and suggested he write a story about it. "I thought it was pretty far out. He talked about things I'd never thought about in the context of a museum. I thought: This is never going to fly at all."

Frank went through city hall "from the top to the bottom," Ruth Newhall remembered, showing around his "Rationale" and talking about what he wanted to do. "Frank had this disarming shyness, humility, and feeling of helplessness," she said. "He has created many illusions in his life, but that was one of the greatest. I won't mention all the people he charmed bringing this dream into reality, but it was a remarkable thing to watch."

In December 1968, Frank formed a small nonprofit board of directors for his project, made up of old friends like Pief (Wolfgang) Panofsky of the Stanford Linear Accelerator Center and Nobel laureate Luiș Alvarez, whom he'd worked with at Berkeley. Louis Goldblatt, the secretary-treasurer of the International Longshoremen's and Warehousemen's Union, introduced Frank to scores of important players on the San Francisco cultural/political scene.

By May 1969, Frank had wangled a dollar-a-year lease on the Palace. In July, he took over a trailer left behind by the KQED annual auction and moved it into the building; the trailer became the Exploratorium's office. While his son Mike and Jackie cleaned out nests left by birds and mice and swept away pine needles, Frank introduced himself to his new neighbors in the posh Marina District, where the Palace, with its peaceful lagoon and park, was situated.

Some neighbors immediately signed on as volunteers. But others worried that hordes of schoolchildren would ruin the tranquility of the park. Typically, Frank tried to see things from their point of view. He wondered whether it was really the kids they feared so much as losing control. For a lot of people, San Francisco in the sixties already felt unhinged. "Normally, you like children," he later reflected, "but if you think everything is going out of control, then you see a bunch of kids going every which way, and it just brings your fears to the surface."

With friends and family all pitching in, the Exploratorium got put together in the manner of an old-fashioned barn raising. Panofsky loaned Frank pieces of an accelerator, with charts explaining how the various components worked. He helped him scrounge other stuff from all over Stanford. "We both agreed that physics was grubby," Panofsky said, "and it shouldn't be pretty and under glass. The visitor should learn that experiments break, and fail, and you've got to fix it. Shops should be part of the museum. Because that is the way that physics is done. Things break. You fix them. You repair them. You change them. You improve them."

Frank went to all the local foundations and eventually got $50,000 from the San Francisco Foundation. Mike put in wiring. Jackie wrote exhibit labels. They hired a single student "Explainer." They cooked up the name "Exploratorium" because "museum" seemed too static, too formal; they wanted it to be a place where people could enjoy themselves while discovering "what's happening in the world."

The museum opened without so much as an announcement. One day in late August, Frank simply opened the doors. "We just watched what happened," he said. People started wandering in. A pair of runners jogged from one exhibit to another. "They never slowed up," Frank said. "They just went back and forth, and finally went through the curling path and out the door again."

Natural Selection

"I really have no qualms about my scavenging efforts. I suppose that one starts off with guilt feelings about both money and sex, but the only way to make any progress with either one is to not worry about the turn-downs."

Frank knew that his first budget was "brazen" (among other things, it called for two million dollars' worth of capital expenditures, which he never got). "He didn't have a penny," Ruth Newhall remembered. "And [yet] it all came off."

He compensated by doing everything on the cheap, using all he'd learned in his years of experimenting and ranching and teaching. "Almost everything that I had ever done helped make the place," he said. On the ranch, he'd learned how to handle heavy equipment. He'd learned to "get unstuck." Mostly, he'd learned how to improvise. "When you're on a ranch that's eighty miles away from a veterinarian, hardware store, you really try and scrounge and find things and put them together," he said. He learned to work long and hard and "to use everything one had to do what one had to do."

He later described the museum as a "weave" that knit the threads of his life together into a tapestry that encompassed everything he liked, and liked to do: "the teaching, the new development, making things myself, working with wonderful people." For someone like Pete Richards, who had known Frank and Jackie as ranchers, Frank's plans seemed consistent with what he'd heard all those years sitting around the kitchen table in the Colorado mountains. "They were trying to create their own world where they could apply their own values," he said. And to a large extent, that's exactly what they did.

Early exhibits came from every source imaginable. "People just wandered in off the streets," Mike said. "Sometimes they said, can I put my thing here, or can I do something. And we said sure, why not?" The place was so vast and so empty that at first Frank accepted practically anything that didn't violate his core principles. Several of the groups that had wanted the Palace for themselves were still waiting in the wings, hoping the project would fail, so Frank did everything he could to claim it. "Frank would say, 'Let's get stuff in here, we gotta make it look like we need the space,'" Mike said. "Everything was fair game. I don't remember any rhyme or reason to it other than that it was interesting."

Frank and Jackie brought in all kinds of equipment from the

ranch as well. A foot-treadle lathe bought at a farm auction was turned into an exhibit that showed how hard it was to generate electricity. A school bell originally from Aztec, New Mexico — used at the ranch to call people in from the hayfields for meals — became the Exploratorium's closing bell. A big jack that they'd used to get tractors out of the mud became a handy multipurpose tool.

Frank also brought ideas from the ranch. He'd been impressed, for example, that newborn calves can stand and navigate a few minutes after birth, whereas human babies take many months to do the same thing. He was fascinated by the idea that the simple act of standing is so complicated, involving feedback from the inner ear and other sensory systems. So balance and motion detection became the focus of many exhibits.

The trips to the junkyard with his Pagosa Springs high school students inspired Pete Richards to build exhibits such as the "Differential," which not only showed how the wheels of a car can spin at different speeds during a turn, but also illustrated general lessons about relative motion and frames of reference.

The process of acquiring stuff was initially "opportunistic," Frank liked to say — like biological evolution. Some exhibits went extinct almost immediately — such as the surplus displays from the telephone company, which Frank thought weren't "honest" because some of the effects were rigged. Others are still around today: the "Touch the Spring" exhibit donated by Lockheed, the Abe Lincoln portrait from Bell Labs, the cloud chamber from NASA.

Edward Condon, a physicist who'd been at the University of Colorado with Frank and was now at the National Bureau of Standards, knew where government surplus property was stashed, and helped Frank to get it. Frank also acquired a collection of stuff from NASA, including a space suit and a film of the first moon landing, which was played continuously. Donations dribbled in: a spark chamber, a loaned Montgomery glider, a forklift.

A gigantic lens that had been used at Lawrence Berkeley Laboratory to observe the paths of high-energy particles became a plaything where people could learn about the effects of refraction. Parabolic mirrors were liberated from the Atomic Energy Commission and put to any number of creative uses. Frank later told another museum director that the trick to making a museum was getting on the list for government surplus property. "You'll find lathes, transformers," he said. "You can stock the whole museum this way."

Frank spotted an automated shoe tester gathering dust in the basement of the National Bureau of Standards, in Bethesda, Maryland, and put it on display too—its patent leather Mary Janes chasing each other around in endless circles. A staff member noticed new traffic lights on Van Ness Avenue, so Frank got the 3M company to donate one and opened it up so that people could study the optics.

Frank saw his role as a filter, a sieve. If something didn't fit, was ugly, didn't teach anything, or would easily break, he'd refuse it. "We've turned away many things, because we do have a sense of what the place ought to be," he said. This process of scavenging and filtering was what gave the place its wonderfully eclectic character. There could be a very finished artist's piece and right next to it a funky table made with two-by-fours and wires held together with duct tape.

He grabbed stuff to build his own exhibits wherever he could find it. When a staff member left her furniture in one of the Exploratorium storage spaces while changing apartments, Frank appropriated her kitchen table for an exhibit base. By the time she found out, he'd already drilled holes in it. (He promised to buy her a new table, and he did.)

Anything involving computers was likely to be rejected out of hand. "There's nothing a computer can do that I want done," Frank said. Everything a computer did was a simulation, and therefore not sufficiently transparent—or even "honest"—by Frank's lights. You couldn't see what was going on inside the box. Besides, he thought computers were both passive and addictive. And new technology—as he'd learned at Los Alamos—could all too easily entrap. "Technology, including the technology of chips, is insidious," he later wrote. "It can carry us where we do not want to go if we follow mindlessly." Lest the obvious connection was missed, he added that the failure of the world to "make enlightened choices about nuclear weapons" had put humanity "on the brink of an unmitigated disaster."

Of all the early exhibits, the most important was clearly the shop. Located right next to the entrance, it was the first thing people saw, and it became a kind of statement: the place was not static. Things were being created. Early on, a graphic artist had wanted to put up a big plastic wall with portholes over the shop. "He didn't know that part of a shop was the way a shop smells when you burn the wood in the saw," Frank said, "or smell the oil from a lathe."

Frank thought the shop conveyed that sense of honesty as well. Like restaurants that let customers see inside the kitchen, an open shop reminded visitors that creating things can be noisy and messy. It's done by ordinary people; stuff doesn't just magically appear ready-made, like the meat in the grocery store or the clothes at the Gap.

The exhibits built there were pragmatic, robust, built to take all kinds of punishment—Jeeps rather than Ferraris. If things broke too easily, or didn't work at all, people would just get frustrated. At the same time, the ability to *stop* working was often built into the exhibits on purpose. Nature imposes limits on all sorts of behavior, and ignoring that is a kind of cheating. The frequency has to be tuned just so or the rods don't resonate; the light focused just so or the images don't converge; the blower has to be aimed just short of horizontal so the balls don't fall out of the air stream.

Testing those limits is what allows us to discover what we call the laws of nature—just as testing limits is what teaches two-year-olds the laws of family and society. How far can you go before something bad happens? How much can you get away with? "People would break something and we'd try to fix it and beef it up so that it couldn't be broken anymore," one shop worker said. "And Frank said, 'Let them break it. They learn something from that.'"

To this day, a sign hangs over the shop, made by Barbara Gamow as a tribute to the common ground between Frank and her late husband, George Gamow, of Big Bang fame. It reads: *Here Is Being Created the Exploratorium, a Community Museum Dedicated to Awareness.*

Here Is Being Created the Exploratorium

"I think that was our first plant torture exhibit."

Gamow's sign perfectly captured the thematic thread that held the Exploratorium together and gave it much of its magic: a focus on perception. Since everything we know is filtered through our perceptual apparatus, all of science hinges on it one way or the other, and the subject offers a naturally interdisciplinary way of getting into almost any subject. At minimum, perception involves physics, neurophysiology, chemistry, biology, and also the technology we use to extend our senses.

Perception also appealed to Frank because it's a subject that puts people at ease, creating a "humanistic atmosphere." No perceptual effect will be the same for everyone, or even for the same person at different times. There are no right or wrong answers. Further, perception was an active field of science in which even fresh discoveries could be made accessible. No one entirely understands how our brains take bits of inadequate and ambiguous information and use them to compose what we take to be the real world. "Although much sophisticated work is currently being done in this field," Frank wrote, "even novices can ask and answer sensible questions."

(In fact, visitors *did* make up their own experiments. One family — after spending time watching how real people appeared to shrink and grow as they walked around a full-sized "Distorted Room" — hit on the idea of bringing their Saint Bernard along as a test case. As expected, if no less bizarre, the dog's apparent size altered as surely as if he had eaten one of Alice's magic mushrooms.)

One of Frank's favorite perception exhibits was a kind of aquarium full of touch-sensitive mimosa plants. He soon discovered that if he touched the tip of the plant with a cigarette, a wave would travel down the length of the plant as the leaves folded in on themselves. So a little heat source was added to the exhibit that visitors could apply to the plants with a wand. "I think that was our first plant torture exhibit," Charlie Carlson, a biologist, said. "I'm surprised the plants lasted as long as they did."

Beyond physiological effects, Frank was interested in what perception could tell us about the social dynamics that seemed to keep the people of the world in a state of continual conflict. Why is it that foreigners often look alike to us? More generally, how do we make judgments about people? If information is ambiguous, how do we decide which interpretation to accept?

Illusions are windows into how the mind works, offering insights into how we determine reality. In effect, they catch the brain in the act of jumping to conclusions: they reveal the perceptual biases that lead us to perceive things we know are impossible; to ignore things that are right under our noses (our noses included); to use information that ought to *dispel* an incorrect conclusion to instead *compound* it; to automatically adapt to just about any stimulus, causing it (perceptually speaking) to completely disappear.

In part, these "mistakes" are caused by the brain's need to take shortcuts, relying too much on some clues (such as skin color) while ignoring others altogether. The writer Morris Berman, visiting the Exploratorium soon after it opened, noted that such filtering makes information, like processed flour, easier to use, but it also takes out the vitamins. And since we tend to screen out all information that is different or threatening, we also screen out much that is interesting and valuable. "Our optical illusions," Berman wrote in an essay Frank distributed widely, "prove to be fairly innocuous compared to our political and personal ones."

Frank loved these political parallels and saw them everywhere. Once we were playing with an exhibit on binocular vision in which lenses sent two different and conflicting images to each of a viewer's two eyes. Because the combination of the two made no sense, some people actually became nauseated when they looked at it. Wouldn't it be nice, Frank offered, if the contradictions and lies of politicians automatically made people "just throw up"?

Animals were no different, and Frank enjoyed discovering parallels between the ways humans and animals perceive (and misperceive) the world around them. One exhibit demonstrated how feather worms would draw in their "feathers" (tentacles, really) whenever someone cast a shadow over their tank (a reaction to the likelihood that a predator was near). But the worms quickly adapted and stopped responding — behavior that in other circumstances could well have perilous consequences. Might this be the same mechanism, Frank wondered, behind the human tendency to adapt to inhumanity or ugliness?

The theme of perception had strong connections even with subjects seemingly as far afield as mathematics and language. The eye's response to brightness is logarithmic, for example, meaning that it "sees" small changes in brightness when actually they are large and increase rapidly with intensity. This is much the same way people "see" the difference between a million and a billion as much smaller than it is — leading them to make poor judgments about everything from national budgets to population growth.

Frank added an entire section on mathematics — not to teach people to calculate, he said, "but to illustrate the extraordinary power and versatility that mathematics has in helping us think about and relate to the behavior of nature." Luis Alvarez told him about logarithmic stacking blocks that teeter impossibly over an edge of thin air; Phil Morrison suggested the exponentially decaying bouncing ball. Both phenomena are great fun to play with; plus, both behave according to a universal mathematical pattern that drives everything from compound interest to nuclear explosions — exponential growth (and its converse, logarithms).

As for language, Frank thought it was inseparable from the

"perception of meaning," as he liked to put it. That's one reason he was so involved in all of the writing that took place at the Exploratorium. He wrote an essay for almost every issue of the *Exploratorium* magazine, including pieces about the power of numbers to explain the universe, the importance of play, and the relationship between a boiling pot of water and a bubble chamber.

Most of all, Frank liked perception because it is the basis of noticing, making us aware of how we often see what we believe rather than vice versa. "The first thing you have to learn about science," the computer pioneer Alan Kay told me, "is that the world is not as it seems." It's hard to know just what will give a particular person that insight, he said. So Frank made a woods of exhibits that essentially all taught the same thing: "Here is something you didn't expect."

"It is impossible to come away [from the perception exhibits] without some sense of awe at the subtleties, complexities and the almost unbelievable reliability of sensory information and processing," Frank wrote. "One frequently comes away with a new awareness that causes one to stare, squint, close one's eye, or cock one's head, in a word, to experience everyday phenomena."

Surprises

"Invariably, as I build these experiments, I observe some phenomena which I have not observed before and which I do not understand, or I find some deviation from the expected result which requires further investigation."

Frank looked at exhibit-building as research. Nature was "full of surprises," so it was impossible to build a good exhibit, or teach well for that matter, unless one was "always reacting to these surprises." Some of the surprises made exhibits more interesting or more beautiful. Others clarified or enlightened. The attempt to teach something almost always teaches the teacher something new. Sometimes these new things are new not just to the teacher; they have never been thoroughly explored by anyone.

So if the exhibits looked perpetually unfinished, that was the nature of science. "It's really often pretty messy and pretty mixed up," Frank said, "and you try over and over again and it sometimes works and other times it doesn't work well." He constantly paced the floor, watching people play with his toys, seeing how and where they were confused or stumped, figuring out ways the exhibits could be improved. New pieces were set up as prototypes, just to see how people would react to them. Frank was rarely satisfied. It was not uncommon for staff members to come to work in the morning and find some filigree Frank had added the night before.

The shop itself was a stew of people and ideas, always in flux, where everyone minded everyone else's business and where the best "meetings" were often held. Frank would stop by after hours to "play" with whatever was new; people crowded around to listen; sometimes they stayed for hours.

In the process, they were schooled in Frank's way of doing (and seeing) things. "Frank had a kind of cockeyed idea of how things work," said Tom Tompkins, the longtime head of the shop and a Saint Francis figure to scores of lost teenagers who came his way. "It's really a rancher's sense; he knows when you can put baling wire on it and make it go." It was also a rancher's aesthetic. Once, a staff member used glue to put together some pieces of plastic, and Frank had a fit. Glue was mysterious, he said. The exhibit builder had to take it apart and use screws, which people could readily see.

Many exhibits grew out of Frank's own everyday noticing of natural effects or found objects. He came upon some plastic rods impregnated with orange and green fluorescent dyes and arranged them into a kind of spiral sculpture. Frank liked it because it was "a pretty thing," he said, but also "because fluorescence seemed like a good way to discuss . . . the interaction between light and matter." And because the ultraviolet light causing the rods to glow flickered on and off imperceptibly, it caused a strobe effect as the rods rotated, and also fan-like moiré patterns. In addition, each rod acted as a light pipe, so that the ends of the rods lit up "quite brilliantly." It was exactly the kind of rich exhibit he liked best.

Mostly, ideas were brought in by people who worked there, like seeds carried to foreign shores by birds. Philip Morrison and his wife, Phylis, spent the summer of 1971 helping out, and left behind a number of perception and color exhibits, including one that showed how white could be "blacker than black" in the absence of light. Other exhibits started off as school projects, and high-school-age apprentices helped out a lot in the early years. The National Science Foundation had wanted to help Frank's project from the start, but at first couldn't find a way to fund a museum directly. However, it *could* support high school kids to work on projects, and so it did.

As the place began to fill, the philosophy began to change. Frank and the staff could think more deliberately about what ideas or effects were fundamental to an understanding of some aspect of science or nature. Then they'd sort out those phenomena that were easier to understand when demonstrated rather than read about and try to find ways to incorporate those effects in captivating, instructive exhibits. Each exhibit became part of a chain of exhibits, all linked to each other in a multitude of ways.

At the time, however, most people were so wrapped up in filling that huge floor space that they weren't aware of the radical nature of what they were creating. "It wasn't until the eighties that I realized we had actually invented interactive exhibits," one exhibit builder told me. "It was really a revelation. Frank created a new way to educate the world. We were living in a world we made up, but we didn't know it."

While the mix of exhibits was eclectic, all shared certain philosophical underpinnings that reflected Frank's values. At heart he believed that other people were basically like him, and the fact that he'd been brought up in a cultured household, studied at the best schools, and knew atomic physics didn't make him special in any essential way. So nothing in the exhibits or the graphics smacked of condescension. Frank never designed exhibits for people who didn't know as much physics as he did. He designed them for himself, for his friends, for the people who worked there.

The staff absorbed this attitude and passed it on to the general public. "You ended up treating the public as intelligent, curious, and thoughtful people, rather than pandering or talking down to them," one staffer said. When a delegation of men in Mao suits came to visit from China, they immediately dubbed it "the people's science museum."

Beyond the rather surprising fact that this "science museum" never talked down to you, the other thing you immediately noticed about the Exploratorium was that it was a kind sort of place. This was deliberate. Frank knew how tough science was for most people in school — how tough school was in general. (When then Governor Ronald Reagan crippled California's public education system with his tax policies, Frank's reaction was "He must have had really horrible teachers.")

So Frank went out of his way to make the Exploratorium as kind a place as possible. He wouldn't allow games, because games have losers. Even the exhibit labels were worded so as not to put visitors on the spot. If a sign told the visitor to look through a prism, for example, it would say, "Notice the colors," not "What do you see?"

He was constantly refining this "kindness component" as he talked to visitors on the floor. During one uncomfortable encounter, Frank asked a visitor what he saw while looking at some exhibit. The man backed away, saying, "Well, what am I supposed to see?" It was precisely the kind of reaction Frank wanted to prevent. When he later told Mike about the experience, he concluded: "That's American education for you."

So Frank tried hard, as he put it, not to make science too sciencey, art too precious, technology too impersonal. He didn't want anything in the museum to convey the message "Isn't somebody else clever?"

The exhibits also had to be honest — sometimes in the extreme. Builders were told to reveal every wire, every connector, even the power supply. And all this authenticity could get expensive. During one discussion about a new exhibit, a staffer suggested using Formica instead of wood for a change. Every year, visitors complained about getting splinters; every year, the wood surfaces had to be sanded and varnished; they got dirty; it was cheaper to use Formica, which you could put down and never think about again. But Frank refused. "No scientist works on Formica!" he fumed. "You've got to be able to bang with your hammer, burn holes with your soldering gun; it's gotta look like a real environment that a scientist would work in. That's the only way people can know what that's like . . . I don't care how expensive it is, it's gotta look cheap."

Alan Friedman of the New York Hall of Science said he was greatly inspired by this attitude. "He had this incredibly pure conception. It didn't matter what extraordinary lengths you had to go to, whatever barriers of safety and practicality and cost. You had to get the real phenomena out there."

All this rough-hewn honesty didn't always go down well with donors; San Francisco society was far more comfortable in proper museums with paintings in gilded frames, the velvet-curtained opera, slick historical exhibitions; the Exploratorium was a little scuzzy for their tastes. Frank knew this made his creation vulnerable, but he was determined to win them over, and eventually, mostly, he did.

He was even honest about the fact that he had very little idea of what the Exploratorium would eventually be like.

In fact, despite the lie he told about his previous Communist Party ties, which undid his career, Frank felt that one of the most appealing and important aspects of science was its intrinsic honesty. "If you are genuinely trying to understand what's going on around you," he wrote, "there's no point fooling yourself, or for that matter, fooling any of your colleagues." Scientists who fabricate data are ostracized; it's one of the tenets that make science special. "I wish it also applied to politicians and advertisers," Frank said, "so that they would ostracize people who willingly and deliberately fabricate data."

He thought reporters or members of Congress who lied or distorted facts should be discredited and never allowed to work in journalism or hold public office again. "Not only can we exert public and social pressure against dishonesty and secrecy, but we can actually legislate against such sins," Frank said. "There is no greater crime against the people and democracy."

Families

"O for God's sake / they are connected / underneath."

Developing intuition about nature requires multiple experiences, not only one or two. "You don't learn about a family just from one member of a family," Frank said. "You learn about a family because you see something about parents and grandchildren, and uncles and aunts, and cousins and brothers and sisters."

It's the same way we learn to recognize that both a Pekinese and a Saint Bernard belong to the general family "dog," even though the animals appear to have little in common. And just as we learn to distill the essence of "dogness" out of many varied examples, so we learn to abstract the essence of ideas, such as resonance or interference, in a way that "sticks" as no dictionary definition can.

This distillation is enormously useful. A water wave, a sound wave, a light wave, and a fashion wave are not obviously all the same thing, but once you see the connection, you can recognize and understand a wave anywhere it might appear—and waves appear everywhere, in social as well as physical phenomena. This remarkable power of abstraction—Frank called it our "ability to discern pattern in a mire of confusion"—reveals that all waves share the same properties, which are easily described through the universal symbolism of mathematics. That math is so general, Frank pointed out, that people learned how to look for gravity waves even at a time when no one was sure they existed.

In essence, all science is about abstracting simple concepts from a messy, confusing world. So even more than a woods of phenomena, the Exploratorium was a woods of ideas. That's why, *Scientific American* editor Dennis Flanagan explained, the Exploratorium was "the only true science museum" — because it was a museum of "concepts."

And so Frank's "woods" contained not just phenomena but families of phenomena — brothers and sisters and aunts and uncles: you'd see resonance in air, water, metal plates, rubber membranes, rods and springs and even light; exponential growth (or decay) in winches, tire pressure, building blocks, and bouncing balls. I remember being surprised and impressed by how eerily similar oscillating systems looked — whether they were made by oscillating electric currents inside a computer or by a felt-tip marker attached to a mechanical arm that drew patterns on paper taped to a swinging board.

Multiple examples convey a sense of overview, a feeling for context, an insight into how disparate things can fit together. And one of the most surprising and satisfying discoveries of modern science is that things do, in fact, fit together. Gravity spills milk and holds the moon in its orbit. Electricity and magnetism — once (understandably) thought to be completely unrelated — turn out to be different aspects of the same fundamental force. The tiny quantum mechanical wiggles in the newborn universe grew up to be the giant garlands of galaxies that drape across the sky today.

In fact, the intuitive connections that people make by messing around with multiple examples of things can be surprisingly deep — even when they think they're simply (or not so simply) playing. Take the surprising symmetries inherent in mirrors, which pop up throughout the museum (and our lives) in endless different contexts. Yet mirrors still have the power to fool us — spiriting images around so stealthily you can't tell where they've been, leaving us grasping at air or believing, despite ourselves, that someone or something can float in thin air, working all kinds of mischievous magic. These same kinds of symmetries underlie the Golden Rule and Einstein's theories of relativity.

"The whole purpose of this place," Frank said early on, "is to integrate different natural phenomena and make them understandable and not as fragmented as they are for the public — and natural phenomena include the things created by man because that's a natural phenomenon as well." Frank loved to quote Muriel Rukeyser's poem "Islands," which begins "O for God's sake / they are connected / underneath."

The fact that the Exploratorium found a home in a huge empty hangar was an accident, but it turned out to be the perfect venue to show off this connectedness. Dividing areas up with walls would have greatly diminished the sense that things are of a piece, that common threads weave through the whole like a theme in a symphony. In fact, Frank often compared the organizational structure of the place to a piece of music that could be heard by different people in many different ways. Even if the listeners weren't aware of the grand ideas or the underlying structure, it was vital for the creators to keep that structure in mind, because an audience can always sense whether the composer "was disciplined in his efforts to achieve the coherence of his composition."

Staff

"This little spaghetti-looking thing right here is called the optic nerve."

By and large, people accreted rather as the exhibits did, with serendipity and opportunism playing big roles. Charlie Carlson was hired (at least partly) because he had a pickup truck. He'd been working with a biologist, Evelyn Shaw, who set up a meeting with Frank. Charlie showed up at Frank's house and found a skinny man pacing around the patio with a cigarette in one hand and a big rum-filled pineapple with a straw sticking out of it in the other. For a moment Frank couldn't decide which to put down in order to shake Charlie's hand. "It was the funniest thing to see," Charlie said. When Frank found out Charlie had a truck, he said, "That's great! How soon can you start?" When Charlie said two weeks, Frank said it wasn't soon enough. "How about Friday?" "You mean tomorrow?" Charlie asked.

Pete Richards was in Baltimore when he got a letter from Frank saying he'd moved to San Francisco and was starting an Exploratorium. "It seemed kind of vague, but also pretty interesting," Pete said. So he joined on, and his wife, Sue, worked in the trailer as an assistant.

Thomas Humphrey, who earned his Ph.D. in physics at Caltech, was working at Fermilab, home to the world's most powerful particle accelerator. At the time, accelerator experiments were becoming huge, comprising as many as a thousand people, and though Thomas loved particle physics, it no longer seemed like fun. Then a friend called and told him that he had a weird idea for him: "There is this strange guy up in San Francisco. He's got this strange place." Thomas said, "Oh, yeah? How strange?" The friend said, "Pretty strange." So Thomas said, "Sure, why not?"

At first, it was a shock for Thomas. For one thing, he was used to being associated with established, prestigious institutions such as Fermilab and Caltech. "And now I'm working in a barn with a bunch of folks that not only do I not know, nobody knows!" he said. Looking back, he still finds it strange that after years of working on "those two acres of asphalt," the "barn" became an internationally renowned institution.

Not everyone had obvious qualifications. Rachel Meyer — now the director of her own museum, and sharp-edged and passionate as ever — was a sixteen-year-old runaway. "Frank would take me to dinner — a sixteen-year-old! — and ask what I thought about things. He really wanted to know!"

Frank didn't care whether someone had a Nobel Prize or even a GED. He wanted people who could think freely, who had ideas. He believed if someone was sufficiently enthusiastic about an idea, that person would learn how to communicate it.

It's hard to figure out what the people Frank chose had in common beyond the fact that they were all immensely curious, open, and engaged with the world on many different levels. Perhaps Mike Templeton, the former director of the Association of Science-Technology Centers, put it best: "He had a deep respect for ideas and a deep respect for people who had ideas."

As much as the staff shaped the Exploratorium, the place changed them. Thomas Humphrey found that the way he thought about physics was completely turned around. "There was this no-bullshit thing Frank would do," Thomas said. "He wouldn't let you hide behind the mathematics, the highfalutin stuff, the jargon. He opened up for me the opportunity to do something which I really enjoy doing, which is thinking deeply about things. I mean, not just knowing them because you do the math well. But really thinking deeply about the behavior of the world. Trying to come to some understanding that maybe is verbal as opposed to mathematical."

The most distinctive staff members by far were (and are) the high school Explainers, forever associated in most people's minds with Darlene Librero, the petite powerhouse who became practically synonymous with the Explainer program and probably turned around more troubled teens than a convention full of psychologists. Darlene came to the Exploratorium as a teenager herself, and soon became indispensable both to the program and to Frank. Thomas Humphrey says that to this day, he considers hiring Darlene as his proudest accomplishment.

Drawn from all parts of the city and sometimes beyond, Explainers are tattooed, green-haired, dreadlocked, multiply pierced, and preppy; black, white, Asian, Latino, and everything in between. They open and close the museum, handle money, mingle with visitors, explain exhibits, report breakdowns, take care of lost children; they are in complete charge of the museum floor and so also serve as the closest thing (though not very close) the Exploratorium has to security guards. Their main job is to help get visitors "unstuck." They have real responsibility for everything that happens, and the experience changes lives.

At the thirtieth Explainer reunion, Ana Wong, now a deputy district attorney, said the experience taught her how to explain things to a jury. A second-grade teacher said that being an Explainer inspired her to take physics classes in college, and she integrates science into all of her teaching. A clean-cut black man with a government job said being an Explainer gave him the confidence that he could figure things out.

"Every time I come in here I get such a special feeling," said another former Explainer. "It's home for me, I feel comfortable, I feel happy. Anybody that means anything to me, I bring here, because this is such a part of me. I'm thirty-two, and when people ask me, what was your favorite job, I tell them, the Exploratorium."

Yet the point of having Explainers was primarily to provide a rich experience for visitors. Adults wouldn't make very good Explainers, Frank thought, because they were too constrained, too worried about giving the wrong answer, too unwilling to guess. He hated how science education so often promoted the myth of the collective right answer. He thought teenage Explainers would "loosen up the whole feeling of learning," because they were learning too. He wanted them to guess if they had to, because that's part of how science really works. Adults wouldn't project anywhere near the excitement of kids who were in the process of understanding science for the first time and couldn't wait to tell someone else what they'd just found out.

The Explainers' racial and socioeconomic diversity was part of the plan as well. Frank wanted to reach out to people who normally didn't visit museums. He hoped the kids would bring their friends and families, make them part of his experiment. And since he was personally interested in people who were different from himself, he went out of his way to make sure he'd meet them. "He was always just poking at things just to see what's going on," a staffer recalled, "and I felt like one of the things he sort of poked at."

Undeniably, most of the Explainers were noisy, messy, and weird — typical teenagers in every respect. Nonetheless, Frank put the Explainers' area (their lockers, a cramped office, a large table often littered with the detritus of too hurried meals) smack in the middle of the museum's executive offices. The scene could be intense and even intimidating, and some staffers went out of their way to avoid walking through the area. Others occasionally lobbied to move the Explainers out of sight (and earshot). Explainers held wild parties (which Frank loved to attend), and a few made bongs in the shop. But they loved Frank, and some even loved Orestes and took him for walks.

As for Frank, he loved being around their exuberance. He loved their slang, their way of thinking. He'd wander into their hangout and say, "Hi, I'm Frank. Whatcha doin'?" Even when they were just goofing around, he knew they were testing things out—personalities, lifestyles, ways of thinking and being—and he thought this made them an excellent floor staff. Sometimes he'd help them with their homework.

Watching Explainers at work in the Exploratorium could be almost as much fun as playing with exhibits. Sometimes they'd just hang around, egging visitors on. "What do you see? A circle? Or a spiral? Here, take your finger. Go like this." They showed people how to figure out "what's going on" when they reached out to touch a metal spring and grasped only thin air. "Here, shine a light on it. See what happens."

They dissected sheeps' brains and cows' eyes.

Suddenly you'd hear the call "Cow's eye dissection! Cow's eye dissection!" coming improbably from a sweet-faced Explainer yelling like a carnival barker. Before long, a half-dozen people would be sitting on benches around a curved counter, and she'd start her spiel: "Here's a cow's eye. It's real. And the reason we dissect it is so you can see how your eye works and what it looks like inside. Does anybody want to touch it before I cut it up?" It looked like something out of a horror movie, but someone always wanted to feel it for themselves anyway.

Explainers would cut off the muscles and fat from the back of the eye, extract the optic nerve, squeeze out the aqueous humor—the jelly-like substance that holds the eye in shape. There'd be "eeewww"s all around, followed by tentative fingers reaching to have a feel. They'd pass the lens around like a jewel, ask if anyone wanted to feel it. They'd peel away the layers one at a time, explaining how the lens grows as we do. At the end of the demonstration, when parts of the cow's eye were scattered about the table, Explainers would often all but beg their audience: "Would anyone like to feel a part of the cow's eye? Please. It may be your first and only time." Before you knew it, visitors young and old would be passing slimy bits around the table.

Formal studies have shown that a four-month stint as an Explainer can substantially affect a teenager's life. And because of the success of the program created by Frank and Darlene Librero, many museums—most of which were once skeptical about involving young people in a serious way—now have analogous programs.

Not least, Explainers helped create the kind of social atmosphere that would get people talking to each other. "The Exploratorium was conceived as a place to teach and learn, primarily because these are things that we all like to do," Frank said. "It is the way we bring up our children, take our friends to the top of a hill to see the view, or call out, when walking through the woods, 'Hey, look, there's a deer.'" He thought people tend to notice more when they're exploring with someone else. If you're walking alone in the woods, you're less likely to notice that special flower or tree than you are if you have someone along to point it out to.

In the name of promoting such social interaction, Frank sometimes went to extremes. For example, he wouldn't put up signs indicating where the bathrooms were because he thought they would give people one less reason to talk to each other.

Certified

"If one writes down a list of all the things schools are currently asked to do, one can only react by throwing up one's hands in despair."

Frank liked the idea that the Explainers would have a chance to serve the community in a way that was connected with what they were learning—to teach what they were studying. Young people spent too many years locked away in school, he thought, unable to make real contributions. Hiring teenagers as floor staff would be a way to let them play some useful role in society "during the long period in which they are being subjected to education."

Besides, having a position of real responsibility in an adult environment provided an alternative way for students to evaluate themselves beyond the tyranny of grades and testing. As it was, learning was almost entirely tied up with the process of certification, predicting how well students would do in jobs or colleges, "guaranteeing" success (or failure) as if they were some kind of product. Society increasingly relied on schools to send young people out into the world with "thorough, albeit not necessarily reliable, labeling." In fact, the whole process of education, Frank thought, was being subverted for that purpose.

Certainly, some type of ongoing evaluation of young people — in fact, of all people — is inevitable. Falling in love is a sort of evaluation, as is hiring someone for a job. Students need to know how they're doing, what is expected of them, whether they're ready to go on to the next step in learning or in life. But they can't do that if they don't know the real norms of adult and peer-group performance, rather than the artificial ones measured in schools.

"If young people are allowed to be around so that they could observe, work alongside and eventually apprentice," Frank said, "then they could accumulate a personal experience record that would be far more reliable to a future employer or next step educational institution than a grade point average."

Making such a system work, of course, meant that people "would have to believe that spending time with young people is not a waste of time," he said. Instead, one of the main functions of school seemed to be keeping kids out of adults' way.

But the fundamental problem with schools, Frank thought, was that they were being asked to do much too much. Knowledge and population were both increasing exponentially; students were more diverse than ever. "The insistence that all education be done in the schools has literally turned the schools into factories," he said.

Part of the solution, Frank thought, was to remove some parts of education from schools. Public education had to become a web of "adjunctive organizations," including museums, libraries, community organizations, and the media, all playing some part in education. So one of the first programs Frank put into place brought teachers to the Exploratorium in order to fire them up about ideas that they could take back to students. Starting with School in the Exploratorium (SITE), for elementary school teachers, the programs grew to encompass middle and high school teachers too; for a while there was even a lending library of small-scale copies of exhibits for use in the classroom, which were much in demand.

Teachers weren't told how to teach. They got involved in activities that left them aching to teach what they'd learned. Building on Frank's Colorado experiences, the programs encouraged teachers to develop materials and curricula that got them engaged and excited. The idea was to turn them into addicts. And the Exploratorium was only too happy to give them a fix.*

The Exploratorium hosted more traditional school programs as well. On many mornings, the place was aswarm with kids on field trips — running around frenetically and sometimes frantically trying to get everything in, sometimes having only the most cursory contact with exhibits. Many people looked at school field trips merely as breaks for teachers, ways to let kids let off steam, devoid of educational content. Frank disagreed. He thought a field trip for a child was an experience something like roaming the stacks of a library, picking up a book here and there. If a child got involved in half a dozen or more exhibits in a two-hour period, well, that was fine.

As the *San Francisco Chronicle* science writer David Perlman observed, "When you first see the kids [at the Exploratorium], it seems like all they're doing is fooling around. They jump on this and they jump on that and they're not really doing any science.

* A good number of the young people who came to the Exploratorium as Explainers, teachers, and high school apprentices are still working there, almost forty years later.

But then you realize: yes, they are." Inside their minds and hands they are storing experiences, weaving and knitting things together, stirring up neural imaginings.

If a woods of natural phenomena has enough sights and sounds to grab and keep their attention, then the kids are bound to get engaged no matter how much random running around they do. One somewhat jaded inner-city high school teacher was amazed at the transformation in his students during and after a field trip. "They went from being little professional criminals to excited kids," the teacher later told me. "Their behavior was flawless."

Funny Money

"The most common misperception is that he was innocent. But he knew exactly what he was doing. He was so sweet and lovable that while you were talking to him he could open your wallet and see how much money you had. And he could get away with it. If you did notice it, you just didn't care."

How Frank got the money to pull the Exploratorium together says as much about him as almost anything else he did. In 1969, he estimated that he'd need around $6.5 million over the next ten years, at which point the annual budget would stabilize at something like $750,000. His calculations proved to be remarkably accurate.

Frank did most of the fundraising himself, continually making the rounds with his hand out; he figured that the person who had the idea was the best one to explain why it was worth supporting. But after that first $50,000 from the San Francisco Foundation, it was hard going.

Ironically, one thing that made it hard to raise money from seemingly obvious sources was the fact that the Exploratorium wasn't in the business of certifying. Neither private foundations nor federal agencies could figure out how to classify his project: it didn't have a permanent collection, so it didn't qualify as a museum; it didn't certify, so it wasn't a school.

One factor that may have helped Frank was his past — and the Oppenheimer name. "He was a moral hero," said the eminent British psychologist Richard Gregory, whose influence on Frank (and on the early perception exhibits) was enormous. "People knew he'd been incredibly badly treated."

As the Exploratorium's reputation grew, the money started to flow more easily. Frank was known around the National Science Foundation as "the world's worst proposal writer," because his plans were "often laughably" vague, according to George Tressell, who at the time was handing out money for the NSF's informal science education program. "A typical Exploratorium proposal would say, 'There's this interesting topic, and we're going to do an exhibit on it, and we don't really know what we're going to do. Please send us two million dollars.' But we always funded them. I don't remember ever turning down a proposal."

Frank didn't really have a fundraising strategy. He just believed so much in what he was trying to do, and he worked so hard at it, that he managed to mesmerize people out of their money. Nevertheless, Frank soon recognized that he needed help. He put an ad in the *Wall Street Journal* seeking a professional fundraiser, and a young mother of twins named Virginia Rubin answered the ad. She'd never been to the Exploratorium, but she'd worked raising money for foundations. Because Exploratorium salaries were so low, Frank had to pay Ginny a higher salary than he was making. It turned out to be one of the smartest moves he ever made.

Together they were unbeatable; people simply couldn't turn them down. If Ginny was the professional, Frank was her "talent." "He was a great showpiece," she said. "His earnestness was so compelling. He was always so worried about the next pay period. He'd be smoking his cigarettes and pacing around. He didn't start at the beginning. He would sort of jump in. He might talk about the relation between art and science, or the importance of not having guards and docents."

Years later, Ginny Rubin ran into an old friend from the Ford

Foundation, who recalled, "Boy, you and Frank, what a show you were. We used to love the two of you walking into the office."

Still, Frank's fundraising was largely scattershot, and he stumbled from crisis to crisis, always landing more or less right side up, but never quite stable. At one point the staff agreed to take temporary pay cuts in order to avoid layoffs. Some turned in time sheets listing only half the hours they actually worked. Some managers delayed paying themselves until funds came in. "They did it out of respect for Frank," a staffer said. It didn't solve the underlying problem, of course. But it did keep the place afloat long enough for Frank to find another pocket to pick.

The one income source he refused to consider was an admission charge (although he did ask for donations). Woods are free, and so should be a woods of natural phenomena. He wanted visitors to wander in and look around and see what his woods had to offer — especially those who weren't particularly interested in science. He wanted people from all socioeconomic classes and all parts of the city. He wanted people to come back, even for short visits.

And come back they did. Children who lived within bike-riding distance came repeatedly after school and on holidays. Adults came back as often as once a month — meaning, Frank concluded at one point, that some 5,000 adults spent 36 hours there each year. "The experience is equivalent to a pretty good course," Frank said, "and is most assuredly a very cost-effective form of learning."

Despite the constant financial crises, the Exploratorium prospered. In 1973, heaters were finally installed, and attendance had grown to 25,000 visitors a month. By 1976, Frank had raised nearly $3 million in grants, and created or acquired 400 exhibits. There were 35 full-time staff members, 35 part-timers, 25 high school Explainers, and 12 college Explainers.

Frank was getting recognition, too. In 1973, the American Association of Physics Teachers honored him with the Millikan Award, which meant a great deal to him.

And at the same time as the woods of natural phenomena was taking root in Bernard Maybeck's crumbling Palace, a parallel story was unfolding in the same cavernous space: the Exploratorium was becoming a gathering place for local (and soon nationally known) artists. In some ways, it radicalized the notion of public understanding of art as much as it did the public understanding of science.

8

A DECENT RESPECT FOR TASTE

"I cannot really say that I have noticed any difference in the way visitors to the Exploratorium behave on sunny and cloudy days. But for the staff and especially for me, and my feeling for you when you come to visit, whether the sun is turned on or not makes an incredible difference. This is because of the Sun Painting. I think it crucially important to have an exhibit of such scale and beauty . . . The exhibit demonstrates light scattering, prisms and mirrors and color, and sunlight. It is a brilliant abstract painting that shimmers and changes as people move in the light path and brush against the Mylar mirrors behind the frosted screen. We have other exhibits of beauty, and without them the museum would be sterile and incomplete; but none are so fine as the Sun Painting."

To Frank, it would have been all but pointless to build a science museum that didn't include art in a serious way. In his eyes, the two were inseparable: "One cannot truly understand nature without also discovering the ways in which it is related to human experience and feeling," he said, "and one cannot appreciate human experiences without learning that they are imbedded in a broad concept of nature."

It was, in fact, impossible to discriminate between art and science at the Exploratorium. When Buckminster Fuller stopped by for a visit, a woman accompanying Fuller sniffed, "I thought this was an art museum. Where's the art?" Fuller waved his arm expansively, saying, "It's all around you."

Early on, I remember experiencing the same confusion myself. I asked Bill Parker — whose "Lightning Balls" have since become standard toys available in many science museums — whether he considered his plasma discharge sculptures art or science. "Neither," he said. "It's nature."

I was also taken aback at first when Frank described particle accelerators as modern analogues of Gothic cathedrals — but of course, he was right. Both are stupendous creations of human minds designed to reach out to the unknown and ask the most profound questions. And as much as scientists create art, artists do science — often adopting forefront technologies or new understandings of visual perception. Artists make intellectual decisions and scientists make aesthetic ones.

So almost immediately after the museum opened its doors, artists started bringing their works to the Exploratorium. The word got out that "anything goes," as Bob Miller remembers it. Bob was living in a dingy apartment in North Beach, experimenting with a rack of prisms and small slivers of mirror, when he heard from an artist friend about a strange scientist who was building some kind of new place at the Palace of Fine Arts. Bob invited Frank over to look at his work. Frank hired Bob on the spot, asking, "How soon can you come over and play?"

The experiment with prisms and mirrors turned into the "Sun Painting," in which a palette of pure color extracted from sunlight creates three-dimensional otherworldly landscapes — a swirling wash of color that spread over the Exploratorium's entrance as webs of pure gold and green and blue and red; wandering through the light felt like swimming in an underwater rainbow.

In keeping with the Exploratorium's philosophy of total honesty, the "Sun Painting" shows its colors only when the light from our local star does. Waiting for the sun to break through the San Francisco fog and bring the sculpture to life could be an endless purgatory of tried patience—and on gloomy days, it didn't come on at all. The experience felt, as Muriel Rukeyser put it in her poem "The Sun Painter," "a good deal like real life."

Many of the earliest exhibits at the Exploratorium were creations of artists. A typical piece was "Bathroom Window Optics," ghostly geometrical light forms that seemed almost alive, transforming in shape and size as one approached. On closer inspection, it turned out that the shapes were made by ordinary Christmas tree lights filtered through textured glass of the type used to diffuse light in bathroom windows and shower stalls. It was so simple, and yet it produced such fantastical effects.

One pivotal piece of (dare I say) serendipity was the donation of the "Cybernetic Serendipity" exhibition from the Corcoran Gallery in Washington, D.C. One piece turned sound into ever-changing images on a television screen. Some people talked or screamed into it, but a pure voice or the sound of a flute could produce quite beautiful forms. One day, a deaf child stayed at the exhibit for nearly an hour, making sounds and watching the patterns. The child was entranced to see that what he was doing with his mouth and throat was having such a singular effect. He stayed there so long he got hoarse.

The show had been put together originally by the Institute of Contemporary Art in London before moving to the Corcoran Annex in Washington. Frank talked the Corcoran into lending it to his nascent museum. When it turned out to be too expensive to ship, the Corcoran's staff rented a van and drove it across the country themselves, and then helped set it up. The exhibit convinced Frank that art could have a major role at the Exploratorium.

The artist August Coppola (brother of Francis, the film director) came by and decided he wanted to build something that had to do with "touch and tactile" senses. So he got a little money from the National Endowment for the Arts and brought in dozens of students, who worked all summer without pay; he persuaded an engineer and an architect to lend their expertise. The "Tactile Gallery," as it became known, is a pitch-dark multistory rabbit hole where people feel their way through a maze of artistically rendered "chutes and ladders" (the metaphor is mixed, but apt) and wind up in a . . . well, that would be like giving away the end of a mystery story.

Tad Bridenthal brought in his "Limbic System," a crawl-into sculpture of infinitely reflecting colored lights, which became the cover art for my article in the *Saturday Review.* (Alas, people tell me that the artist later either accidentally dropped or deliberately threw the sculpture off the back of his truck.)

The physicist Jan Pusina remembers being a "poor struggling artist" when he walked into the Exploratorium and met Frank. He built a "Multiplied Glockenspiel," and Frank was delighted to discover it created "artificial harmonics."

An Englishman stopped by on his way to Australia and left his "Light Form," a ghostly impressionist sculpture created by light reflecting off rapidly revolving brushed-metal plates. A student at San Francisco State University brought in a "Polaroid Projector," which created shifting, brightly colored patterns out of plain white light using only two plastic Polaroid filters, cellophane tape, and an overhead projector.

Jackie Oppenheimer discovered Ben Hazard's "Pinball Machine" — a mesmerizing installation that exploited polarization to create an ever-changing landscape of colored forms — in the Oakland Museum. Doug Hollis created an Aeolian harp that perched on the roof over the entrance like a spindly singing insect, long threads attached to speakers, greeting visitors with a low drone, as if the building were humming to itself — rather like Frank himself.

This infusion of art wasn't restricted to physical stuff. The Exploratorium was a three-dimensional blank slate, a magnet for all kinds of innovators. The publisher Stewart Brand had parties there to raise money for the *Whole Earth Catalog*. Artists and musicians used it as a venue for sometimes wild experiments.

One musician, in preparation for a concert, spent the afternoon in the shop cutting up tiny strips of mirrors. He clamped them onto a frame, put transducers on selected ones, and stuck them on the glass. Then he hooked them up to a sound mixer and installed an amplifier with humongous speakers. As the concert began, he shined lights on the mirrors. He hit the mirrors with a tiny mallet to make them vibrate, sending out screeching tones. Finally, he turned up the volume as high as it would go and hit the mirrors harder and harder until they started breaking. By the end of the performance, nothing remained but shattered glass.

Another group of musicians put choirs at opposite ends of the huge cavern; they wanted to experiment with the two-second delay created as the sound waves made their way from either end of the building, but since the choirs couldn't see each other, it was very hard to conduct.

"The place was almost empty," Frank said. "There was just this thousand-foot-long space. The music was just incredible. One person, Dinwitty, wrote a special composition just for the space. Then there was Carlos Carvajal, he had a small dance group here. Holden had a concert. That just kept up."

(Of course, being Frank's place, the Exploratorium had its fair share of unplanned events. Someone brought in a nineteenth-century projector, with hand-painted slides and a kerosene lamp inside. One day it caught fire. "There were great flames," Frank recalled. "All the public visitors stood around and looked at it, waiting to see what was going to happen, assuming it was just another exhibit." Frank ran to grab a fire extinguisher. "But for the visitors, it was just a normal thing happening in the Exploratorium.")

By 1976 there were short films on weekends, a program put together by Liz Keim (and still going strong, with Liz as curator). Liz came to the Exploratorium as a weekend receptionist and quietly became part of Frank's family. Frank hardly ever missed a screening. "He would see a film and tell me why it really worked in terms of the rhythms of the music and the image," Liz said. "And if anybody ever laid music over a film that wasn't intrinsic, it would infuriate him."

Frank invited poets to read, painters to paint. In the San Francisco Bay area, the Exploratorium become known as one of the major arenas for new music, new film, new art. He asked his friend Muriel Rukeyser to come and help with writing explanations: "How does one explain lateral inhibition in the retina of the eye, or the way in which light waves can cancel each other to produce darkness from light?" he asked. "How does one imagine electricity?" He thought a poet could come to his rescue. But it wasn't so easy — and was much more interesting. "It turns out that the communications of the poet do not necessarily pop out automatically to the uninitiated any more readily than do those of the physicist," Frank said. "Muriel could help, but she frequently had to start from scratch along with the rest of us."

What usually happened was that people asked Muriel to read. And so Frank and Muriel put on a series of "Readings on the Forefront of Science and Poetry." Writers would read their poems, and scientists would read whatever they had written, and then people would talk about the similarities and differences between the ways scientists and poets use imagery, how they perceive meaning, how they describe newly discovered phenomena. "At the leading edge of experience in philosophy, science and feeling," Frank mused, "there is inevitably a groping for language to translate the insecure novelty of noticing and understanding into a precision of meaning and imagery."

Frank attended most of the sessions. He noted that when science was young, before jargon had become standardized, scientific writing was not so "dehumanized" as it is today. And while he tried to find differences in the ways physicists and poets communicated, he could not. "Both could evoke expressions of caring, of imagination, and of passion. Both could be either starkly descriptive or intensely polemic."

During a discussion after one of the readings, Frank complained that many of the young physicists he encountered didn't seem to really care about the answers to the questions they raised in their research. One of the poets was surprised at his comment. Like most people, she thought scientists weren't *supposed* to care. "What a strange misconception has been taught to people," Frank said. "They have been taught that one cannot be disciplined enough to discover the truth unless one is indifferent to it. Actually there is no point in looking for the truth unless what it is makes a difference."

A Decent Respect for Taste

"Why do we admire children when they build a pattern of objects with great symmetry, but then refuse to understand that when they 'ruin' it with a misplaced object, they, in fact, have made an aesthetic decision?"

Art and music had always been central to Frank's life. His flute went with him everywhere, whether he was in the mountains, on a boat, or in the office. You'd be in the midst of a conversation and suddenly Frank would be gone; you'd be answered with a sweet snatch of Bach or Purcell. He encouraged the staff to play as well, putting on "talent shows" where staff members would play the piano or display their photography or paintings. He encouraged my early efforts at the flute, and later — awful as I was — we'd play together while his second wife, Milly, accompanied us on piano. I remember telling Frank (only partly in jest) that if I could learn physics and the flute, then surely I could learn just about anything. Perhaps I would try my hand at painting. Frank shook his head. "Painting is *hard*," he said.

When I visited Frank at his home in Sausalito, I bumped into art everywhere, from the Picasso drawing in Frank's office to the silly wooden quacking duck hanging from the ceiling. By the time I met him, most of his art collection had been sold — some to support the family during their years of exile — so that only some Picassos remained. Still, few people walked into Frank's house without being stunned by the Blue Period *Mother and Child* hanging without fanfare in the living room. When Frank was teaching in Colorado, a graduate student went to visit him at home. "That's the best Picasso imitation I've ever seen," he said. He was shocked when Frank responded matter-of-factly, "It's not an imitation."

In Frank's mind, aesthetics had a place in the most mundane things. He railed at objects that didn't "feel nice"—usually if something was frustrating or aggravating in some way. Vacuum cleaners were high on his list. "They are hard to steer and they make that awful noise. It's because nobody cares what it is like to use one."

He hated buying batteries in a plastic package that you had to cut open with a knife. "It makes that awful crinkly noise." Batteries should be sold loose, he insisted. "And frankfurters too! I don't like them packaged. They are all sort of shrunk that way." One of the few museums he didn't like was the Guggenheim in New York, because you could walk only one way, and even if you went against the stream, you had to go in a particular order.

At the same time, he took great aesthetic pleasure in things others might find decidedly distasteful. When Frank visited my home in Port Washington, New York, during an oppressively muggy summer, he immediately pronounced, "What a glorious day!" At his insistence, we took a leaky rowboat out to our ancient O'Day Mariner, Frank in his business suit, water up to his ankles, humming as he bailed water. He stood on the bow of our boat with his suit coat open to catch the nonexistent wind. Reveling in it all.

Frank's often peculiar "aesthetic" once played a role in an uncomfortably close encounter between Orestes and me. It happened one weekend when Frank and Jackie and I drove to Bodega Bay, on the northern California coast, to look at some property they'd bought. Orestes came along as always, taking up most of the back seat of Jackie's broken-down Plymouth Barracuda, while I squeezed myself into the other corner.* Orestes smelled bad even

* I don't hate dogs, really; I have a big black Lab myself; Orestes was an entirely different matter.

in the best of times, but then Frank and Jackie thought he should get out for a run, so we stopped the car, and the dog ran out and disappeared. A half hour later, he returned, encrusted in cow dung, and resumed his sprawl on the back seat. I was disgusted beyond words; Frank found the episode funny — a broad comedy of the "city girl meets country dog" variety.

He later wrote an essay about the subject of smell in which he admitted liking all manner of odors, including acetone, ether, manure, dirty socks, even a "whiff of skunk." He found it remarkable that every part of the human body — ears, mouth, underarms, feet — had its own strong smell. And he was disturbed that advertisements implied that smells like sweat were offensive. It was part of a "concerted attempt to sterilize human experience," he wrote, "part of the set of rules that say we are not to raise our voices, not to argue about religion, not to revolutionize politics, not to be awake when we should be asleep or drowsy when we should be awake or ever miserable or mischievous.

"We can now be odorless when we smell. But have you ever cringed as you rushed to your car in the rain? I have! Then with a flash of insight I say to myself, 'What the hell?' and I walk slowly with my face skyward, letting it be washed by the rain."

He concluded the essay in a way that almost suggested a connection between the use of deodorant and the mindset that led to the use of the atomic bomb: "We might know each other better and more humanly if we ignored the deodorants that are foisted on us and stuck to plain soap. It is silly to scoff at inventions, but it is equally foolish to let them carry us where we do not want to go. We do not need all the different kinds of umbrellas that we have invented. They get out of hand and instead of protecting us from nature, they isolate us from it and so from each other."

For a man who almost always wore a suit and tie, whose tastes were so refined and whose manners were so gentle, this disdain for coverings and filterings of all kinds was oddly out of kilter. But it was consistent with his insistently inconsistent character. He was both cultured and crude, airy-fairy idealistic and intensely practical, on a cloud and down-to-earth, smooth and elegant and rough-and-tumble. Somewhere along the line, the filters had fallen off, the internal Caution signs we all carry around ceased to work — or perhaps, given what he'd been through, they no longer seemed to matter. Either way, the result was that his sense of aesthetics could be all over the map.

Frank's aesthetic pervaded every aspect of the Exploratorium, right down to the doorknobs. "It may actually be better to make a doorknob square, but those big heavy round doorknobs feel nicer," he said. "They're really lovely. So we use them." He even made an exhibit out of an enormous ball bearing, which he simply put out on the floor with a sign that read "Some machinery feels nice."

The fact that children could run around the museum at will was part of Frank's aesthetic too. And years later, when the Exploratorium was forced to charge admission, he came up with the idea that each entry ticket should buy a six-month pass. Though he didn't want to charge admission at all, this solution, he said, "felt nice." In fact, most of the decisions he made about the Exploratorium, he later concluded, were based on "things that I'd like to happen to me."

For aesthetics to become deeply ingrained, Frank thought, a "decent respect for taste" should be inculcated in the very young — even when the choices might upset their parents or teachers. He believed, for example, that children should be allowed to choose their own clothes from the earliest ages, no matter how outlandish the outfits. One day at lunch, in a rather fancy restaurant, he encouraged my then four-year-old son to play with the silverware and — hey, why not? — the food. Frank thought children weren't even encouraged to believe they could convey something important in the paintings they did in school.

"We need to look at children with new eyes," he wrote. "Why do we self-righteously ignore (and even berate) children's intense discrimination among textures and tastes of food, or object to their enjoyment of the feel of food on their hands and faces? Why does a group of adults invariably laugh at children when, as two-year-olds, they begin to move in response to music? Why do we refuse to recognize that knocking down a just built, teetery structure of blocks is a fine example of an order-disorder transition?"

As an example of such early aesthetic sensibilities, he often told the story of a four-year-old girl he watched uttering "a shriek of delight after watching a rather spectacular disorder-to-order transition." A boatman at a lake in Golden Gate Park was preparing to move the public rowboats from the docks to a shed in the middle of the lake, where they were stored at night. "He started by untying them from the dock and tying them together in an impossible looking, random mess of every which way boats," Frank wrote. "The four-year-old looked on with increasing anxiety. Finally, the boatman attached his putt-putt to one of the boats and took off. The fifty or so boats broke out of their tangled web and followed him in two lines that formed a beautifully curved symmetric 'V.' It was at that point that the four-year-old burst forth with her shriek of aesthetic delight."

As for the toys children were normally expected to play with, Frank thought they were "an abomination," because "they do not feel nice." Frank thought that toy stores should stock their shelves with "commercial, industrial and military surplus items." He himself vividly remembered the pleasure of being five years old and "sinking submarines" by hurling "a large, marvelously built and balanced screwdriver" at tin cans.

He also remembered with delight a multiwire egg slicer he discovered in a kitchen drawer. "Such slicers still exist, but they were stronger and tougher in my day," he wrote. "They made wonderful music as one twanged the five or six strings that were each somehow under slightly different tension. I still twang egg slicers whenever I find one, but I am usually disappointed. They do not feel or sound as nice as the one in our kitchen used to feel and sound."

It was the aesthetic of such toys, he believed, that helps children develop taste — helps them recognize what is nice, what is beautiful, what is humanly desirable. He wondered whether toys were an introduction to art. "Toys that are well conceived and not just junk may be the prime movers that induce people to strive for a nicer and nicer world in which to live," he said.

Frank himself loved to give people toys — not surprisingly, very nice ones. He gave my son a stopwatch and a microscope-spyglass combination before he was six years old. He gave the perception researcher Richard Gregory a variable prism. "He loved ingenious gadgets," Gregory remembered. Many of these toys he made himself — a necklace of ball bearings, a pendulum toy. I treasure the brass top he turned for me on a lathe: Frank designed it to "sleep," as he put it. Normally, a top precesses as it spins, tilting tipsily as it swirls, but Frank's top spins perfectly upright until it runs out of steam and falls over flat. He also made me an "earthquake-proof" flute stand with a huge hunk of lead worked into the wooden base.

One of the reasons Frank valued play so highly is that it is one of the few activities that explicitly encourages people to rely on their sense of aesthetics. In play, you try something because you like it. You do things that are pleasing, that seem "right" in one way or another, that reward the senses. This "aesthetic feel," he believed, was a critically important human quality. Taste was an intellectual tool as valuable as logic, empathy, or common sense.

A Matter of Urgency

"It's through familiarity with the arts that I think we will make the kinds of changes that make life stay human."

A respect for aesthetics, Frank thought, should be a central part of sound decision-making. He didn't think it would be out of place — though he admitted it would be impractical — if Congress, unable to decide on a difficult matter, took a recess to visit the National Gallery for guidance. "Art," he liked to say, "is not valid merely to decorate our surroundings with statues in the plazas of skyscrapers, any more than science is valid because it provides the conveniences of electric shavers."

When people said, "We need more art," Frank complained, they tended to say it in the same tone of voice they used to say, "We need more trees." True, both art and trees make our surroundings more pleasant, but artists also make discoveries about nature and human nature that are on the same level as the discoveries scientists make. And in the same way that we can make better decisions about global warming if we know what scientists have discovered about the earth's changing climate, so we can make better decisions about human affairs and environments if we pay better attention to what artists have learned.

Frank worried a lot that the arts were undervalued, and that aesthetic considerations were largely ignored whether people were designing schools, supermarkets, bridges, or "topless dance joints, nuclear weapons, and homes for the aged." If money was tight, aesthetics was the first thing to go.

"We're in terrible trouble because of that," he said. Historically, places that respected the arts and based decisions on aesthetics were also places where "better things happen," he argued. If art were considered more important, he wrote to David Rockefeller, "many of the things that now shock or degrade people's sensitivities would not be tolerated."

Of course, as a physicist Frank had learned to trust aesthetics. Scientists often try things because they "smell right." They believe in theories because the mathematics behind them are "beautiful," even when they contradict evidence. Like artists, scientists develop an eye (and ear) for nature, a sense of what is true and what is not. Laypeople, too, should be encouraged to rely on their aesthetic sense to guide their decisions, Frank thought; if a certain course of action or behavior struck them as "ugly," then it probably was.

Artists and scientists, Frank liked to say, are the official "noticers" of society — those who help us pay attention to things we've either never learned to see or have learned to ignore. Artists of all ages and in all lands have traditionally sensitized people to nature through their poetry and painting, sculpture and drama, and, less obviously, through their music. Without art, "one even ignores what people's faces are like," Frank said, "but by seeing paintings of people's faces you begin to look at them again, and I think that the same thing is true of science. You look at the sky and you see the stars, and it is just an amorphous mass; but suddenly somebody talks to you about it and you see that some stars move with respect to other stars."

He gave artists credit for teaching us great human truths. Without art we might not have recognized the universality of the feeling between mother and child, he said, or the emotion between man and woman.

"If you don't know how to notice, you can't do anything well," Frank said. "You can't even relate to people well." You can't tell if someone is angry or amused or hurt, or if the weather is about to change and maybe you should get an umbrella. You'll miss that guy lurking in the shadows, and Saturn shining overhead.

Frank wondered why urban planners didn't look at paintings in order to learn how to design cities; why architects didn't look at Cézannes to design cafés; why people didn't look at portraits to find meaning and wonder in the transformations that occur in aging faces and bodies. Why didn't people realize that paintings enable us to find pattern and structure in scenes that would otherwise seem shapeless, amorphous, and emotionless?

So above all else, the Exploratorium was a place that encouraged the kind of everyday noticing that helped people develop an eye and ear and feel for the social and physical universe around them — an almost artistic sensibility.

Visitors to the Exploratorium certainly build up intuitive feelings for physical phenomena as much from artistic works as from "science" exhibits — whether the subject is wave mechanics or the nature of light or fluid dynamics or the quantum properties of matter. To this day, when I imagine stars being born from swirling interstellar clouds, I think of Ned Kahn's "Whirlpool"; when I think of exotic bits of matter coming into being seemingly out of nothing, I see his "Visible Effects of the Invisible." Most of my intuitive feel for light and color and shadow and reflection comes from Bob Miller, and there is a lot of the spinning black hole in Doug Hollis's "Vortex."

Even in terms of process, Frank pointed out, artists and scientists work in similar ways. They both start by noticing patterns in space and time, trying to make sense of them, rearranging them, and then linking patterns together in ways no one had thought to do before. They make sketches with equations or charcoal. They elaborate and synthesize. "They end up with a composition which means more than what they started with," Frank wrote—melodies and theories. In essence, they make patterns of patterns that reveal new insights. Their compositions, theories, and other works separate relevancies from trivialities; provide a framework for memory; reassure by creating order out of confusion.

Of course, all people spend much of their time perceiving and making sense of patterns; even animals do it (the dog knows exactly what follows the fetching of the leash). Frank once told me that when he can't see a pattern, he gets "miserable." But artists and scientists spend their whole lives looking for patterns in nature, and so perhaps learn to see more than the rest of us.

To Frank, artists were people who looked at human experience in the same way astronomers looked at the sky through telescopes. Just as astronomers collect, codify, interpret, and communicate what is known about the stars, so artists collect, codify, interpret, and communicate what we know about human feelings.

The reason we need this knowledge so much, he argued, is the rapid pace of change. If things didn't change, then perhaps education could simply be a matter of learning to conduct business and follow directions. But everything in nature changes. People inevitably change the world in which they live. They change themselves. And as people(s) change, at some level there's always a worry that we might lose some of that indefinable and extraordinary specialness that makes people human. And who can define that essential nature of humanity we so want to preserve? Who can tell us (or remind us) what is fine, what is beautiful, what is important, in humankind? Frank claimed that was the role of artists.

Decisions about how to adapt to inevitable changes are based, by necessity, on what we believe is possible. Science tells us what is possible in the physical realm, and in doing so, gives us a basis for action. If we don't know that it's possible to make antibiotics, for example, we won't learn how to protect ourselves against disease. In the same way, Frank thought, art tells us what is possible in human experience. What's more, it tells us how we *feel* about the various possibilities — or at least how an individual artist feels, and therefore one way it is *possible* to feel. "If you don't know those things, you are not going to make good decisions," he said.

And just as technological inventions help us cope with changes in the external environment, we need "heightened social and emotional awareness and invention," Frank said, to cope with changes in the human environment.

Alien Territory

"There are two things that people [are surrounded by and] avoid trying to understand. One is music, and the other is electricity."

If science seems unfamiliar territory — a realm where strange characters do obscure things with complicated machinery — at least it is territory that is socially acceptable to steer clear of. Art is another matter. Because art is "culture," it is a realm where everyone (everyone who is not a barbarian) is expected to be at home.

Yet museums and art galleries can seem less than welcoming to outsiders. Some are musty and mausoleum-like, places of reverence where you can almost smell the incense and holy water, final resting places for artifacts of long-dead worlds — as remote and untouchable figuratively as physically. Others are antiseptic and sterile — dead in a different way (they often seem to maintain this aura even when the artworks themselves are playful). The appropriate emotion in either case is awe. One treads silently and cautiously, avoiding at all costs public displays of ignorance or confusion or emotion. (And oh, the stares you'll get at the symphony should you commit the faux pas of applauding between movements!)

For most people, museums fall into the same category as church or school — places to be revered more than really enjoyed, places to learn and appreciate rather than to play, places of silence and solemnity. In a very real sense, museums are preserves that keep precious objects safe from people.

So for me, nothing was quite so unorthodox as Frank's take on museums, which he loved in the same passionate, meddling way he did everything else. For one thing, art, according to Frank, was an absolutely appropriate venue for play, and also for breaking rules. Going to museums with him turned my long-held notions upside down, changing museums from repositories of things to adventure parks. You should race around, talk loudly, laugh, and touch just about anything when the guard isn't looking.

On one occasion when Frank was visiting New York, we went to see an exhibit at the Whitney Museum of Duane Hanson's uncannily lifelike sculptures of ordinary people: the woman laden with shopping bags, the camera-toting tourists, the janitor. Suddenly I noticed that Frank wasn't there. It took me a few moments before I realized that he'd hidden himself by freezing into the backdrop of the art, posing as a Hanson sculpture. He wanted to see if anyone would poke him to find out if he was real. He got the idea because he himself had just poked a very still, but very alive, security guard — thinking the guard might be a statue. He was so amused, he wanted to try the same experiment on himself.

Frank well understood that for many people, art was every bit as intimidating as science. And not only was art itself alien territory: artists — like scientists — were often seen as strange characters who do unpredictable, outrageous, sometimes dangerous things. If scientists were logical robots, artists were creative freaks whose work had nothing to do with our daily lives.

One of the reasons for this attitude, Frank thought, was the fact that the work of both artists and scientists was increasingly inaccessible to the general public, increasingly removed from ordinary experience. Neither the public nor students were expected to understand the latest discoveries in physics. And yet people *were* expected to understand contemporary art even if they had no background in the arts whatsoever — and that made them feel confused, stupid, and distrustful. In school, children were often exposed to "modern" art divorced from context or history — given no tools to help them "read" a poem or a painting or really listen to a piece of music. Art was too often taught as science is — as a "nonexperiential and hollow mimicry of what artists (or scientists) are publishing at the forefront of the field."

Artists shared some of the blame for this situation, just as many scientists shared some of the blame for the impenetrability (to the general public) of much of science. "Contemporary artists," Frank wrote, "tend to either sneer at people who cannot extract meaning from their works or, alternately, deny that their works have meaning, insisting that they should be appreciated as meaningless aesthetic experiences."

So Frank hoped the Exploratorium might do for art what it did for science — make people comfortable and involved enough to participate in the process, to ask themselves questions like "What if?" What if I tuned the frequency as high as it could go? How would changing a single element in a painting alter its impact?

One quasi-didactic attempt to make people ask questions of art — and one of my personal favorites — was an installation exploring the use of balance in Saul Steinberg's drawings. The drawings were enlarged and mounted, but an element would be deliberately taken out and instead put on a plastic overlay, which the visitor could add or remove to see how balance was achieved — not just through composition, but also through meaning.

This was dramatically illustrated by a sculpture that towered over the exhibit entrance. A long horizontal platform was perched on the top of a thin, pointy pyramid, rather like a seesaw. On the left side were the enormous figures 5¼ + 2¾; on the right was a tiny numeral 8. Although the numbers as objects were grossly out of balance, and should have tipped the seesaw in a crashing tumble to the floor, the *intellectual* balance was perfect—effectively and completely counterbalancing (so to speak) the visual effect.

Frank also tried to bring a certain comfort level to the arts through programming. One of the earliest series of events was initiated by the flutist Leni Isaacs, now the public affairs director for the Los Angeles Philharmonic. "Speaking of Music," as Leni called it, presented all kinds of music, from classical to avant-garde, in an informal setting that encouraged people to ask questions about anything from the subtle intricacies of harmony to the reason a violin was shaped like an 8.

It wasn't long before art became seamlessly incorporated into just about every aspect of the Exploratorium. Thomas Humphrey (an artist as well as a physicist) was already working closely with the San Francisco Art Institute (one of the works he created for it mimicked Marcel Duchamp's iconic *Nude Descending a Staircase,* using a strobe light). He was soon asked to design a course, which he called "Perception in Art and Science." When Thomas left temporarily in 1978, Frank and Rob Semper, another physicist, took over the class. Frank would dream up demonstrations that were "incredibly simple and incredibly insightful," Semper said. "It was the most fun I had with Frank." Artists became involved in the school at the Exploratorium almost from the start.

And artists in residence became a regular presence at the museum. One of the first of the invited artists, and the one who gave direction to the program, was Pete Richards, the neighbor of the Oppenheimers who grew up in Blanco Basin. Pete was just getting out of graduate school with a degree in sculpture when Frank asked him to help create the Exploratorium. In his art school, "all of the work we were doing was totally introverted," Pete said. "It was all about Who am I? What am I trying to do? And the work we made had to look like 'art.' Then I walk into this place and here's these gizmos made out of two-by-fours and gaffer's tape and surplus motors, stuff that looked like it should fall apart at any moment, yet it communicated some really interesting ideas . . . My art changed drastically after coming to the Exploratorium . . . It was all about noticing things and trying things out and developing a real interest in the way people interact and respond in public situations. I was experimenting more like a scientist."

Richards started experimenting with tides, an exploration that culminated in his "Wave Organ," which sits at the end of a jetty in San Francisco Bay, not far from the Palace of Fine Arts. Pipes leading into the water create a panorama of sound that encircles you, a symphony conducted by the bay itself as it swells or sinks or chops or calms in response to the waves and tides and weather. "I like the way it's got a direct link with the cosmos," Richards said. "The way it behaves is directly related to the position of the moon to the earth and the earth and moon to the sun, and what you hear is really the result of that relationship."

Hardening of the Categories

"It has something to do with physics."

The sensibilities of several artists were so in sync with Frank's aesthetic that they played a major role in shaping the Exploratorium. One of these was Bob Miller. Impossibly tall (about six foot seven), eccentric, and uncannily creative, Bob complemented Frank perfectly, and together they made a wonderful team, eternally playing in the intersections of physics, perception, reality. One of Bob's sculptures, for example, is an optical illusion in which a concave object appears to pop out and follow you as you walk by. The sculpture is made up of an inside corner of a box that appears to turn into a cube, and so Bob calls it his "Far Out Corner." At one point, Frank encouraged him to patent it. They were both amused when the U.S. Patent and Trademark Office maintained that he couldn't patent an effect that existed only in someone's mind. As if there's a work of art or science that doesn't! The office finally relented, and Bob got his patent.

Another of Bob's pieces came to the museum from an art fair. It was a box of silver Christmas tree balls stacked together in a way that made for curious optics; infinitely reflecting light actually made the edges of the balls look black. Bob stuck the box on a stick and put it in a planter for the show. Later, he noticed a mother and child stopping to look at it. When the child asked what it was, the mother dragged the boy away, grumbling, "It has something to do with physics."

The insight behind many of Bob's creations — and the lesson that lodges in your head after spending time with them — is that no one ever sees anything *but* light, and then only as it emerges from whatever surface it last scattered from or traveled through. Everything else is imagination and projection. As far as your brain knows, light has no history; even if it's been bent and spun around several times over, your brain assumes it's coming at you in a straight line. So you "see" the image *behind* the mirror even though you know full well there's nothing there. Your brain places the image exactly where it would be if there were an object behind a *window* instead. Because we believe that reflections are real (and why not? all we're ever seeing is light), it's no stretch, in Bob's "Floating Symmetry," to stitch together one-half of a person plus the person's reflection and "see" that person fly.

Like Frank, Bob was an acute noticer, so if you sat with him in a restaurant, for example, the table became a big toy box of things to experiment with — glasses and silverware and white cloth and colored lights and shadows. People would stare at the goings-on at our "kids' table" and, more often than not, come by to see what was going on. The one sad part about such outings — or at least those without Frank present, especially after Frank got sick with leukemia and then lung cancer — was that we often drifted into discussions about what would happen to the Exploratorium when Frank died. What we were really worrying about was what would happen to *us* when Frank died.

Another artist whose work is deeply entwined with both Frank and the Exploratorium is Ned Kahn, whose tornados and clouds and wind sculptures have earned him a worldwide reputation (and a MacArthur Foundation "genius" award). Ned specializes in making the invisible visible, creating sculptures that capture the complexity of the nonlinear dynamics behind the way water droplets organize themselves into clouds, stars into galaxies, birds into flocks, neurons into thoughts, motions of rocks into avalanches, electrical impulses into heart attacks. His work draws people into what he calls "cloud time" — the internal life of fog — and alters their perceptions so they begin to see kinetic sculptures in windblown trees or in bits of garbage that twirl into tornadoes in the corners of buildings.

Ned came to the Exploratorium right out of college as a shop apprentice, and for the first six months he had almost no contact with Frank. Then one day the head of the shop came to him sheepishly and announced, "Well, Ned, we had a meeting, and I'm really sorry, but we sold you to Frank." Frank was already pretty sick at that point, and he needed an assistant. "And they were all kind of expecting me to freak out," Ned remembered. "But I was just so excited, because I hadn't really had any connection with Frank."

Ned and Frank started spending several hours together almost every day. Ned would wander into Frank's office, where the two would go through the drawers in a metal rollaway cabinet. "It was full of weird little Frank things, strange artifacts he had collected over the years, all kinds of bizarre optics and irises and crystals," Ned remembered. "It was a box of wonders. I'd say, 'What's this, Frank?' and he'd say, 'Oh, that's from . . .' and it was like pieces of some atomic physics thing. Everything had a story."

They'd fool around with half-baked ideas Frank had for exhibits — most of them made out of string or cardboard. He'd tell Ned what he found interesting about whatever it was, and suggest that he make an exhibit out of it. "We were amazingly prolific, the two of us," Ned said. "We just cranked stuff out. It was a great collaboration. But the best part of it for me was that I got to spend all this time with him.

"I was full of all these burning questions about the physical universe," Ned said. "And a lot of these questions were on really basic things. Like I remember asking him what electricity was, like when you turn on a light bulb, what was actually running through that wire? And he'd spend hours and hours trying to give me an inkling. I kept asking him, 'But what is actually going through that wire?' And at a certain point he said, 'Well, nobody knows. We know how to do stuff with it, and we know the effects that it has, but what is actually going through the wire, no one really knows.'

"That was an earth-shaking thing for me. The whole time I was in college, my whole life up until then, I thought I didn't know what was going through a wire or all these other basic questions because I hadn't taken the next class. So hearing that from someone like Frank, who was the smartest human I had talked to up until that point, was mind-blowing for me.

"And that was a major influence on me," Ned said, "because I got interested in the edges of what's knowable . . . those phenomena that are so complicated and intricate that they're physically unpredictable."

Frank also learned science from artists. One example he talked about a lot was Doug Hollis's "Vortex," a large, clear glass cylinder filled with water that swirls into a sinuous aqueous tornado. Watching it, Frank noticed that the tornado doesn't go all the way to the bottom but has "a little fine fuzz"; that things twirl around at different speeds in a way that can't be entirely explained by conservation of angular momentum; that complex ripples embellish the undulating form in unexpected ways. He talked about being inspired to try to calculate some of these effects.

"I mean, you begin to think things out just by watching a thing like that, whereas I've never thought of watching a bathtub or thinking about a tornado," Frank said. "So [there were] all these things I'd never thought of until I saw that exhibit, and it was done by an artist."

Aesthetics and the Right Answer

"I just don't like the idea that there's no right or wrong in art."

One reason Frank thought that people didn't take aesthetics

seriously was because they didn't think art had right answers in the same way that science did. If physics was seen as a tyranny of right answers, art seemed to have no right answers at all.

Physics students spend most of their time solving problems for the "right answer," Frank noted. Most textbooks listed the answers for, say, the even-numbered problems; students who can't find the answers for the odd-numbered ones feel guilty and stupid. Physics is taught as a "right answer" subject, while its metaphysical implications are ignored "along with the creative nature of scientific activity." Art students, on the other hand, are rarely told that "right answers" are also important to artists. "In the popular view," Frank said, "no one looks to art to provide any answers at all."

But Frank thought there was every bit as much validity — and as many "right answers" — in art as there was in science. Just because artists deal with more complicated subject matter, such as human feelings and emotions, it doesn't mean that you can change a line on a Picasso and not ruin it. The works of artists were valid, Frank thought, because just like theories in physics, they led to the discovery of things that existed in nature but that no one had yet perceived.

During one long tape-recorded conversation about validity in art, Frank and I argued at length about the idea that art could be judged valid under much the same circumstances science is — that is, if a work of art somehow predicts the existence of phenomena that might be found in nature but have never been seen before. This could be anything from the colors in faces to abstract shapes in a painting that also appear in the shadows of buildings or hills.

And what of the performing arts? I asked. What about the ballet? Frank thought perhaps the corresponding natural phenomenon might be "the wonderful sense of freedom from earthliness" that is often reproduced in dreams. So even if the art doesn't correspond to a real human experience, to be valid it ought to correspond to a *plausible* human experience. "I think that's at least something we can speculate about," he said. "How to test that, I haven't the slightest idea."

Not having the slightest idea didn't stop him at all. Frank was always poking at art the same way he poked at people and other natural phenomena. In an introduction to the *Exploratorium* magazine on the subject of color, he digresses from discussing the physics and physiology of perception to the "experiential, emotional, and aesthetic components of color. Why do we say we are 'feeling blue'? Whence the term 'mood indigo'? Why Royal Purple? Artists talk of warm and cool colors or those of foreground and distance. Colors can be bright and gay or soft and soothing. In fact, many contemporary artists are experimenting with color divorced from a context of form. They create large juxtaposed canvases each with a single uniform color, or alternatively, canvases with precise narrow stripes of contiguous brilliant color which seem to send conflicting sensations to the brain."

When you think about it, it's something of a wonder that we don't all ask such questions constantly. By all rights, we should naturally be wide-eyed noticers: stopping to sniff things up close; staring and turning things over to see the hidden side; snuggling up to experience and nature and art and poking at them for all they're worth. Was the problem all those years of being told what we were supposed to see and how to get the right answers? What if we don't even know the questions? Then we can't help feeling a little squeamish and uncomfortable — like sitting at a formal dinner table and not having a clue which fork to pick up for the salad.

Frank's Exploratorium brought to art the same sense of comfort he brought to science. The result was a decidedly un-museum-like museum. And yet, of all the awards Frank eventually received, none meant as much to him as the Distinguished Service Award from the American Association of Museums, which he won in 1982. "When I started developing a science museum," he said, "there was no organization whatsoever that thought of science centers as part of the museum world."

9

THE MAN WITH THE GOLD-RIMMED GLASSES

"How would Orestes like a nice bath?"

By the time I met Frank in 1972, his Palace of Delights was a vital, humming, whirring, dazzling operation — full of flash and spunk. Frank liked the cover story I wrote about the Exploratorium for the *Saturday Review: Education* — a publication that lasted a little over a year and dissolved into bankruptcy soon after my article appeared. He invited me to lunch. To my utter amazement, he offered me a job writing explanations of exhibits and phenomena for the museum. For reasons I will never understand, Frank had decided I was good at physics — a subject I didn't particularly like and had never studied, save for a single elementary course in college.

Despite serious misgivings, I said I'd give it a try, on the condition that I take virtually no salary at first (little did I know how hard it would be to get a paycheck out of Frank in years to come).

As I was soon to learn, I wasn't alone in my inability to say no to Frank. Donald Kennedy — president emeritus of Stanford University and until recently the editor of the journal *Science* — told me that when Frank called him asking for help in putting together the museum's nascent biology section, he was impossible to turn down. "He was so persuasive, he could have made Tom Sawyer paint his fence," Kennedy said. "It's hard to imagine that someone as sweet and mild-mannered could have so much stealth drive."

Needless to say, a twenty-six-year-old didn't stand a chance.

After the *Saturday Review* went bankrupt in 1973, my family returned to the East Coast, and so I essentially commuted to San Francisco off and on for what turned out to be the next thirteen years, until Frank's death.

Frank would pick me up at San Francisco International Airport and drive me in his beat-up blue Dodge Dart — the flooring and dash torn to shreds by Orestes — past the city, over the Golden Gate Bridge, through the rainbow tunnel, sharp right at the Spencer Avenue exit, past the old firehouse, up and down the Sausalito rollercoaster streets until we came to the house on Cable Road, and Frank gunned the car into a nearly vertical position in a driveway so steep, just stepping out of the car could make you dizzy. A long, narrow staircase led from the garage to the house; once, Orestes got his tail caught in the door, and as he bounded up the stairs into the hall, wagging it furiously, he splattered blood over the white walls in patterns worthy of Jackson Pollock. Later, when Jackie got sick with cancer and was too weak to manage it, the stairway was outfitted with a lift so that she could ride up from her wheelchair. (Before the lift was installed, Frank had asked a hefty Exploratorium staff member to help him pick Jackie up from the hospital. He'd strapped two pieces of wood to an office chair to make an improvised litter to carry her up the steep garage stairs. "She was not light," the staffer remembered. "I was lifting weights at the time, and I thought, 'Can he do this?' But he made it all the way up. I felt it was an honor that he'd asked me. He could have had an ambulance take her.")

Usually I slept in the bed in Frank's office, a cozy room with a Picasso on the wall, sculptures and mobiles by local artists, and his flute. It was a lovely, airy, light-filled house, with sliding doors that opened onto a deck overlooking the bay, where we'd have drinks and watch the fog roll in and fuss over Jackie's begonias. In the morning, Frank would pad around in his worn plaid bathrobe, light a fire in the kitchen, make toast or eggs with green chilies. We'd pile into the car for the drive to the Exploratorium, and I'd be sorry I'd bothered to shower, since I shared the passenger seat with Orestes and arrived covered in slobber anyway. ("How would Orestes like a nice bath when I come out?" I wrote in a letter before one of these trips. And in another: "Love to Jackie and that stupid retriever.")

Sometimes I'd arrive to find the house filled with guests, and I'd have to sleep on the couch. Phil and Phylis Morrison* would come from Cambridge, Bob and Jane Wilson from Chicago.

While Frank had been making his way back into academia at Colorado, Bob Wilson had left Cornell to build Fermilab in Batavia, Illinois. Since Wilson was a sculptor as well as a physicist, he conceived of Fermilab as a work of art, planting his works everywhere on the lab's 6,800-acre grounds. He modeled the main administration building on the Beauvais cathedral in France, with gently curving glass walls that swoop up toward the open sky. There is a Möbius strip in a pool on top of the auditorium, a staircase shaped like double-stranded DNA, a capacitor constructed like a tree. The utility poles look like the symbol for pi.

In so many ways, Frank and his friend Bob were two of a kind. "There was this irrepressible warmth between these two guys," Jon Else remembers. "They were like a couple of little old ladies,

---

* Crippled by polio, Phil was by this time getting around mostly in a wheelchair, and I'll always think of him as E.T. riding off into the moonlight, since that's the image PBS adapted to promote his TV series, *The Ring of Truth*. An elfin figure, Phil was an Institute Professor at MIT and still doing work in astrophysics when he died in 2005, at the age of eighty-nine — one of the last surviving Los Alamos scientists. Phil was one of the first to visit Hiroshima after the explosion. His subsequent efforts to control nuclear proliferation made him a target of the FBI, subjecting him to constant harassment and almost costing him his job. Still, he never gave up, and became an expert on both conventional and nuclear armaments (one of his last books is titled *Reason Enough to Hope*). Phil was also a key figure in initiating the Search for Extraterrestrial Intelligence. He wrote beautifully and prolifically, including regular columns and reviews for *Scientific American* and the book *The Powers of Ten*. Phylis, a teacher, collaborated with him in many public education efforts.

just yammering on and on. [One time] they were talking about resonance, and there was this little dog in the Exploratorium, and it was wagging its tail, and they were trying to sort out the resonant frequency of the tail and the dog's body."

Dinners in the Sausalito house were lively affairs, with frequent company, a fire going in the living room, and good food. We'd sit at a round table with a lazy Susan holding fresh crab and salad and bread; it was perfectly suited to conversation, which was loud and lasted long into the night. It was during one of those evenings I learned that the atomic bomb blast at Trinity turned the desert to glass — a vision that's never left me.

If I stayed over the weekend, we'd sometimes drive to Bolinas, or to San Jose to visit Frank and Jackie's daughter Judy, now a successful pediatrician. On one such trip, we had to pull over frequently because Jackie was suffering so much nausea from her cancer treatments.

I loved Jackie a lot. She was tough and solid, straightforward and funny, and intolerant of pretension. I thought I wanted to have a face like hers when I got old; it was creased and crisscrossed like a map of her long and often hard life — wrinkles in time. At the Exploratorium, she was my editor, relentlessly nagging me about dangling participles, misplaced modifiers, and run-on sentences. She laughed at my mistakes ("complimentarity" for "complementarity") and argued persistently (and unsuccessfully) to do something about my long subtitles (for example, *Facets of Light: Colors and Images and Things that Glow in the Dark*).

Despite Frank's infidelities, he was extremely devoted to her. When he was invited to visit China — a lifelong dream — he insisted that Jackie come along. And when the federal agency sponsoring the trip refused — even after Frank offered to pay — he decided it wouldn't be worth it without her. One thing I'll never understand is how the two of them fit into that small double bed with Orestes.

The only unpleasant experience I remember during all those visits with Frank was the day I made the mistake of going sailing with him and some friends on San Francisco Bay. An hour into our sail, a dense, soupy fog cloaked sea and land alike, and only the low and persistent horns warned of freighters bearing down on us, it seemed, from every direction. The rest of us wanted to head back, but Frank wouldn't have it, reveling at the helm in his colorful wool cap. I spent the rest of the trip below, fearing for my life, downing little bottles of whiskey.

Apple Pie à la Mode and Wine

"The fifteen years I was around Frank, I was never bored. Not even in his trigonometry class."

On weekdays, Frank and I would usually go to the Exploratorium together while Jackie took her equally beat-up 1967 Plymouth Barracuda — Orestes' teeth marks all over the steering wheel, his scratches obscuring the view out the window, the carpets duct-taped in a vain effort to repair the damage. Meanwhile, I'd hold my breath as Frank wove back and forth between lanes on the Golden Gate Bridge, using both his hands to demonstrate some scientific principle and always looking straight at me — never at the traffic.

Frank lived his life, as he liked to say, on the edge of chaos. As one who tagged along for part of the ride, I can tell you that it was an exhilarating — if often exhausting — place to be. Everything became a bit of an adventure, even the most mundane chores. One day we stopped to have a mechanic look at his car, and for some reason, the sight of the hood gaping open, waiting for attention like a patient in the dentist's chair, struck him as funny. "Say aah," he said, giggling, to the puzzlement of the mechanic.

Anything could happen when we got to the Exploratorium. Bob Miller and I might start arguing over something such as, say, centrifugal force (I suppose most people not only don't argue about such things, they don't think they're arguable; but we often did). Frank would wander by with pencil and paper, and soon he'd be engrossed in tiny illegible calculations and then go rummaging around in his office for all manner of miscellaneous stuff he kept in the drawers of his desk and we'd start playing around and then we'd be out on the floor experimenting with gyros and such — by this time, with maybe half a dozen or so staff members tagging along — and we'd fool around with the exhibits and the visitors would get involved and before you knew it, class was in session! And then, also before you knew it, people would feel satisfied that they knew what they wanted to know, at least for now, and would sort of drift away as haphazardly as they'd joined us.

That might go on all day. Once, a Xerox repairman came to fix the copier, and a couple of us standing around got interested in how it worked. An hour later, dozens of people (including Frank) had gathered, and the repairman was still explaining.

For lunch, we'd crowd into Frank's car and go to some local dive, like Joe's on Chestnut Street, or Upton's on Lombard. Upton's was his favorite; he'd pull right up in front of the diner and park on the pavement even though there wasn't a parking spot and order apple pie with chocolate ice cream and a glass of wine.

The routine didn't change even when VIPs were around. "There were a lot of times that we had special funders and the idea would be to go to lunch," an executive from the development office remembered. "We'd go out to the parking lot and try to head the party away from Frank's car. But it would never work and we'd end up stuffing everybody in the back and go careening up to Upton's . . . I wondered if they gave us money because they didn't want to have to go out to lunch with us again." (Real dignitaries didn't get spared either. When John Kenneth Galbraith came to visit, a staff member who showed the distinguished economist to Frank's office was horrified to find instead "that disgusting dog lugubriously licking his manhood.")

No matter what we were doing, Frank was always experimenting. We'd be sitting in a restaurant, and suddenly he'd get an idea and start scribbling on a napkin. He'd make a gyroscope out of pats of butter and a bread plate to explain precession, or demonstrate how to find the center of gravity of a spoon.

When Judy was growing up, Frank would sing to her and play the flute in a minor key while acting cheerful and bouncy. He wanted to see if the sad emotions so often evoked by music in a minor key were innate or culturally transmitted. (Since not everyone who sang or danced with her went along with the experiment, he never got an answer to his question.)

At a staff member's wedding, a musician was playing an electronic flute; Frank marched right up between numbers and made her take her flute apart to show him where the microphone was. For another wedding, he bought a music album as a gift but wanted to listen to it himself, and by the time the bride received it, the wrapping was off, the package undone. Curiosity always got the best of protocol.

During "workdays," Frank could often be found in the Exploratorium parking lot, poking a stick down his car's gas tank, say, or otherwise trying to diagnose a problem. Sometimes he'd come inside in such an absorbed state that he'd leave the car running.

Even the tiniest things would delight him. Once, the physicist Richard Feynman offered to introduce Frank to Arnold Beckman, of Beckman Instruments, a potential funder. Feynman and Frank had to drive for several hours, and when they were halfway to Pasadena Feynman said, "Frank, at this point, if you get into the middle lane and stay there, even if the road narrows and widens, you'll never have to change lanes." And Frank said, "No kidding?" And sure enough, he stayed in that lane and was thrilled by his discovery.

"Frank had an intensity and energy about everything he did," Phil Morrison remembered. "That was his most characteristic trait. He was always trying to formulate things in another way, even if it was just the way the wind was blowing."

Curiosity not only won out over protocol, it also trumped fear. One staffer never quite got over being with Frank during an earthquake that had everyone scared out of their wits. He was in the graphics trailer when he felt the quake, "a long, serious one," he said. The trailer was shaking, and when he looked out, he saw every engineer, every expert in the shop who knew about the structure of buildings, running outside as fast as they could because they knew the big concrete slabs on the ceiling were bound to fall. It was very scary to be inside, and everyone panicked. Everyone except Frank, of course, who stayed inside. "There was Frank, just looking up with this really goofy, bizarre smile on his face," the staffer said. "He wanted to watch it all happen."

Understanding Is a Lot like Sex

"[The Exploratorium] represents the hope that people will be convinced that the world, including the private world, is understandable."

Frank wasn't the sort of wizard who hid behind a curtain; he invited us behind the curtain, and behind the curtain was the world. It was a world as exotic as Oz and as (extra)ordinary as the air. Through his eyes, we all became three-year-olds again, looking at old things afresh and finding universes we'd never known existed — bursting with curiosity about anything and everything. His vast knowledge and unstoppable enthusiasm were the ruby slippers that transported us to magical places, and we took them with us even when Frank wasn't around. Because once you got hooked, you couldn't take them off even if you wanted to.

Frank often wondered why most people weren't as curious as he was, especially young people. Some unknown circumstance of society had rendered people "artificially celibate," he thought. After all, curiosity was common among many animals. If you put a block of salt in the forest, he liked to point out, cows will always find it, though it looks like any other rock. Cows are explorers. They'll walk all over the place and lick any rock they see sticking out from the grass. Eventually they discover the rock that provides the salt they need.

"If cows are innately curious, how is it we have managed so thoroughly to suppress curiosity in ourselves and our children?" Frank asked. "We're told not to play with wall plugs, not to talk to strangers, not to stick our head out of a car to feel the wind. We are told, 'Curiosity killed the cat.'"

Along with curiosity, people seemed to have lost the conviction that the world is understandable at all, perhaps because so many aspects of modern society seem incomprehensible. As Frank liked to point out, most of us are in daily contact with at least as much that we do not understand as were the early Greeks or Babylonians. And yet we do not, in general, ask questions because we have no context to help us absorb the answers. "We end up in the paradoxical situation in which one of the effects of science is to dampen curiosity," Frank said.

People simply hadn't had the experiences that would get them asking questions on their own. They used cameras, Frank noted, but never held a simple lens to make an image on a wall. They used electric toasters, but never connected a wire across a battery and watched it get red hot. They may have spent hours watching ocean waves, but never observed the way waves pass through each other, bend around corners, or bounce off cliffs. No one had pointed out such things or encouraged such play.

This seemed strange to him in light of the endless curiosity that drives most scientists — even in the face of failure. Because scientists regard their work with so much passion, they tend not to lose hope when experiments don't work or calculations turn out to be unsolvable. True, they may fall into "the deepest, outrageously discouraging gloom," Frank said. But due to their prior experience of even minor successes, or perhaps because they have witnessed the successes of others, they overcome the gloom. They try another approach. "In the long run, it's the successes, sometimes trivial, other times earthshaking, that egg us on and keep our curiosity alive," he said.

To Frank, curiosity and confidence in the ability to understand things were not just niceties. They were fundamental human needs — and fundamental for survival. As he liked to put it: "There's a lot of practical fruits to understanding, but it's like sex. There are practical fruits to sex, but nobody would say that's why you do it, normally."

The sex analogy,* like so many Frankisms, turned out to have more behind it than might have been immediately apparent. In truth, the two activities do share more than superficial similarities. Understanding, as much as sex, is required for the propagation of culture — perhaps even of life.

And once you've gotten a taste of it, you can't stop wanting more. Knowing that a solid rock is mostly empty space, that the atoms that make us up were forged in stars, that space and time are elastic — these are as thrilling as any science-fiction movie. Hollywood's gazillion-dollar special effects are child's play compared to what the universe provides for free: black holes that pinch off space-time, quantum particles that can be here and there at the same time, the existence of perhaps ten dimensions of space. Learning enough to play in this sandbox (and you don't need much) can be unreasonably rewarding.

On the other hand, being surrounded by things we don't understand is distinctly uncomfortable — even creepy. When everything seems unconnected and unexplainable, we are like children sitting in the dark, listening to strange noises that could be monsters trying to get in the window as easily as the scratching of branches on the pane. Anything can happen, and what does has no apparent rhyme or reason. Knowing why things happen, and also that the most amazing array of phenomena can be understood with a few simple ideas, brings with it a long existential sigh of relief — as I was rapidly learning.

(And fear, of course, is the enemy of curiosity: fear of getting lost, of wandering into dangerous territory, of appearing stupid, even fear of being a nuisance.)

---

* More and more, I have seen a version of this quote attributed to Richard Feynman, usually in this form: "Physics is like sex. It may give some practical results, but that's not why we do it." Since he and Frank were friends, it's likely they batted around such an idea, but Frank's version is fundamentally different. First, his focus is on understanding in the broader sense of something that is essential for survival, not just physics. Second, Feynman's version implies that the reason physicists do what they do is because it's fun (surely that's part of Frank's meaning also), but Frank also meant it as a fundamental part of human nature. So, in short, while they may have said something very similar, or picked it up from each other, the message of Feynman's quote is more flippant and humorous; Frank's had a more serious meaning.

"Understanding brings a sense of order, a reduction in uneasiness, and a definition of beauty," Frank wrote. "Understanding simplifies. Things that did not seem to fit together become part of a whole. Through art and science one becomes aware that each thing that happens around us or is felt within us need not be a separate thing to be separately learned and remembered but can belong to an interrelated family."

Human beings are naturally driven not only to learn but also to tell what they've learned to others. This obsessive sharing is, in a very basic sense, what we call culture. We share what we've read or overheard or seen at the movies or bought at the store as much as we share our political views, our children's successes and sorrows, our illnesses, loves, sex lives. We share family histories and customs and beliefs. "Whether one calls it teaching or education or just being human, I don't know, but it seems to be something that most people want to do very badly." Frank marveled at the fact that people seem to have a compulsion to tell others the smallest things: "I have to cough" or "I have to take a leak."

Prehistoric peoples shared stories around campfires, painted on walls of caves, dreamed up theories of the universe which they passed on by word of mouth. "If it were not for this drive to share what is known with others," Frank said, "we probably wouldn't have any culture at all; and without culture, we two-legged creatures would not have survived. The desire to teach and learn must be a very basic instinct."

Frank freely admitted that he would tell what he knew to anybody who would listen long enough. He compared the pleasure of communicating to a pianist who has just mastered a difficult sonata — and has an irresistible urge to play it over and over.

Of course, in some sense, the reason sex is essential for survival is straightforward: if people don't procreate, no babies are produced to preserve the gene pool and carry on the culture — and the whole thing just stops. The argument for the central role of understanding is perhaps more roundabout.

It does seem pretty clear — assuming that survival depends crucially on the contributions of the human mind — that a reduced level of enthusiasm for understanding can lead to disasters of potentially catastrophic proportions. When we don't use our own heads, we are forced to rely on other people to make decisions for us. So when we hear about stem cell research or missile defense systems or claims that quantum physics can help us make money, we have no choice but to fall back on figuring out whom we can trust — and often the most persuasive people are neither unbiased nor well informed. In that case, "our decisions about social, political and economic matters are inevitably based entirely on the most appealing lie that people dish out to us," Frank said.

In addition to the obvious advantages of using one's head, however, understanding affects survival because people who feel they don't understand things often give up, lose interest, succumb to apathy (often misunderstood, or rationalized, as coolness); in other words, they stop participating. This itself is understandable, and I'll wager that few people on the planet haven't resigned themselves at some point to simply not "getting it" — whether the subject is physics, art, music, politics, or love. Sometimes it all just seems too much.

The price, however, is accepting the fate of the frightened child in the dark. "Without this conviction [that they can understand things] people usually live with the sense of being eternally pushed around by alien events and forces," Frank wrote. And why get involved when ultimately others will decide anyway?

Such wide-scale dropping out deprives society of the wisdom and good sense of most of the populace — just when we need them most. And when enough people give up, the whole mechanism of government changes. Participatory democracy becomes oligarchy. We relinquish our power to the few, and far worse, eventually lose faith in the value of our own ideas. "The intellectual apathy that I am told now exists among young people," Frank wrote, "may have come about because these youths have never been convincingly taught that . . . they can make a difference."

The kind of understanding Frank talked about had nothing to

do with filling people up with information. "A lot of guff has been written about how democracy needs an informed citizenry," Frank wrote, "as if this were all by itself going to make everything work well, and I don't believe that." Informed people can come to very different conclusions. Edward Teller and Robert Oppenheimer were probably the world's greatest experts on hydrogen weapons in the 1950s, and yet their views on what should be done with them were polar opposites.

Rather than an informed citizenry, what was needed was a citizenry composed of people with the ability and confidence to figure things out for themselves — to detect lies, check facts, and stop feeling they had to rely on received wisdom.

This isn't an easy thing to accomplish in a society where even the income tax system seems as impenetrable as quantum mechanics. Yet success in understanding just a few things has an enormous payoff in terms of the willingness — and perhaps more important, the desire — to understand almost anything. The idea that such a thing might be possible, that you just might be able to understand the economy if you tried, or AIDS, or how to download music onto your computer, was a gift that just kept giving. "It was an enormously liberating thing to discover as a grownup," Jon Else said, after spending a year, off and on, in Frank's company.

When Frank couldn't understand something, or get somebody else to understand something he wanted to convey, no matter how trivial the subject, he'd slam down his cane, furious. One time, we were having one of those frustrating "who's on first" conversations about going to the grocery store; Frank yelled at no one in particular that he could understand why miscommunication sent countries to war when even friends had problems using language to convey the simplest ideas.

For Frank, promoting understanding was the highest kind of civic duty. It annoyed him that camera manufacturers, for example, didn't provide small lenses with every purchase, along with a brochure about lenses that encouraged people to fool around with them. "Unfortunately, there is no sense of obligation by the makers of cameras to help people feel that they are masters of what they are doing," he complained.

The Lady with the Lamp

"Telling someone to 'think' is about as effective as telling them to 'wiggle your ears.'"

What Frank was addicted to most was creating other addicts like himself, and as a pusher, he was fanatical. A new staff member would come to the Exploratorium to work, say, in the bookkeeping department, and Frank would drag him out onto the museum floor for hours to play with optics. "I remember when I first came here," recalled an accountant. "Frank asked me if I found optics interesting. I said, 'No, not particularly.' Then he said, 'I want to show you something.' Now I love this stuff."

You could never talk to Frank about "demystifying science" because he didn't believe you could take away the mystery even if you wanted to. He called science "the search for the ever-juicier mystery," and he wanted everyone around him to join in the fun. I think he succeeded in part because we'd never seen anyone who let curiosity so completely off the leash; for the most part, it was our first lesson in the irresistible pleasure of allowing passion to prevail. "More than anything else he just always seemed to be interested in stuff," said Charlie Carlson. "He would stop by and poke his head in and wonder what was going on and talk about stuff. You could show him almost anything, and he would say, 'That's really interesting.'"

And the interest didn't die when the conversation did. You would think a conversation was over, but he'd come back weeks later and say, "You know, I've been thinking about such and such and . . ." Or he wouldn't bother with the transition, leaving you momentarily lost. He'd just pick up the thread and expect you to know he was following up on something you'd discussed a week or so before.

When Frank got the itch to try something, practical considerations couldn't stop him. At one point, he wanted to create an exhibit on pheromones (chemical sex attractants), so he and Charlie drove to UC Davis to watch male moths, propelled by an overpowering urge to mate, fight furiously to follow the scent of a female pheromone in a wind tunnel. Frank was entranced. He decided that the Exploratorium should give visitors bits of the pheromone and have lots of moths flying around inside the Palace. He envisioned people wandering about the exhibits followed by little moth clouds.

Danger was no impediment either. Frank wanted to demonstrate electrolysis by running a current through water to break down H_2O into oxygen and hydrogen and show that when the gases that made up water were separated, you always got exactly twice as much hydrogen as oxygen. It was a challenging exhibit, most of all because separating the water molecules required a very large current. And while Charlie was working on it, there was an explosion that destroyed the apparatus.

"It rocked the whole place," Charlie said. "There were visitors, and I was standing there trying to figure out if I was still whole or parts of me were missing. I did not want to go through another explosion like that." Six months later, Frank came back to Charlie, wondering why the exhibit wasn't finished. "One explosion and it gets you stopped on these projects," Frank scolded.

Frank never minded a bruise or two in the interest of discovery, and he didn't understand why others should either. So when some staffers worried that a spinning water tank they were building might thwack a visitor in the head and cause an injury, Frank spun the tank as fast as he could and stuck his own head in its way. *Thwack*. "Not so bad," he concluded.

Frank's addiction to understanding was infectious all by itself, but he also had a knack for passing on his madness, and it was something he'd thought and written about a great deal. For example, he had concluded there were certain things he could not teach, such as how to solve problems. "Telling someone to 'think,'" he said, "is about as effective as telling them to 'wiggle your ears.'" The ability to think is a result of experience, not a skill that can be imparted directly.

It also doesn't work simply to allow someone to imitate what you're doing. The effect of that approach, he said, "is about the same as if I had shown a small child how to saw a piece of wood by standing behind him and making his arms move the saw. The child would know what is required of him, but he still could not do it himself."

What Frank could and did do was give people that small taste of success that whets the appetite for more. Even understanding something very simple, he believed, was "contagious and addictive."

"The first thing I try to do as a teacher is to get my students to understand so clearly — some phenomenon or device, such as the twinkling of a star or the ring of an electric bell — that they understand that understanding, like eating or making a basket during a ball game, is satisfying and fun," he told the Pagosa Springs PTA in 1957. "If I can succeed in making understanding seem like fun, then I believe that the student will want to understand many things, that is, he will become curious. If I can [accomplish that], then perhaps the course I am teaching will have the effect of enriching his whole life. It may also make him a more useful and sympathetic person."

Adults were more resistant than children, as Frank learned from all the people who came to the Exploratorium and told him, "Gee, I wish science had been taught that way when I was a child." They seemed to think the opportunity was lost and that they couldn't start over, but Frank believed they could — and that once they had experienced the pleasure of understanding, it would be self-perpetuating.

There was nothing more addictive than feeling "I can understand that," and no worse feeling than being stuck. "I find no matter what the situation, if I can think of something to do next, I'm usually optimistic; it's only when I can't think of the next step that I get really gloomy."

The responsibility of a teacher, therefore, was to help people get unstuck. If a student stayed stuck, it was the teacher's fault. Or, as Frank said when my super-smart stepdaughter failed a physics course at Harvard: "Harvard should be ashamed of itself." It was also the teacher's responsibility to get students intrigued in the first place — and Frank was always delighted when he found the one little thing that would tweak someone's interest.

Part of the trick was to know what students already knew, and what they could master, and when. Frank seemed to grasp where people were even before they did. No matter how low their initial level of knowledge, he never talked down. He never mistook an unfamiliarity with terminology (an unfamiliarity with anything) as ignorance or stupidity. "He never made anybody feel like their question was not well thought out, or that you didn't have enough background," said Darlene Librero, the guiding light of the Explainer program, "or that if you had taken this or that course, you wouldn't ask that question. Even if you said something in beginner form, he didn't see you in a beginner place. He made everybody feel like their brains were working just fine."

Sometimes meeting people on their own terms required rethinking basic assumptions about teaching and learning; for example, rethinking what it meant to "make sense" — an idea often misconstrued by teachers, Frank thought. For one thing, what makes sense to the teacher isn't necessarily what makes sense to the student. "What we call making sense seems to be the accumulated longtime effect of nonsensical understanding," Frank wrote. "Greek sculpture, Shakespeare and Newton now seem to us very rational. Jackson Pollock, E. E. Cummings and Feynman do not. And yet we can understand them all. We can understand the way in which light travels through a vacuum, although it doesn't make sense to say that there is a wave with nothing that waves. We can understand the uncertainty principle although it does not make sense to [the extent] that if we know where something is well enough we cannot possibly find out how fast it is going."

In fact, very little of modern art or physics "makes sense." When teachers don't give students credit for being able to deal with nonsensical things, they eliminate much that is interesting about science. A better approach, Frank thought, would be to show students how to understand science in the same way we help them understand and enjoy poetry.

One of the most effective methods for creating learning addicts, Frank decided, was to turn them into teachers themselves. It was one of the reasons he so encouraged the kind of discovery that comes as "something of a surprise, a triumph." Because an idea you discover yourself is one you want to tell others about over and over again.

And so it didn't always matter if the teacher got everything exactly right. Frank liked the teenage Explainers at the Exploratorium to go right ahead and offer explanations even if they were a little off the mark. They were experiencing the joy and confidence of imparting something to others; just as in science, the "right answer" wasn't as important as the process of thinking things through.

As a consequence of the confidence he gave people, they often tried new, and sometimes unreasonable, things. Frank's one-time secretary Ester Kutnick, who is also an artist and photographer, offered to fix a friend's typewriter, though she knew nothing about how the machine worked. "I remember thinking: Use your powers of observation. Use logic. See if you can see what's not working. And I fixed her typewriter. And my friend said, 'How did you do that?' And I said, 'I really don't know. I just kept looking at it and poking at things.' But I had that confidence not to be afraid to try and fail, to keep looking until I found something that made sense to me."

There is one story Frank loved because it perfectly encapsulated what he thought a little familiarity could accomplish. A woman told him that after visiting the Exploratorium, she went home and repaired a plug on a lamp. It was important to the story that nothing in the Exploratorium dealt even remotely with table lamps or plugs. Rather, the incident showed that the woman had become comfortable in a previously alien part of her world, at home enough to explore it. "She must have felt that since she had made sense of some of the difficult ideas in the museum," Frank wrote, "she could make sense of other things as well."

There's no way to tell how many people Frank affected in this way. But a letter from Frank to a Fred Duncan in El Paso, Texas, written in 1984, makes you wonder: "Dear Fred," the letter reads. "Thank you for sending me the two ham cans that show the intriguing caustics formed by the sunlight. I have not put the escargot dish onto a turntable yet, but even just oscillating it back and forth by hand it will do things. Best wishes, Frank Oppenheimer."

Increased Neural Activity of Imagining

"I've become absolutely obsessed with the idea of seeing a particle accelerator."

For me, hanging out with Frank was a real awakening in every sense. In fact, what happened to me (and oh so many others) is best summed up in a letter that a visitor wrote to Frank after he noticed some perplexing changes in himself following his excursions to the Exploratorium. Is it possible, the visitor asked, that the exhibits on electricity were setting up currents in his gold-rimmed glasses? How else to account for the "increased neural activity of imagining" he experienced for days after his visits?

Frank wrote him back in all seriousness: "No, those currents wouldn't last that long, but that was one of the points of the Exploratorium; to increase imaginings, and notice things."

It certainly worked on me. On my many airplane trips between San Francisco and New York, I'd take a little piece of Polaroid material and rotate it to watch the sky darken and brighten and make colors appear in the stressed plastic windows. In one letter, I wrote to Frank that the windows of my L-1011 appeared to be prisms, making rainbows around the clouds. I was apparently carrying a small magnifying glass,* and when I looked through it, the

* I say apparently because it's not something that I remember, but something I refer to in a letter.

colors disappeared. Asking him why that might be, I allowed: "It's very strange and magical."

At home, I soon realized that I could find just about every phenomenon I was learning and writing about with the help of a small prism, a magnifying glass, spoons, mirrors, door peepholes, and the like. I showed anyone who would stand still long enough how any surface becomes a mirror if you look at it from a shallow enough angle; how to make (polarized!) blue sky and orange sunset in a glass meatloaf pan by adding powdered milk; how to put red, blue, and green flashlight beams together and get pure white, and then make shadows of complementary colors on the wall.

I bought a bunch of magnets (ostensibly for my young son) and was amazed, I wrote Frank, "at how strongly you can 'feel' the 'shape' of the field around one magnet using another. There's something really eerie about being able to detect this strong 'thing' which has a very defined shape with a tool, but not with your finger . . . It makes you wonder how many other things our perceptual mechanisms can't detect."

I wanted to know why "accelerators are like cathedrals" and told him, "I've become absolutely obsessed with the idea of seeing an accelerator . . . What strikes me so much about this sort of physics is that it is exactly what most people take for science fiction."

Of course, I also struggled, frustrated at the books I read that "seem to delight in telling what things are without ever hinting at why." Sometimes, the more I read, the more I learned how much I didn't know — how many utterly magical things I'd never seriously thought about before. I marveled that magnets could "stick together" right through the heavy fabric of my denim jeans. If electricity and magnetism were two sides of the same coin, I asked Frank, how could this be?

Some of my questions, I've since discovered, weren't so stupid as I'd thought. I recently showed a prominent astrophysicist friend one of my letters to Frank which posed a series of questions about the relationship between electricity and magnetism, and he said they were very hard questions — the answers still not entirely understood.

My "lessons" took a range of forms, including letters, phone calls, and conversations in San Francisco or while wandering around New York when Frank visited. Many of them came in the mail, on audiocassette tapes that Frank made in response to things I'd written or questions I'd asked. Some of these tapes have sat in my desk drawer for fifteen years or more. Listening to them today truly gives a taste of Frank's insatiable and infectious addiction to addicting me (as well as others).

He would stop mid-explanation to suggest experiments I should try. When we were talking about angular momentum, he suggested I let my car idle and try to gun it to see if it would lurch in the direction opposite the rotation of the motor. He stressed the importance of play, courage, and guesswork science. One of his favorite expressions was Mark Twain's quip "There is something fascinating about science. One gets such wholesale returns of conjecture out of such a trifling investment of fact."

If, in passing, I mentioned a "law of nature," Frank would start ruminating: What *is* a law of nature, anyway? When Galileo rolled balls downhill and observed that the acceleration is constant, he did not discover a "law," Frank insisted; he discovered a pattern of behavior. "One talks about these patterns as if they were laws that have to be followed," he said. "But it isn't. It's just something he noticed."

Frank also frequently stirred emotions into the stew. In one conversation, he talked about shadows in his typically crazy, wonderful way: "You pour spaghetti into a colander; the colander creates a shadow for the spaghetti but not for the water; in this case, you may be interested in the spaghetti. But if you use a tea strainer, then the strainer provides a shadow for the tea leaves but not for the water, and you're interested in the tea." So much for physical shadows.

Then he told me to be sure not to downplay the importance of shadows as emotional refuges, "places to hide, to have a private life"; shadows may appear flat, but they are, after all, three-dimensional; the shadow we call night sticks out four thousand miles from Earth at midnight.

Into the Woods

So what?

The best times were those I spent with Frank at the Exploratorium—alternately playing in Frank's personal Oz or trying to decipher his illegible scribbles on his office blackboard or wondering at his marvelous demonstrations. Once, he took a bright red Life Saver and smashed it with a hammer to show that anything you grind up finely enough turns white, and for the same reason ocean foam is white: the smashing creates millions of tiny mirror-like facets.

Like Frank, I get as excited today about the ideas I learned during this period as I did when I first heard about them twenty-five years ago.

Take, for example, the idea of transparency, something we all think we understand. A clear piece of glass is transparent, but a concrete wall is not. My first inkling that this was (sorry) transparently wrong was a beautiful exhibit called "Convection Currents," which revealed how water (and even air) can cast a shadow. The exhibit comprises a thin aquarium with a movable rod inside, illuminated by a concentrated source of light. When you heat the rod, it makes the water just above it rise in beautiful sheets and swirling eddy currents that build and grow. You "see" these patterns by the shadows they cast on a white wall behind the tank. The denser, colder water acts as a molten lens that directs the light into certain spots and away from others; the shadows form where no light gets through.

The idea that you can create a shadow with something transparent struck me as pure magic. I tried putting my clear eyeglasses in front of the light, and sure enough, even the clear parts cast shadows.

And that was not half of it. The concrete ceiling of the Exploratorium was perfectly transparent to radio waves, cosmic rays, television signals, neutrinos. A clear window, however, was opaque to ultraviolet light—which explains why you can't get a suntan indoors, and also why astronomers have to send spacecraft above the atmosphere, because even thin air is opaque to many wavelengths of light. And as Frank noted: "The nothingness of a vacuum is surely finely textured, yet it is opaque to sound; there is no way that we can listen to the thunder of erupting sun spots."

Frank's "props"—as he sometimes called the exhibits—were extremely important in all of this. It's one thing to know that visible light and heat radiation are the same, but quite another to feel the warmth of an electric heater in the same spot that you see its image in a parabolic mirror; one thing to learn about rotational inertia, but quite another to feel the strangely powerful force that prevents you from tipping a spinning wheel—or if you do manage to tip it, how forcefully it can push you around in response. Frank, after a theory by Ernst Mach, called this force "pushing off the stars."

I've co-taught a number of astronomy courses, and felt enormously handicapped because I didn't have something like the Exploratorium's "Solar Spectrum" to show students that you really can see what ingredients go into making a star. This exhibit consisted of a simple spectrum of the sun, spread out into a long rainbow; here and there, it was etched with thin, crisp black lines—like a bar code. Each dark line was a place where a particular frequency of sunlight was missing because it had been gobbled up by atoms on the sun's surface. And since each kind of atom absorbs only particular frequencies, the lines spelled out clearly and precisely which elements made up its atmosphere—just like a fingerprint. (The same lines can reveal what the star is made of, how fast it's moving, and how hot it is, among other things.)

To me, this felt like hearing secrets from the star firsthand, tuning in to the songs of its component atoms as directly as if I'd put my ear to the surface. Most remarkably, before the accidental discovery of this so-called Fraunhofer spectrum (named for Joseph von Fraunhofer in 1814), people sensibly assumed that there was no way one could ever find out what a star was made of. You couldn't travel to a star, take a sample, and bring it back to the lab. But as it turned out, nature had provided a Rosetta stone that translated the language of stars into plain English (or math, as the case may be), allowing astronomers to find out almost anything they wanted to know about stars (and other astronomical objects, such as quasars, black holes, and supernovae).

In terms of learning physics, the best thing I got from all this playing around with "props" was intuition. Physics is really a very simple subject, with almost everything, on some level, related to a few simple ideas; if you know how waves behave and interact, how things move and exchange energy, how to think of dimensions and symmetry, you've basically got the subject licked. (Almost everything in physics is also easy to spell, which remains one of its charms for me to this day.)

When waves move in and out of step, for example, they make interference patterns, which create (to name just a few examples) the colors of soap bubbles, opals, oil slicks, and butterfly wings; the hovering images of a hologram; the patterns in picket fences and moiré silk; the subtle throbs that tell you a musical instrument is out of tune. Rosalind Franklin used interference patterns produced by x-rays to "see" the spiral structure of DNA; your 767 en route to Maui relies on laser-produced interference patterns to tell the plane's inertial guidance system how you've turned; astronomers use the interference of radio waves to see matter falling into the throats of black holes. Soon physicists will be using interference patterns to catch gravity waves — the ripples of space-time itself. Ghostly subatomic particles called neutrinos change their identities as they move through space because they have a complex internal structure that lets them get "in and out of tune" with themselves!

In the Exploratorium, interference bloomed everywhere. One of my favorite examples was a small, rather obscure exhibit that few people visited but that allowed you to build a hologram, one point at a time, by focusing laser light as it diffracted around an interference pattern; it stood next to an enormous circular interference pattern, or zone plate — as tall as a building — that focused sound in exactly the same way.

Frank often rode the Sausalito ferry to work, and noticed an interference effect while walking back and forth on the deck. As he paced, the combined sounds from the boat's two engines produced audible beats, and the pattern changed depending on how fast he walked. He turned this into an exhibit called "Walking Beats."

With enough experiences like this tucked away, almost anyone can begin to understand complex phenomena "in their bones" in a way that no amount of book learning could ever duplicate. This isn't all that surprising, perhaps, given that what we call understanding, especially intuitive understanding, is often just a matter of getting used to things through repeated and intimate exposure.

I have to admit that not every idea Frank presented immediately appealed to me. With some, it took a while for me to see what the big deal was. For example, I couldn't imagine why he had so many exhibits on resonance — the phenomenon colloquially known as being in tune. A resonator is anything that vibrates at some natural frequency, like a spring or a pendulum, and when two resonators are in tune, their sympathetic vibrations vastly magnify each other's response (think of an opera singer shattering glass). The Exploratorium had endless variations, and at first I thought it was overkill.

But resonance, as I soon realized, is one of those ideas that explains an enormous range of familiar phenomena. The rings of Saturn have gaps because the tiny particles that make up the rings are flung out of those places where the orbits of the planet and its moons are "in tune"; apples are red because molecules absorb light that resonates with their natural frequency, and the pigment in apples absorbs all the frequencies of sunlight except red, leaving only that color to be reflected to your eyes. Soldiers marching over a bridge have to be careful to break step because if the rhythm of their marching resonates with the natural frequency of the bridge, it will start to resonate too, wriggling wildly like a frenzied snake, leading to catastrophic collapse, as happened at Tacoma Narrows, in Puget Sound, in 1940.

Characteristically, Frank also brought in the human dimension: "I guess that's true of people as well as piano strings," he mused. It's a familiar phenomenon: when people are out of sync with each other, an uncapped toothpaste tube can get magnified enough to wreck a marriage. I wound up writing my first *New York Times* "Hers" column about the way relationships often turn on resonance—but it was really an excuse to play (intellectually) with this newfound treasure.

Words

"I see no reason for a museum to cater to the fact that many people have been put off language by the way it is so deliberately used with dishonesty in commercial and political life."

Frank loved language almost as much as he loved natural phenomena and considered it a central part of doing science. After all, when scientists first discover new phenomena, by definition those phenomena don't have names; they have to be anointed. "Dark energy," for example, is now thrown around at physics meetings and in newspaper articles as if the term had always been part of the lexicon, but in fact, the phenomenon that "dark energy" describes—a repulsive energy of the vacuum that appears to be pushing galaxies apart—wasn't accepted as real until a few years ago (and some cosmologists don't accept it to this day).

Someone had to come up with "quark" and "electron" and "gene." But how to communicate an idea before those words exist? Like the science writer, the scientist has to resort to imagery built on words from everyday language. More and more, Frank worried, scientists talk mostly to each other, speaking mainly in jargon. He wanted to put the imagery back. He liked to quote the father of quantum theory, Niels Bohr: "When it comes to atoms, language can be used only as in poetry. The poet, too, is not nearly so concerned with describing facts as with creating images."

Frank recognized that translating scientific concepts into ordinary words didn't sit well with everyone, that it "scandalizes some narrow-minded scientists." He also recognized that the task was extraordinarily hard. "It takes many attempts and the best talent we can find." He hired, and then quickly discarded, more science writers than I care to remember.

Still, he refused to be put off either by the challenge or by the familiar complaint that people these days no longer like to read. If people didn't read, it was because words didn't speak to them. Words didn't need to be intimidating to convey information, or slick in order to be exciting. Fancy words often conveyed no information. By way of example, Frank brought up legislation that had required ingredients to be listed on medicine bottles. The disinfectant spray Bactine had an "unenlightening" label that read "Alcohol 3.17 per cent, Methylbenzethonium chloride, isooctylphenoxypolyethoxyethanol and chiorothymol." "Why should not this information be given in a pamphlet explaining which ingredient serves as bacticide, which as fungicide and which as deodorant?" Frank asked. Such a pamphlet could explain why these particular organics are effective, painless, and commercially profitable.

He thought most museums — art and science alike — had far too little in the way of signs and labels. This was a good thing for me, as I was initially hired to write many of the explanatory materials at the Exploratorium. And signage was only the start. It's difficult to pause and reflect in a museum the way you can at home on your own time, so Frank developed an array of take-home materials, most of which, at one time or another, involved me in minor ways.

I thought of my job as writing paths through Frank's woods, helping to point out connections and generally guide visitors through what might have seemed a chaotic environment. So we wrote extended signs on the exhibits that included sections on what "To do and notice" and "What's going on?" and, later, "So what?" "These were greatly improved when Jackie Oppenheimer was in charge of graphics," Frank noted in one progress report. We wrote "catalogues" of exhibits, with Jackie in charge of design.

Frank loved to worry over the meaning of words — mostly to arrive at a better understanding. On one tape we were talking about creativity and somehow drifted into debating the relationship between creativity and nobility, and what "noble" meant. Is it noble to paint a picture for money? Was Napoleon noble?

"I don't think you can be evil and creative," Frank said. "Why is that so?" I asked. "I'm not sure it is," Frank answered. "[Then] I don't think you can just say things like that," I told him. "No," Frank answered, "but you can have a good discussion."

He wrote a whole essay on the meaning of the word "map," arguing that even a novel could be a map; that a lens is a map because it creates a correspondence between points on a face or a tree or a building with points on a photograph or your retina, just as a biographer maps the events in a life onto a page in a book, and so on. I think of maps much more broadly because of this.

In fact, use of language was one of the realms where people really noticed that Frank had quite different standards than most of us. In the smallest ways, he made sure that nobody used words as Humpty Dumpty did — to mean "just what I choose it to mean." Words were precision tools, not to be taken lightly. A development officer at the Exploratorium incurred Frank's wrath by writing a thank-you letter to a donor expressing gratitude for a gift. "It's impossible to explain . . . [how grateful we are]," the letter began. "It's not impossible," Frank said. "It's only very very difficult."

He could be literal ad absurdum. When the writer Pat Murphy started working at the Exploratorium,* Frank asked her what title she'd like, and she answered that "writer/editor" would probably be appropriate. He responded that writer/editor meant "writer divided by editor."

To some extent, almost everyone who worked on writing with

* Pat not only passed muster with Frank; she stayed around for more than twenty years.

Frank eventually internalized his views, carrying a little "Frank the editor" on their shoulders like Jiminy Cricket, a crabby conscience whispering disapproval of fuzzy thinking, embellishment, and bullshit. "I remember one time I sort of got carried away by my own writing, and the content was slowly seeping out the sides," said Thomas Humphrey. "It was, I think, an article on equilibrium. And Frank wrote some notes on the page that absolutely took the wind out of my sails. And I realized that I had written seven consecutive content-free sentences. And I felt somewhat embarrassed by what I had done. And so I took it as kind of a way of being, to really try to find that honesty."

I think of Frank's respect for words whenever I hear reporters and politicians unthinkingly toss around the term "weapons of mass destruction" to describe chemical and biological weapons as well as nuclear ones — a grouping that would also put sparklers and exploding stars in the same general category. Certainly, biological weapons could cause thousands of casualties, especially in places where public health services are poor. Chemical weapons could kill hundreds and cause great panic, but even so, people can protect themselves to some extent. But only a nuclear weapon can vaporize a city in an instant. And there is no conceivable protection. "Subsuming these three types of weapons under the rubric 'weapons of mass destruction,'" as Phil Morrison pointed out, "approaches the disingenuous."

In the end, Frank probably changed the way I think about words as much as he changed my relationship to the world of physical phenomena. He got me to constantly ask, What does it mean? Whether the "it" is some weird property of the universe or a word on the page.

So by the time we got around to writing a book together, based mainly on Frank's unpublished work, I was already as deeply addicted to Frank's way of thinking about the written word as I was to his ideas about science and art, and especially the way all three were connected. When I started reading his letters and essays, however, I also realized that there was a deeper level to his Palace of Delights—one I'd managed to mostly pass over: politics. All the time Frank had been exiled on the ranch, he'd been thinking and writing about how ideas from science could be used to change the way people feel about nature, about each other, about war. With enough understanding, he thought, people could create the kinds of social inventions that would protect us from one another, just as vaccines protect us from disease. The more Frank and I worked and talked together, and the more I read, the more I realized that, at its core, the Exploratorium was a political institution.

10

THE SENTIMENTAL FRUITS OF SCIENCE

"The Manhattan Project had a political motivation as well as a technical one. And so does the Exploratorium."

During Frank's years on the ranch and then at the University of Colorado, he spent a great deal of time trying to make sense of how science (or at least his cherished view of it) had somehow gone astray — either abdicating its proper role in society or being shoved out. He wrote dozens of essays and letters to editors and gave speeches when he could as a way of thinking out loud about the issues that so troubled him. Frank wasn't concerned with weighing the obviously positive contributions of science (such as antibiotics) against the obviously sinister ones (such as the atom bomb). The greatest inventions of science, he concluded, had little to do with either and everything to do with changing the way people *felt* about things — their place in the natural world, their relationships with each other, their potential for change.

Science, in Frank's mind, was at heart not so much about "practical fruits" as "sentimental" ones — ways of thinking that can be applied, indeed are applied, to almost every conceivable realm of our existence. The need for racial equality (as well as the Golden

Rule) reflects the same principles of symmetry behind Einstein's relativity theories; tools for reconciling conflicts emerge naturally from quantum mechanics; both democracy and progress depend on accepting the fact that everything in the universe is constantly changing.

Ideas such as these offered some hope, Frank thought, for buying time until humanity learned to cope with its seemingly insatiable thirst for self-destruction. Science had already produced ways of understanding that reached far beyond what most people considered its natural boundaries, and if some of that understanding could permeate the popular culture, perhaps people would discover "social inventions" that could save us from the worst of ourselves.

But if we stopped trying to understand things, "we'd all be sunk," he often said. "The only thing we could do then would be to clobber each other and push each other around and coerce each other." Frank built the Exploratorium because he felt the need for understanding—and in particular scientific understanding—was an urgent political matter. Underlying all the fun, it was a political answer to a political problem.

To me, Frank's ideas were as resonant as they were unfamiliar. Like most people, I had always regarded physics as the study of inanimate things—mere stuff. Sure, the history of physics is the story of people and the things they did and discovered, but the thing itself—the science—was only a set of ideas about objects and forces and how they behave. As far as human nature was concerned, physics was irrelevant. Its realm was matter, not mind. It would be absurd to characterize its fruits as emotional; "sentimental" seemed sillier still.

I soon learned that this separation of science from feeling was quite a recent development; until a century or so ago, physics was known as natural philosophy. The physics of one era, as an earlier generation of physicists noted, becomes the metaphysics of the next. Our very notions of right and wrong, of human nature and human potential, of fairness and progress, are deeply embedded in our beliefs about how the physical world works. These philosophies become the invisible underpinnings that shape perceptions, affecting how we look at everything from social arrangements to economic priorities: it makes a difference whether or not you believe that things can change and evolve; whether you believe that point of view matters or that people have free will. It matters what you fear.

But somehow science had lost this connection with culture — the body of beliefs that subtly yet powerfully guides every aspect of our lives. This was a strange thing, Frank thought. When we study ancient cultures, we always look at their conceptions of nature, their technologies, their views on the origins of the universe and the relationship of people to both the physical universe and each other. Even the Bible begins with what is essentially cosmology. "It would be inconceivable to study the Greeks without taking into account their ideas about the natural world — much of it embodied in mythology," Frank wrote. "Native American cultures are often described primarily in terms of people's relationship to nature."

In modern society, however, science was taught primarily as a vocational subject. He thought this was one of the reasons science was so unattractive to people, "because it has not succeeded in changing the way they look at themselves."

And just as science had molded culture in the past, Frank hoped that the discoveries of physics — those philosophies embedded in the equations as clearly as Newton's clockwork — could be applied in the service of making the world a nicer, fairer, and, most of all, safer place. "The basis for social change as well as technical change is understanding how nature behaves and how people behave," he said, "and if one can promote that, if we can create confidence in that, then there's some chance that we won't blow each other up and that we can have a decent society."

Of course, he well understood that such a conflation of church and state, so to speak, wasn't always well tolerated in his profession. Many scientists frowned on the idea of applying the strictly defined methods of physics to the ill-defined realm of human feelings. "They fear getting into the quicksand of human relationships," Frank wrote in the foreword to my first physics book, *Sympathetic Vibrations*. Many scientists worried (and with some reason, I think) that laypeople would not use analogies wisely. Physicists "are quite willing to talk about how the exchange of virtual photons between two electrons can account for the attraction or repulsion of these charges," Frank complained, "but they are much too timid and unsure of themselves to point out that the exchange of unspoken words between two people can account for their mutual attractions and repulsions."

Still, Frank thought it was both tempting and inevitable to extend scientific thinking into human realms. "If one has a new way of thinking, why not apply it wherever one's thoughts lead to?" he asked. "It is certainly entertaining to let oneself do so, but it is also often very illuminating and capable of leading to new and deep insights . . . Most of the physicists that I know talk with me and with each other with full awareness that the way in which we think of the physical world profoundly shapes the way we think of the human and ethical worlds. For them physics is a part of culture and philosophy."

This same line of thinking pops up in talks Frank gave beginning in the postwar years and especially during his time in exile and while creating the Exploratorium. When he advised teachers in Budapest in 1965 on how to approach science education, for example, he took a distinctly humanistic tack. "It must be part of the task of science education in developing countries to teach science in such a way that it appears not only as a tool for the service of humanity but that it also can help *to make people more aware of their humanity*," he told them (emphasis mine).

Soon after I started reading his essays and speeches, I told Frank he should try to get them published, and I set about typing up a descriptive inventory. Eventually Doubleday gave us a contract, on the condition that we rewrite them together, and to my embarrassment, our endless (and not very successful) drafts litter the

Frank Oppenheimer Collection at the Bancroft Library (along with all my letters scolding Frank for being so far behind).

Frank died before the book was finished, but even today, revisiting these pieces, I am moved by his passion, inspired by his courage, and exhilarated by his optimism. If I had my way, every politician, policymaker, and advertiser would read and take to heart what he says about the "ethics of coercion." And his ideas for coming up with social inventions to protect us from the harm we can do each other seem more necessary now than ever before.

A Thing of Infinite Promise for Human Values

"The spirit of truth and the highest human welfare are inseparable."

These ideas of Frank's did not emerge from a vacuum. When I met Victor Weisskopf of MIT for the first time, I was stunned by how much he sounded like Frank, and soon I discovered that this was because they were both intellectual children of Bohr and Einstein and Schrödinger and Max Born and so many others.

Robert Oppenheimer wrote and spoke about such notions often, although never with the charm and clarity of Frank. "In the days of the founding of this republic," Robert writes in his book *Atom and Void,* "politics and science were of a piece. The hope that this might in some sense again be so was stirred to new life by the development of atomic energy."

Occasionally, Robert could be eloquent. "The discoveries of science . . . have changed the way men think of things outside its walls," he wrote. Also: "It is my thesis that generally the new things we have learned in science, and specifically what we have learned in atomic physics, do provide us with valid and relevant and greatly needed analogies to human problems lying outside the present domain of science or its present borderlands."

But the conviction that science has something to say about human values is centuries old at least. In his introduction to *Newton's Philosophy of Nature,* John Herman Randall writes: "Newton himself . . . and even those who attacked him . . . would all alike have been amazed at the more recent contention that natural science has nothing to do with 'values,' that it can and should itself remain 'value-free,' and that those seeking a direction for human life have nothing to learn from our best knowledge of the nature of things.

"Even a little science," Randall concludes, "is a thing of infinite promise for human values."

Francis Bacon widely promoted the idea of applying science to the human good in the seventeenth century; he proposed building a "house of Exploration," uncannily like the Exploratorium, where science and technology would be made available to the public for hands-on investigations. "There is indeed [such a house]," Richard Gregory wrote in his book *Odd Perceptions* almost four centuries later, "across the Atlantic, the Exploratorium in San Francisco . . . This is Bacon's dream come true."

When I met James Heckman, the economist and Nobel laureate who had been so inspired by Frank as a high school student, he told me about a little red book that Frank carried with him everywhere. It was *Science and the Moral Life* by the philosopher Max Otto, and its influence on Frank is unmistakable. Ethical neutrality in science was perhaps possible when science had little power, Otto notes. But that was clearly no longer the case. The survival of the world may well depend on the extent to which scientists reclaim their role as citizens.

In fact, Otto's little book is full of the kinds of ideas that so appealed to Frank: faith in common people, especially the young; the power of one person to make a difference; art as a way of knowing; the importance of "kind wisdom"; the idea that objectivity does not mean neutrality; the conviction that the world is not a given, but is ours to make up; and most of all, the duty to remain "in battle" against economic inequality, war, and injustice. "We *must,*" Otto writes, "or the higher interests of life are doomed."

Like Frank, Max Otto argued that it's perfectly appropriate to extend scientific thinking into the social realm. Not in the sense of "emptying of human life into test tubes," as he put it, but rather in the sense that "the dependable, the objectively testable kind of thinking which is the rule in the natural sciences should be put to work in the great laboratory of man's search for a good life — the good life richly and profoundly conceived."

Otto's book spoke directly to Frank's broad view of science and his distress at seeing it hijacked for evil or trivial purposes; like Otto, he wanted to nurse it back to ethical health. "Somehow we must keep trying to do better at arranging matters so that all the wonderful things that can happen between people will continue to occur," Frank wrote, " . . . without at the same time allowing every newly discovered possibility for horror from becoming probable."

What follows is an attempt to distill the essence of Frank's thinking on science and feeling as laid out in his many writings during his years of exile and reemergence. Somehow or other, I believe, he managed to incorporate just about all of these notions into the physical and social structure of his Exploratorium.

Science and Fear

"The reason why all mortals are so gripped by fear is that they see all sorts of things happening in the earth and sky with no discernable cause, and these they attribute to the will of God."

Perhaps the "sentimental fruit" that most obviously touches human emotions is the effect of scientific understanding on fear, and it was something Frank had experienced firsthand. As a child, he'd had a bronze figurine of a dog, and he attached certain rituals to its care that had to be followed precisely or bad things would happen (he might miss his bus to school, for example). The day he realized that the dog was powerless and the rituals meaningless, he experienced an enormous sense of relief. Science — in the sense of knowledge — had freed him from superstition, and it's a lesson he never forgot.

A similar lesson grew out of the acute fear of blood poisoning

Frank had when growing up. Even a simple infection from a scratch on his arm was cause for alarm (and in fact could easily have been fatal). Such fears disappeared virtually overnight when antibiotics became widely available. It made such a difference in his own sense of fear that he began to imagine what it must have been like when people feared just about everything. Before most illnesses were explainable, sick people often felt they were being punished for some mysterious reason or visited by evil spirits. Lightning, earthquakes, floods, plagues, birth defects, and famines were all seen as retributions for unknown sins, or portents of catastrophe. Not knowing what to do to protect themselves from these assorted terrors, people resorted to rituals ranging from ceremonial dance to human sacrifice.

Thanks to scientific understanding, we no longer have to "rack our souls" trying to figure out what we or somebody else did to bring on these disasters. We can't control lightning or predict where it will strike, but we can put lightning rods on our houses and be careful to avoid standing under tall trees during thunderstorms.

But when people don't know what to do to protect themselves, they'll understandably try anything — which is what makes fear so dangerous. Irrational terrors typically lead to frantic, incoherent, and usually ineffective forms of action. Thus fear can be used to manipulate people, making them lash out at the wrong things. As Frank put it, "People are more sensible when not frightened of each other or of nature."

Fear can provoke normally harmless people to behave in violent ways. Most cases of police brutality can be traced to frightened police officers; dogs and horses detect fear in people and become much more dangerous to those who are afraid of them than to those who are not. Fear of pain in childbirth can increase the actual pain, and fear while driving or rock climbing can cause accidents. Widespread fear of a stock market crash can lead to panic behavior and economic disaster.

Vague fears are the worst. If you're afraid of elevators, you can always take the stairs. If a car is coming at you, at least you know what to do. But if you're afraid of insane people or leaders of nations who behave in unaccountable ways, then it's not clear what to do. Frank admitted that groups of boys on dark city streets and insane people left him at a total loss. And while he believed that at least he had learned to commune with drunks, he thought his behavior in their presence was more like a "rain dance" than a rational response to the situation.

Rain dancing seemed an apt description of much of U.S. foreign policy during the Cold War and Vietnam War years — and perhaps especially today when the enemy, terrorism, is more diffuse and elusive. In our fear and confusion, we do many different, often incompatible things — waging war while brokering peace, killing civilians while dispensing humanitarian aid. "We rely on the balance of terror to deter war," Frank wrote, "but continually increase the quantity of this terror." Our actions were about as effective as those of prescientific societies that engaged in all sorts of bizarre behaviors as protection against disease, infertility, or plain bad luck.

Even the most rigid rituals (like Frank's meticulous attention to his bronze dog) don't help much, because we never know which part of the ritual is decisive, so we're afraid of leaving out some critical aspect. We only know that somehow we have so far avoided some kind of disaster — whether missing a bus or starting a nuclear war. But because our behavior has included so many contradictory elements, as Frank put it, "we can only keep adding to our ritual without daring to abandon any part of it, since we have not the slightest notion which parts are effective."

People generally fear most what they don't understand, and what they don't understand most is other people. Yet scientific understanding has done little to help us deal with our fear of other people — especially categories of people that seem at various times in our history to pose particular threats: Commies, or Arabs, or groups of boys on dark city streets.

We fear war above all else because it's what we understand the least. For Frank, the problem wasn't so much that he himself didn't understand war as that he thought nobody did. If you knew that *somebody, somewhere* understood it, at least you could have confidence that you *could* understand it yourself if you took the time and trouble. Since no one seemed to understand war, Frank admitted he was very afraid of it.

He saw the world as increasingly like a saloon in an old-time Western: "people crouch behind the furniture whilst silent desperados eye each other for the least sign of motion. Today fear is the cocked hair-trigger of our silos that could start the futile agony of a World War." If only we could begin to understand people as well as we understood lightning, he thought, perhaps the danger could be brought under control.

Unity, Change, and Progress

"When I go to New York and I see new buildings go up, or when I see scratches on a rock and know that the rock came from a glacier, it reminds me that the world is always progressing. Or at least, it is always changing—and that leaves the possibility of progress open."

A related "sentimental fruit" that Frank thought went a long way toward mitigating fear was the increasing sense of unity in the cosmos that science had wrought—the knowledge that nature was not as complex and confusing as it once had seemed; that many seemingly disparate things were, in fact, connected.

Before Newton discovered his universal theory of gravitation, people believed that the heavens and Earth were separate realms, ruled by different forces. What an enormous simplification it was to see that apples fall to the ground for the same reason that planets circle the sun. What a satisfying sense of connection there was in knowing that Earth and heaven weren't divided between "us" and "them," imperfect and ideal. Stars, moons, people, begonias, Ford Explorers—all are ruled by the same forces, all created from the same elementary particles. "How horrible it would be if they all seemed disconnected to us," Frank wrote.

Biology had taken a similar course. Above all, Darwin's theory of evolution taught us that we are all part of a large and vastly extended family tree, related not only to each other but also to dinosaurs, insects, bacteria. Our modern understanding of genetics all but erased the biological basis of class and racial distinctions. "This new insight of science makes it much harder to believe that other people are really, fundamentally different from ourselves," Frank concluded.

If an awareness of the unity of nature became pervasive, perhaps the enormous gulfs that divide and frighten people could be more easily bridged, Frank thought. "Surely this understanding has enabled us to think more deeply about ourselves as a part of nature, and not as a separate kind of being," he said.

This often surprising and awe-inspiring unity, in fact, is one of the main qualities that separates science from pseudo-science. In pseudo-science, it doesn't matter whether any particular explanation fits with what is already known to be true. That's what made it so "unpleasant" to Frank, and "conversely makes science so wonderful."

Science had also profoundly affected "the nature of our humility," Frank thought. In ancient times, people and the planet they inhabited were seen as both separate and special—literally the center of the universe. Today we know that Earth (and everything in and on it, including us) is made of regurgitated leftovers: the ashes of long dead stars. We bask in the light of an ordinary second-generation sun, which itself orbits an otherwise insignificant galaxy. In cosmic terms, we've fallen a long way.

At the same time, however, we've gained enormous power. In the Earth-centered universe, human beings were the powerless and imperfect playthings of all-knowing, all-powerful gods who controlled everything from plague and pestilence to the course of weather and love affairs. The only way to ensure survival was absolute obeisance to authority, unflinchingly good behavior before gods, church, and rulers. There was no sense of an individual's power to make a difference. There was no such thing as a self-made man or woman.

So what we lost in cosmic status we gained in the ability to control our own destinies. "No longer do we believe that everything that happens in the universe is merely for the benefit of people," Frank wrote. "Yet we also know that we have the power to destroy much of it—perhaps destroy humanity itself. This is a strange and frightening thing."

There was one "sentimental fruit" Frank firmly believed was the basis of democracy, and indeed all human progress: the discovery that everything in the universe is constantly changing. In Aristotle's universe, everything from the positions of the stars to the status of slaves was fixed forever in space and time. Today we know that the solid ground we stand on shifts beneath our feet. We find fossils of ancient sea creatures in the Himalayas, evidence that the earth's highest mountains once sat at the bottom of the ocean.

We know that our atmosphere was once a deadly poison; that the air we breathe today is manufactured, daily, by plants. We ourselves evolved from a long and unlikely line of ancestors, including fish, singled-celled organisms, and stars. The universe itself evolved, perhaps originally from some mysterious speck of primal nothing. Energy and matter have a history.

Even now, everything continues to be recycled. At any given moment, stars are dying by the millions, many going off in fantastic displays of fireworks that would make the universe sound like a giant vat of popcorn if only sound could travel through empty space. At the same time, stellar nurseries all over the cosmos are pregnant with newly emerging stars. We have gone from a universe where everything was predetermined to a universe where everything evolves, which means that the social and economic status of people can change too—the idea at the heart of democracy.

With the acceptance of change, we have become accustomed to the notion of progress, "or at least progression," Frank said. The abolition of slavery would have been unthinkable in a society that didn't accept change as a matter of course. "All of us act with the conviction that the future will be different from the present or the past," Frank wrote, "and that society is not necessarily permanently imprisoned by our present faults and limitations."

The Fundamental Things Apply

"My description of people must have a certain symmetry to it . . . It must be independent of where I look or how I am living at the moment."

Most impressive to me among these "sentimental fruits" — probably because it was the most surprising — was the idea of symmetry as the guiding force behind most laws of nature; that is, the idea that the fundamental things are those that do not change, no matter what. In physics, a symmetry always describes a change that doesn't change anything, like rotating a snowflake 60 degrees. This is the basis of Einstein's relativity: it's the fact that the speed of light never varies that makes space and time elastic. No matter how fast you run toward or away from a light beam, it makes no difference in the light's measured speed. Of course, weird things *do* happen to both space and time, but that's because the speed of light turns out to be more fundamental than either of them.

The same holds for other surprising symmetries — that energy and mass are different forms of the same essential stuff, for example — an idea neatly captured in Einstein's famous equation $E=mc^2$. Since matter and energy are one and the same, each can be transformed into the other: the sun changes two dozen ocean liners' worth of mass into energy every single second.

So if we want to look for truths that are deep and valid under all circumstances, we have to look for things that don't make a difference. In fact, these notions of symmetry are already enshrined in various laws as measures of fairness. If you divide some resource into two parts, and the division creates two equal pieces, then it shouldn't make any difference who gets which piece. The outcome, either way, is perfectly symmetrical. The same is true of the Golden Rule. If you "do unto others as you would have others do unto you," then it shouldn't make a difference whether you are on the giving or receiving end.

Frank thought understanding such symmetries could guide us in creating a better society. Among the differences that appeared not to make a difference were race, nationality, appearance, economic status, and disability. Whatever categories people might fall into, everyone feels pain, makes art, experiences awe — whether they call it religion or something else.

So civil rights, for example, were necessary for more than just humanitarian reasons. They followed from principles of symmetry. "The reason we need civil rights," Frank said, "is because of the fact that the transformation between white to black skin doesn't change a lot of things . . . And the fact that those things are invariant has consequences for the way we should behave."

This notion that unchanging fundamental truths could lie hidden beneath such apparent differences was perhaps even more powerfully developed in quantum mechanics. In the subatomic world, mutually exclusive descriptions of the same phenomenon can both be right. Light is a wave, but it is also a particle. Electrons are particles, but they are also waves — as are all other particles in the universe. What determines whether light is a wave or a particle is the way you measure it. Set up your experiment to look for waves, and waves are what you'll see; the same is true of particles. The answer you get depends on the question you ask.

On the face of it, nothing can be a wave (which extends into space) and a particle (which is concentrated in one place) at the same time. But this is how nature has arranged things. The fact that the human mind hasn't evolved to grasp such a dual thing as a "wavicle," as some have called it, is beside the point.

The human realm is full of such seemingly nonsensical dualities. Most people, Frank noted, have no problem reconciling the fact that they are "merely flecks of nature — ashes and dust" with the intense feelings they have about themselves and their loved ones. "We are not perturbed that our mental image of ourselves includes both the view that human life is the most precious thing in the universe, and that the destruction of life on earth would make a negligible difference in the universe," he said. "We can be passionately in love and yet realize that nothing is going on but some complicated chemistry."

Simply knowing that such dualities were necessary for an accurate description of "inanimate nature" made a profound difference in Frank's thinking about human problems. "I can be reassured that thoroughly contradictory ethical statements need not, either one of them, be a heresy when applied in the appropriate contexts," he said. He didn't mean to imply that there was no difference between right and wrong, but only that the opposite of one truth need not be totally devoid of merit, and so perhaps scientific understanding could "mollify some of the fiercest intellectual battles of the present and the past."

He thought that if other people understood these ideas, it would allow them to "relax a bit" about the contradictions that complicate their lives and often tear people apart — as well as lead to policies that might deal with such conflicts more sensibly. The abortion debate and end-of-life issues are good examples. Despite how the press and pundits try to frame the arguments, it's not contradictory to believe that life is precious and also that abortion and euthanasia may be necessary and even desirable in some situations.

Bob Miller once asked Frank how he could accept the fact that the universe was essentially purposeless. Frank didn't see why a universe without purpose and a universe with purpose were contradictory. "Why can't there be islands of purposefulness in a purposeless universe?" he asked.

Prediction, Control, and Understanding

"Perhaps some wonderful new social invention would appear if only we had an inkling of why it is that people enjoy listening to music."

Given the success of science in changing the way we think and feel — in lessening fear and irrational behavior, in opening the door for change — why haven't we made more progress in building a fair and peaceful society? Why haven't the social sciences made the same kind of progress as the physical and biological sciences have? In Frank's view, it was because social scientists misunderstood how physical science worked, and what factors made it so successful. As a result, they tended to focus on the wrong things.*

For example, social scientists, like most people, seemed to believe at some level that the power of science came from the fact that it could predict and control things. The truth is that it can predict and control very little. Take a simple scenario: a ping-pong ball bouncing on a table. Can you predict where it will land? The answer is no. There are too many variables: a dent in the ball, an unexpected sneeze, a tremor in the hand — all can change the outcome. Or what if the ball lands on a mousetrap?

Frank's responsibilities during the Trinity test of the first atomic bomb included predicting whether the radioactive cloud would spread along the ground or rise in a tall mushroom. His calculations told him unambiguously that it would rise. But that didn't stop him from plotting escape routes.

Of course, the ability to predict is not limited to science alone. Parents can accurately predict that the lack of a nap will produce a cranky child. A dog can predict that a walk is in store when it sees its owner go for the leash. People predict rain by watching the flight of swallows. "I think our professional modesty in this respect ought to be greater than it often appears to be," Frank said. If anything, the belief that science is good at making predictions caused many people to lose their faith in it. "I think that people got so disappointed that the scientists lied to them, or that their predictions didn't come true, that they don't pay attention. So they don't take science into account at all, or very little."

The misconception that science has some special predictive power is partly rooted in the language of science itself — specifi-

* Needless to say, the social sciences, especially psychology, have advanced significantly since Frank was writing, but in general his points are valid. For example, while the idea of "multiple intelligences" is widely accepted, IQ scores are still used as the standard measure in many contexts.

cally the way scientists use the words "the theory predicts." The phrase doesn't mean that the theory predicts the future. It means that if the theory is right, it can make predictions about what will be found if one looks in a certain place in a certain way at a certain time. Einstein's theory of general relativity predicted the existence of black holes. They have been found. But nothing in his theory was predicting the future. Black holes have always been there, waiting to be discovered.

In this way, scientific understanding has been able to "predict" the existence of some remarkable things. For example, once visible light was understood to be an electromagnetic wave, Heinrich Hertz predicted that similar, longer waves must also exist—and so discovered radio waves. "But this kind of prediction is not soothsaying," Frank warned. "We are not being oracles."

One thing scientists *are* often able to predict quite well is what can and cannot happen. Global warming is a case in point. We need to know what can and cannot happen so we can take the appropriate action. So if scientists say global warming might happen, and that its effects might be severe and irreversible, then it's probably a good idea, at the very least, to monitor climate patterns carefully.

Since the ability to predict implies the ability to control, Frank thought, many people become uneasy about science, especially when it's applied to the human realm of psychology or social arrangements. But despite the great advances in science, we actually control very little. Newton's discoveries about gravity have given us no control over the orbit of the moon or earthly tides. We have light after the sun goes down not because we've stopped Earth from moving but because we invented electric lights. We talk about controlling insects, but as Frank pointed out, where they exist, they fly about as naturally as ever. All humans have been able to do is exclude them from a few places.

On the other hand, we can control things without any scientific understanding whatsoever. "We can control the flow of rivers," Frank said, "but this control is clearly not uniquely the result of scientific understanding; beavers build dams."

Rather than control or prediction, the power of science to help humanity has come mainly from its success in making us immune from the worst threats posed by nature and human nature. A house provides a safe haven from the cold and rain without controlling the weather. An immunization protects a child from a disease without controlling the disease itself. A court of law is designed to provide immunity against injustice. These inventions, like a cell wall, provide a safe harbor where life can go on and creativity can blossom, Frank said. They "promote a freedom to develop rather than repress it."

But either way, it was clear that the success of physical science owed very little to either prediction or control, and everything to understanding.

Social science would make much more progress, Frank thought, if it became more concerned with the quality of understanding rather than "the clarity of the crystal ball" — predicting, say, how people will vote in an election or who will do well in college: "We attempt to develop stable economies although we cannot properly account for the phenomenon whereby a unique currency can be used to purchase items of incommensurate value such as a shovel, a pickle or a Picasso."

A classic example in social science of how the prediction (and control) "cart" was often put before the understanding "horse" is the case of IQ tests. IQ tests are used routinely to predict what people are capable of achieving, though our understanding of intelligence is tentative at best. Intelligence is no doubt more complicated than a physical concept such as motion, Frank pointed out. And yet it took centuries to come up with proper laws of motion because people were confusing many connected but distinct concepts — for instance, velocity, acceleration, and momentum. Why should intelligence be any different? "Certainly intelligence must be much more divided up than motion," Frank said. "And yet they lump it all together."

Human beings and human societies are far more complicated than even the most esoteric physics, so it wasn't surprising that "social understanding has lagged behind the physics of the atom,"

Frank said. But that shouldn't stop us from trying our hardest, because it was the only way to create effective "social inventions" — something Frank thought should be as widespread and readily accepted across the globe as the current acceptance of inventions such as electric motors, milling machines, and lasers.

Science and Social Invention

"Just as present technology had to await the explanations of physics, so one might expect that social invention will follow growing sociological understanding."

When the focus of science shifts from understanding to forecasting, what is lost is the potential for the kinds of new discoveries that "make it possible for people to improve their lot," Frank said. These discoveries are the raw materials that become the basis for invention. Just as understanding disease led to vaccines, so Frank believed that understanding human social and political behavior could lead to the invention of "inoculations" against other threats to human welfare, including war and injustice. He often mentioned the family as such an invention; also democracy, the courts and the Fifth Amendment, the United Nations, the Marshall Plan.

In fact, he thought the Marshall Plan was such a great invention that the same strategy should be expanded into a system of what he called (only partly tongue-in-cheek) "goodie drops." In effect, the United States had bribed European nations into becoming strong allies through our generous postwar rebuilding effort. Similarly, he thought, we'd do better to bribe our enemies by dropping food on them rather than dropping bombs.

While the "goodie drop" idea never caught on among policy wonks, I personally managed to use it to great effect. When my daughter was an infant, the teenagers next door practiced heavy metal far (and loud) into the night. The previous occupants of our house had called the police to protest — and got a rock thrown through their window in response. Following Frank's idea, I baked cookies for the boys and then asked them nicely to turn down the volume. Remarkably, it worked—so well, in fact, that I began to bribe their nasty dog with biscuits.

Another of Frank's schemes focused on devising weapons that a global police force might use to contain dangerous regimes. He wanted something that would be both humane and effective. At one point, he came up with the idea of spewing radioactive fission products over a city to make it uninhabitable (only after, of course, warning people to get out). It wouldn't win a war, but it might discourage countries from arming. Frank himself called this idea a "dumb thing."

A less destructive one we both rather liked, the inspiration of a young student, was dumping Jell-O on our enemies to incapacitate without seriously harming them. (Talk about shock and awe!) The terrors that threatened the world were so serious, Frank thought—and so impervious, it seemed, to conventional solutions—that "we really have to begin to think somewhat outrageously." (The Jell-O attack would also provide the victims with something to eat afterward, another attractive quality.)

A social invention can be a very simple thing—anything from a stop sign to a handshake. The popular vote, public education, labor unions, families, and taxation are all social inventions. Frank was particularly taken with those that protected human rights and promoted democratic institutions—for example, the notion that a suspect was innocent until proven guilty, the prohibition against unreasonable search and seizure, the checks and balances built into the structure of the U.S. government. He liked the veto power because he thought it was a great way to get people who distrusted one another to sit down and talk without fear that the outcome would be unacceptable to either party.

Frank's main worry was that the raw materials necessary for social inventions were drying up. You could get only so far by recycling the same ideas. Without fresh ones, "we become stuck," going around and around the same paths. It was the infusion of such new raw materials that made the scientific revolution possible.

And in order to come up with inventions capable of protecting us from the worst of the modern world — a world with instant global communications, the constant threat of terrorism, exploding populations, and widespread ignorance — economists, political theorists, psychologists, and other social scientists would have to "discover truly new things about people and society," Frank said. What we needed was a "renaissance of creativity" in these realms.

He thought of these ideas rather like recessive genes. Most genetic mutations, like most new ideas, produce no great changes. But when the environment alters, and living things have to adapt to new situations, those recessive genes may make the difference between survival and extinction — as they did when the oceans receded and some water-dwelling species were able to adapt in ways that allowed them to live on dry land.

Many problems in science are solved by drawing on just these kinds of "recessive" ideas. In physics, solutions to modern mysteries often turn on scientific or mathematical inventions that have been lying around for decades. In fact, it's rare that you can solve a complex problem merely by spur-of-the-moment thinking. In the same way, Frank argued, the framers of the Constitution drew heavily on political ideas that had been developing in Europe. And so it was today: "In contemporary America, the government, in times of crisis, has more than once turned to the stockpile of ideas to be found in the academic world."

Part of what made discovering truly new things possible was providing a way for people to explore nature without any thought of practical outcomes or getting something done. It can be difficult if not impossible, to really understand something at the same time that you're engaged in doing it, Frank pointed out more than once. He believed it was this very difficulty that gave rise to the "invention" of the Sabbath as a day of reflection. Frank once testified at a congressional hearing concerning a proposed conference on the humanities that such a Sabbath was essential if people were to avoid making mistakes — because it was the one time they could reflect on the purpose and value of their work.

In addition to time, people needed an environment set apart from messy everyday reality. This ability to isolate "fragments of some natural phenomena" in a laboratory, or by highly selective observations, was extremely important in science but almost impossible to achieve in most areas of human life. "The establishment of separate research environments in which people are paid just to find things out has been a key element that has made science flourish," Frank said.

People had, in fact, already invented an institution well suited to providing both the necessary isolation and a place for stockpiling tentative understandings: the university. A university was a place where people could feel free to try out all kinds of ideas without fearing the consequences of bad ones. Even far-fetched notions could be probed and tested and generally kept on hold until they were needed. The university was a receptacle for the "fund of knowledge and theory which may prove to be indispensable to society as it develops and finds itself confronted with undreamed-of problems," Frank wrote.

Yet these days the reservoir seemed "woefully low," he said, perhaps because the critical role of universities had not been fully understood. For universities to produce reservoirs of new ideas, it was essential that no one working within their walls feared facing "personal or social consequences" for simply accumulating ideas. The creative process that goes on in academic settings often involves pulling theories "out of the thin air of the mind in order to see what happens to them when they are criticized and developed by others," Frank said. "Ideas at this stage are tentative and fragile things." Getting the most out of universities meant people would have to resist the temptation to demand quick results or "immediate service" to society. The "golden eggs" that a university brings forth keep coming only so long as the institution remains autonomous and free.

The more or less idealized university Frank described did not, of course, reflect some of his own experiences with such institutions, but he loved universities nonetheless, and saw great hope in them.

In more ways than one, he had just such a university in mind when he created the Exploratorium.

Science and the Ethics of Coercion

"Advertisers who promise a better sex life if we use their toothpaste don't really help us understand how to have a better relationship with people."

In the absence of understanding, the only way to change the behavior of people is through coercion, a generally ineffective tool—especially in the modern world. The symmetry of power brought about by advanced weaponry meant that there was no longer a safe way for people to coerce each other. As Einstein famously put it, whatever weapons World War III might be fought with, World War IV will be fought with sticks and stones. "Attempts to provide a safe passage to the future by means of the coercive manipulation of people," Frank said, "will be as futile as would be reliance on calming the ocean in order to cross it in a ship."

Coercing is not the same as overpowering, as Frank took pains to point out. When the United States bombed North Vietnam, it was obviously not trying to coerce the people who were killed, though they were certainly overpowered. "There is no way of getting them to the conference table," Frank said. Instead, the real purpose was to coerce those who were still alive to do our bidding by threatening them with actions we had the power to turn on and off at will.

Coercion, in fact, need not depend on violence at all. You can just as easily coerce people by withholding (or granting) foreign aid, affection, or even grades. The essential quality that separates coercion from persuasion is that the coercer has complete control over rewards and punishments. Coercion, in other words, requires only power—but not the knowledge necessary to change somebody's mind.

In contrast, persuasion always requires understanding. To persuade someone to stop smoking, to stay in school, or to refrain from embarking on a nuclear weapons program—all require the knowledge to make a convincing argument. "Persuasion involves explanation," Frank said. "When we are persuaded, we feel that we have understood and that we have made up our own mind."

In the absence of understanding, coercion becomes the only way to get people to do what you want. So if people stop trying to understand things, Frank argued, a coercive society becomes inevitable. In a coercive society, there's no particular *incentive* to understand, because understanding isn't all that valued or useful.

The real problem with coercion is that it's mindless. And in an age of atomic weapons, the combination of unlimited destructive power and often complacent ignorance adds up to an assuredly deadly combination.

Frank thought he was witnessing a profound change in American society in which coercion was replacing persuasion as a primary method of influencing people, and that this, above all, explained why so many people had given up trying to understand things. "Politicians who tell us to build stronger armaments because the Russians are coming don't help us to understand why the Russians might be coming or what, besides arming, we might do to stop them," he wrote. Advertisers who promise popularity by peddling a particular brand of beer don't help us have better relationships. "The claims and counterclaims are so mindless that we eventually give up trying to understand why poverty persists or why there is inflation and unemployment or why the U.S. auto industry is in trouble."

Frank acknowledged that coercion was necessary in some cases, but that when a whole society became based on coercion, it was doomed to failure. Where learning was not revered, little new was learned. Coercive societies, like coercive individuals, could not cope with change: "Excessive coercion inevitably breeds disasters, not because it is so cruel, but . . . because it negates the essential value of the understanding that is gained through art and science."

Coercion became immoral, Frank concluded, when it threatened to preoccupy society to the extent that understanding was discouraged or precluded. U.S. foreign policy had become immoral primarily because it was consumed by coercion. It was almost entirely based on giving or selling other countries what they wanted — whether it was food or fighter planes — in exchange for good behavior. If they didn't do as we wanted, then the message was clear: "We won't sell or give you anything and will do everything we can to make life miserable for you."

He saw the same thing happening domestically — in education, for example. Learning had become so completely based on rewards and punishments (usually in the form of grades) that teachers were forced to direct their teaching to tests; students were forced to concentrate their learning for tests. There was no room left for asking real questions and therefore producing new knowledge. "The exclusion of understanding that accompanies the use of coercion can ultimately reduce the rate of invention to zero," Frank said, "a reduction which then occurs precisely when there is a growing need for the discovery of new mechanisms by which people can protect themselves from one another."

He was angered particularly by the implicit coercion involved in lying and twisting the meaning of words. "The coercive uses of language especially as they become amplified by modern delivery systems are almost as discouraging as the coercive uses of bombs and their incredible delivery systems," he wrote.

Lying can induce someone to buy a product, elect a politician, support a war. Such lies can seem like part of the process of persuasion, Frank said, but the situation described by the coercer is entirely under the coercer's control. And when promised rewards or punishments don't materialize, the fabricator merely escalates the lies. "Eventually, both the coercer and the coerced become wholly preoccupied with the process, the former with inventing more and more absurdities and the other with desperately trying to decide how to act."

Widespread deception, Frank believed, was also behind the much-discussed political apathy among the general public. If people couldn't believe what they read or heard from politicians and the news media, why *should* they pay attention? The whole idea of rational argument becomes irrelevant when lying is pervasive, he said. "I believe that large sections of this so-called inattentive public would come to life in a situation in which they didn't feel they were being fooled and lied to all the time."

Frank used persuasion to great effect in his own life, insisting that everybody understand every decision he made. As one colleague remembered: "The key to Frank's thinking was that he would explain why something was being done, that it would make logical sense to people. So they could accept it, or handle it. They would agree to it even if it was not in their best interest."

It was an attitude he passed on to those who worked with him. Thomas Humphrey remembers Frank advising him on the proper role of a supervisor: "He said it should only be on rare occasions that you have to tell someone what to do. If it was the right thing to do, it ought to become evident to other people, so there is no need for autocracy. And that still kind of lives with me."

Perhaps the most widely quoted Frankism (at least by me) grew out of his conviction that coercion harms the coercer as much as or more than the coercee. I first heard him articulate this philosophy during a dispute over parking spaces at the Palace of Fine Arts. Most of the building is occupied by the Exploratorium, but a portion belongs to the Palace of Fine Arts Theater. At one point, Exploratorium staff members complained because the theater people were using spaces specifically designated for the Exploratorium. One person suggested (only partly in jest) that the Exploratorium retaliate by taking parking spaces that belonged to the theater.

Frank's response was "The worst thing a son-of-a-bitch can do is turn you into a son-of-a-bitch."

This philosophy, he thought, applied equally well to foreign policy and other realms. Too often, the best of people (and countries) allow themselves to be made horrible in their efforts to control outside horrors, and it's a losing proposition all around. Frank argued this point at some length in a letter to Norman Cousins, then editor of the *Saturday Review,* saying that in its policy toward Cuba, the United States was behaving out of character. "I have always been somewhat on guard because I realize that the worst thing that an unpleasant person could do to me is to make me unpleasant, and the most unwanted effect of any set of circumstances would be to change me into the kind of person I really do not want to be," he wrote Cousins. "But now this very disaster is occurring, not just to me, but to the group of people that I am one of and want so much to be proud of."

After recounting some of his experiences with Cubans retrieving his cosmic ray equipment and reviewing recent Cuban-U.S. relations, Frank concluded: "Everything the Cubans have done has provoked us into acting meaner and more spiteful . . . The kind of coercion and tit for tat blackmail that we have increasingly resorted to is clearly despicable. We are a decent nation; let's try to act like one."

As Thomas Humphrey later reflected, Frankisms like this had a special impact because they held a mirror up to ourselves.

If Frank's ideas live on after his death, it is at least partly due to the fact that he put so much time and effort into persuading people of the value of those ideas. Unlike coercion, which becomes ineffective as soon as the coercer loses power, persuasion leaves a permanent mark.

When the year 1984 loomed near, Frank suggested the Exploratorium celebrate the occasion by honoring Rachel Carson. The idea didn't make much sense until he explained that George Orwell's *1984* changed the way people think — much like Carson's *Silent Spring*. Whereas Orwell's 1949 book woke people up to the authoritarian abuses of power (inspiring the term "Orwellian"), Carson's eloquently written book drastically altered people's perceptions about the indiscriminate use of DDT — and thus laid the groundwork for the environmental movement. By honoring Carson, Frank said, we would be honoring the power of a single book to make a significant difference — through pure persuasion.

11

THE ANARCH

"At a certain point, we started to realize that we were getting famous."

For roughly a year, beginning in the summer of 1981, I sublet a flat in the Marina District just blocks from the Palace. For the first time, really, I became a full-fledged employee. I got my own key to the Exploratorium — a huge thrill that never wore off. The very idea that I could walk into that vast cavern of marvels at any time of day or night made me feel unspeakably lucky (and not a little important). I also got a typical Frank title. Needing some kind of professional designation for a meeting or a résumé, I don't remember which, Frank suggested I call myself "Exploratorium Expositor." (At least it was better than "writer divided by editor.")

But much had changed, both in Frank's life and at the Exploratorium. In March 1980, Jackie had died of cancer after a long period of deterioration. Frank was sad and lonely and his house seemed terribly empty. "He was absolutely devastated by the death of my mother," Mike remembered. It was clear that some of the old spark had left him.

Jackie's death was a huge loss to the museum as well. She'd always had her own spheres of influence — the store and the graphics department, both of which Frank was indifferent to at best.

She was quiet but powerful, afraid of nobody, and intolerant of even the slightest whiff of uppityness or disloyalty to Frank. Frank relied on her judgments, especially about people.

For many months after Jackie died, Frank kept her clothes in the closets, her stuff around the house. He talked about her often and would say, out of the blue, "Gee, I really miss Jackie." Usually people would change the subject, unnerved to be reminded of the dead. But Frank thought that was precisely the point: being reminded of her was the only way he could keep her alive a little longer.

Then one day, upon returning from a trip to Colorado, he announced that he'd gotten married. Frank was as giddy as a kid. A couple of staffers saw him in the shop, working on a lathe, and for once he wouldn't tell them what he was making. "Something" was all he'd say. The something was a ring for Milly Danielson, his new bride.

Milly was an old friend of Frank and Jackie's from Boulder, and Frank told Mike that marrying her felt like a kind of "continuum." Warm and tough and funny, she was a social activist who was more than up to welcoming the whole eccentric gang; somehow, she even put up with Orestes. Among the additions Milly brought into Frank's life (and mine) was a grand piano, which she moved into his living room. She was endlessly tolerant of my inept flute playing, so I got to do something delightful that was previously outside my experience: sit around making music, just for the hell of it.

The Exploratorium had changed too. By the late 1970s, Frank decided he needed more floor space and started planning a mezzanine that would add thirty thousand square feet. The money was finally raised, and the Exploratorium closed down in late 1979 while a partial second floor was built. When it was reopened in May 1980, Frank was relieved to find that the somewhat fancier construction (and carpeted floors) of the mezzanine didn't seem to constrain the behavior of kids. "Fortunately they feel free to run there just as they run every place else in the museum," he said.

By 1982, the museum had more than five hundred exhibits, indoor restrooms, a theater, and a restaurant (though Frank still preferred the pretzel cart, or Upton's, for lunch).

Accolades had begun to pour in. *Look* magazine called the Exploratorium "the most innovative science teaching mechanism on the planet." MIT's renowned Victor Weisskopf insisted that Frank's creation was, in fact, "the ONLY science museum." "There is simply no other museum of science in the whole world that can be compared with this institution . . . It is an institution whose exceptional value cannot be overestimated, especially since nothing like this exists anywhere else."

The museum's influence spread. The Association of Science-Technology Centers commissioned a show of Exploratorium exhibits that traveled to museums around the country. By the early eighties, the Exploratorium had published three volumes of "cookbooks," detailed instructions on how to construct exhibits, which were used by science museums worldwide; the publication program (including the magazine and my catalogues) was going strong. The Exploratorium began to receive serious grants for the specific purpose of replicating its programs elsewhere.

Frank, too, was receiving some measure of recognition. He was awarded the University of Colorado Distinguished Service Award, the Caltech Distinguished Alumni Award, the American Association of Museums Award for Distinguished Service, the Milliken Award from the American Association of Physics Teachers, and others.

"At a certain point," Ginny Rubin, Frank's development director, told me, "we started to realize that we were getting famous."

In March 1982, PBS aired Jon Else's *Nova* documentary *Palace of Delights*, a moving tribute to Frank and the world he'd made up. Frank became a minor celebrity, and visitors began to stop him on the floor and make a fuss. Sometimes Frank himself seemed to have a hard time believing how well his unlikely experiment had succeeded. He'd enter the Exploratorium through the shop door and look around at all the enticing "props" that filled the once dead space with light and sound and motion. He'd shake his head in that Howdy Doody way of his and smile like a kid who'd just gotten away with something outrageous — something far more outrageous than taking apart a player piano or becoming a Communist or catching cosmic rays or birthing a calf or building an atom bomb.

And when people came up and thanked him for creating such a seriously magical place, as they often did in those days, he just beamed.

Democratic Centralism

"He was like the leader of a commune."

During the year I spent as a full-time staff member, I got to see how Frank managed his "woods of natural phenomena" up close for the first time. I'd never really seen him in action as an administrator, and it turned out to be quite a show — a ten-ring circus with disaster stalking every tent. Actually, "administration" doesn't come close to describing it. The only thing that does is a phrase cooked up by electronics guru Larry Shaw: the Exploratorium, he said, was an anarchy, and Frank was the anarch.

To be sure, Frank was a funny kind of anarch in that he wielded absolute power (an absolute anarch?). He determined the rules (or lack thereof, in his case). He chose the staff, made final decisions about exhibits, and drew up the budgets. The rest of us followed along while Frank led the charge: "That way!" Sometimes it was a very gentle "That way." Other times, you had to hold on for dear life.

"It was sort of like in the Westerns when you see this horse taking off and this rope trailing behind it and somebody grabbing on to the rope and being dragged along in the dust," one staff member said. "That's the way you usually felt when you were around Frank." The Exploratorium was Frank's crusade, and there was never any question about who was in the saddle.

His intensity was such that some people stayed out of his way, but if you fell within his orbit, you were pulled in, no matter what the occasion. Even if he just called you into his office to celebrate something, say, and asked you to grab some Styrofoam cups from the coffeemaker so everyone could have a drink from the whiskey bottle in his bottom drawer, there was no refusing. Everyone took their shot.

Frank was very like a child, and had all the bad temper and willfulness that went along with never growing up. He could be dismissive, even mean. But he was often joyous and uncommonly kind. When he came through the office playing the flute, everyone loved it — partly because it was so beautiful, but also because it meant he was in a good mood.

Like most children, Frank was also unpredictable, and unabashedly so. He could go on vacation and not bother to tell anyone where he was going or when he was coming back. You never knew how he would react to what you told him, which gave even the smallest interactions a certain charm. "He had a kind of pixie quality," said the historian of science Bruce Wheaton. "You could never quite tell where a conversation was going. Sometimes you had to chase after him to catch up" — a continual theme. Occasionally he would disappear for hours at a time into a little room above the theater, accessible only by ladder.

Despite these eccentricities, or perhaps because of them, Frank had no problem getting people behind him. For one thing, he had a genius for convincing people that they had a lot more influence than they actually did. Staff members could do whatever they wanted — so long as it was OK with Frank. To make sure it was, he'd butt in anywhere, right down to the lost-and-found box and the exhibit labels. A new graphic artist joined the staff in the early 1980s and was surprised to find that the art department was still using rubber cement instead of wax for paste-up. She mentioned this to someone in passing, and the next thing she knew, she heard from Frank: "What's this about wax?"

Part of his meddlesomeness was just plain curiosity. He was bursting to know what people were up to, and he wanted to be a part of all of it. But his attention to seeming trivia was also quite deliberate, part of his management philosophy. "The way the place is emerges from decisions made about many details," he said. "The general view is that an executive should not pay attention to all these details. I don't think that's the right way. It doesn't work when you are doing experiments and it doesn't work in art."

Beneath Frank, the administrative structure of the museum (such as it was) was almost entirely flat. "There were three levels at the Exploratorium," said Tom Tompkins, the head of the shop. "There was Frank, there was everybody else, and then there were the kids who cleaned the fish tanks." "Everybody else" included a huge range of people from physicists and department heads to shop workers and Explainers, but Frank didn't make distinctions in terms of rank. Credentials were largely irrelevant, and titles almost nonexistent.

He dealt with outsiders in the same democratic (or anarchic) way. As the Exploratorium grew more famous, Frank started to get a lot of visitors, some of them big-name muckamucks from science or business or government. If he thought they were interesting and excited about the place, he could, on a moment's notice, spend an hour taking them around the museum. If he thought they were phonies or had nothing to offer (which was not infrequent), he didn't give them the time of day.

In dealing with each other, the staff took Frank's cue, and everyone felt more or less on an equal footing. Everybody did everything. If there was garbage on the floor, Frank picked it up — and he expected everyone else to do the same. "This was not a popular activity," the biologist Charlie Carlson remembered. He was certainly not prepared, after his first staff meeting, to be given a broom and told to start sweeping.

Most of all, Frank wanted everyone to be a teacher, whether they were working with schools or creating exhibits or writing exhibit labels; he wanted everyone to be an inventor, whether they were involved in fundraising, accounting, or curriculum development. It was not uncommon for a secretary or an accountant to wander into the shop and idly ask, say, what it was like to melt metal, and this would start a two-hour conversation that might eventually turn into a project of some kind—exhibit or artwork or written material. "We have lowered artificially imposed barriers not only between the arts and the sciences," Frank said, "but also between activities of the hand and the mind."

The administrative staff, in fact, played an important role in exhibit-building. When a prototype was near completion, the builders would go grab someone out of the office to play with it—preferably someone who knew little about science and technology. In this way, the exhibit developers got to see what worked, what needed fixing, or what may have been forgotten. But it was also part of the general gestalt that everyone should be included in everything. Nobody wanted to feel left out, and usually they didn't have to be.

The staff accommodated themselves to this total lack of boundaries, almost always ready to get roped into anything. Whether you needed a table built or an electronics problem solved or a sign written, someone was happy to help. If you had questions about basic science, you could usually find someone who'd pretty much drop everything and go at it full throttle. More often than not, these discussions went off in all kinds of interesting directions. I once asked one of the resident physicists what I thought was a simple question about the source of the aqua color in the waters off Virgin Gorda, and the subject turned out to be so complicated and interesting we made a whole magazine out of it. (I also wrote a column for the *Washington Post* about it.)

Several of the scientists said they found the administrative arrangement not unlike that of a large physics experiment. One physicist, Elsa Feher, said she immediately felt right at home. What Frank had created, she said, was "an academic kibbutz." Rob Semper had much the same reaction. In physics, what's respected is ingenuity and getting results, he said, and it was the same at the Exploratorium. And just as in Ernest Rutherford's lab, people were expected to find what they wanted to do, and then do it.

It was a system that didn't so much encourage creativity as require it. The result was that when someone created something or solved a problem—whether it involved exhibit-building, writing, or accounting—they wound up with something that was original and also really their own.

The Exploratorium became the center of people's lives, which was exactly what Frank wanted. He wanted people to feel as if they could work there forever, that this was their home. "This place belongs to you," he told more than one staff member. He said it belonged to everyone who built it.

Staff meetings felt like family affairs. Once a week on Wednesday night, someone would buy beer and pizza and everyone would sit on dirty old couches around one of the big fireplaces in the Palace. It was never a meeting to inform people about decisions that had already been made. It was a meeting to discuss what was going on and to make decisions. "It would be like: 'Gee, we're totally strapped for cash nowadays. What are we going to do about it?'" one staffer remembered. "'Would anybody mind taking a twenty percent cut? Who would that severely impact?'"

After the mezzanine was built, the staff met in the new library around a big conference table, but the feeling of the meetings didn't change. One staffer remembered a long discussion about how to solve a persistent fly problem. Some people argued for hanging flypaper. Others were concerned about killing living things, no matter how unheathful or pesky; were there alternatives? "The thing that was so remarkable was that everyone was so passionate about everything," the staffer said. Everyone knew that, in the end, Frank would decide. But he listened carefully to everyone. "He was respectful."

It occurs to me now that this may have been the kind of "democratic centralism" Frank had once hoped to find in the Communist Party. Whether that's true or not, he reveled in it.

The Edge of Chaos

"What's wrong with that? That's science."

The Exploratorium was an undeniably chaotic environment; such paths as there were through the woods diverged in many directions; there were no detailed maps and no easily definable goals — no beginning and no end to the journey (or any particular activity, for that matter). There were no right answers and no particular piece of knowledge or experience that people were supposed to come away with. If someone was using an exhibit upside down, that was fine with Frank. If visitors came to the wrong conclusions, perfectly OK — as long as the conclusions were their own.

Truth be told, Frank loved that people used exhibits in the "wrong" way. One of his favorite examples was the "Bernoulli Blower," a truncated rubber highway marker that funneled fast-moving air into a powerful, upward-moving stream. Because of Bernoulli's famous principle (fast-moving air creates less pressure than slow-moving air), a ball could get "caught" in the air stream, and you felt a force if you tried to pull it out. People sometimes tapped the ball to watch it oscillate, experimenting to see how far out it could go before gravity grabbed it; others would throw the ball into the stream or cover part of the hole where the air went through. But people also did lots of things that had nothing to do with Bernoulli. "Girls let their long hair stream up in the air current," Frank reported with obvious delight. "Kids hold their T-shirts over the orifice and let the air stream cool their bellies. Some people play catch with the ball." All no problem.

Once Frank noticed a middle-aged woman pointing out some star-like holes in the ceiling of the Exploratorium to her companion. The holes were made by seagulls that had punctured the black painted surface of a skylight with their feet. (The surface had been painted to darken the area where the optics exhibits were located.) The woman casually commented to her friend that she supposed if she knew more, she'd understand what "those little lights" meant. It didn't matter that the "lights" were seagull footprints. What mattered, said Frank, was that the woman and her friend "were perfectly happy as they went on to play with other exhibits."

Beyond promoting a feeling of being at home in the world, such experiences also gave people the gift of real discovery. It was not the typical "discovery method," where things are rigged so that people discover only what they're supposed to. Whether the discovery was "right" didn't matter, but whether it was authentic surely did. "Personal discovery, whether it occurs through art, through science, or just by wandering around the city or the country or a museum, brings far-reaching satisfaction and personal consequences that are vastly greater than knowledge which is just handed to you or told to you," Frank said.

This feeling of ownership, many surmised, was behind the surprising lack of vandalism in the Exploratorium. Because the "trees" in Frank's woods could be used for things they were never designed to do (much like real trees), it was almost impossible for visitors to get bored. "As long as people can invent something to do with the exhibits," Frank said, "they don't feel that they have to destroy them."

Given that the goal of the Exploratorium was to build intuition rather than acquire specific pieces of knowledge, a tolerance for letting things (and people) bounce off the walls was essential. Intuition-building can't be made to happen. It has to be allowed to happen. Despite the temptation to lay down clear paths from one thing to another—to attach a tag stamped "dog" to every Saint Bernard and Pekinese—connections have to forge themselves actively and independently and invisibly, each neuron reaching out to the next, etching the pattern into the structure of the brain.

It works, but having the patience to simply let it happen turns out to be extremely difficult; because the process is invisible, you have to trust what you can't really see. You have to be willing to let go to a degree that is contrary to almost any traditional standard of what it means to teach and learn, which no doubt explains why the Exploratorium has never been duplicated. You have to give people unlimited choices about what to do and when and how to do it, how long to linger and when to leave. Too many museums and schools fear that, given too many choices, people will get distracted and not learn, Frank thought, but instead, such channeling seemed to have the opposite effect.

There was a period, he recalled, when some architects decided that classrooms should be built without windows — probably because they felt that students would be distracted by what was going on outside. But the feeling of confinement in windowless schools interfered with learning rather than promoted it. "If the students are not permitted to *decide* whether or not to listen to the teacher, they got less out of the teacher," he said. "Furthermore, if the teachers can also look out the window they may see something happening out there that was interesting and relevant to what was being talked about in the classroom, especially in science classes."

As in exploring, the ability to get off the path often leads to stirring discoveries and connections. It gets people hooked. "Often it is precisely as a result of such aimless exploration," Frank wrote, "that one does become intensely directed and intensely preoccupied."

Frank acknowledged that this extreme open-endedness could be disconcerting even for the staff — never mind outsiders. "When staff members are frustrated by our visitors' tendency to this kind of 'Brownian motion,'" Frank wrote, "I urge them to look back and remember how many different kinds of patterns and circumstances in their own learning were wonderful like the variety of my mountain walks."

One Frank story that has become the stuff of legend because it captures his philosophy so well was filmed by Jon Boorstin, and concludes his documentary *Exploratorium,* which was nominated for an Academy Award in 1974.

In the film, a clot of staffers is gathered around Frank, playing with an exhibit prototype, and as usual everyone is throwing in their two cents on what works, what doesn't, and how it might be improved. One exhibit builder worries that people won't get the right point, and observes that all too often, people make connections between phenomena that have nothing to do with each other. As an example, he tells of a kid he saw sitting at a resonator, looking at a meter, and seeing a friend playing with a harmonics wheel. "Hey, you know when [you] did that right, I got all the way up to fifty over here," the kid shouted to his friend. In

Boorstin's film, the exhibit builder says with exasperated disbelief, "They tied the two exhibits together!"

And Frank responds, with some irritation: "What's wrong with that? That's science. For them that's science. They're figuring something out. They're not just getting somebody else dishing it out for them. There's enough in there so that they actually made a connection themselves. Now, they don't go on and do experiments to see if it's right, but that's hard to do in a museum. But at least they saw something and figured it out. And got something that nobody has given to them, something that was just their own."

"Often Frank seemed so fuzzy," Dave Perlman of the *San Francisco Chronicle* remembered. "But then, when you thought about what he talked about, it wasn't fuzzy at all. People were going to perceive things; some kind of understanding would come to them as they manipulated things or watched something happen; [there was] a pile of sand and you pushed a button and the table vibrated and suddenly you saw things you'd never thought about sand before — even though you might have walked on the beach a million times and watched your own footprint. How does that happen? Why does the sand trickle down this way? Or form a ridge in a high wind? Things you don't ask any questions about, suddenly you change your view of those phenomena . . . He had it right and he could express it in a way that a large mass of people could understand and accept."

Of course, there *was* an enormous amount of structural scaffolding to the Exploratorium, but it was primarily conceptual. As the physicist Bob Wilson noted, there was a "carefully thought out pedagogy; there is a coherence to the whole of it." Frank spoke of his mission as "trying to organize experiences into ideas."

A Public Beach

"How does one evaluate the educational outcomes of play?"

The people who came to wander in Frank's woods were as deliciously diverse as the stuff and experiences they found there. It was the same "audience" one might find at a public beach, Frank liked to say. One very elderly woman stopped by to tell Frank she'd been waiting seventy years for such a place. Artists used it as inspiration for poems, music, films, architecture; teachers of perceptually disabled children used it to sensitize parents. People used it to teach theology, psychology, art, neurophysiology. The mix of races and ages and income groups was a refreshing change from the rest of largely segregated San Francisco. "We were a place people knew where they could get in and have some kind of experience, have somebody to talk to, and it was safe," one staffer said. Public beaches were free, and so was the Exploratorium.

Any attempt to define the Exploratorium's audience struck Frank as offensive; he felt it could only lead to further narrowing, diluting, filtering.

"There is one piece of oft-repeated advice to which we have not paid the slightest attention," Frank said. "Over and over again I have been lectured at by exhibit designers with the statement 'You have to decide who your audience will be.' We do recognize that it is essential that neither I nor the staff are bored by our exhibits, that we learn something as we make them and that we enjoy showing them to people, especially our friends and colleagues, over and over again. The best physicists I know usually learn something in the place and are also delighted by some of the things they already know because the presentation is novel, illuminating or beautiful. But as far as we can determine there is no age limit, no training limit nor any cultural limit to the range of people who use and enjoy the place. Preschool groups and old-folks-home groups come and come back. Mentally or sensorily retarded groups repeatedly come and special classes for gifted students use the place as the basis for a variety of projects. It is one of the few formal institutions that attracts teenagers."

Just as Frank resisted any attempt to define his audience, he also fought against the increasing pressure to evaluate what people learned there. While it's clear that intuition is central to both science and art (as well as to just plain getting around in the world), it's not something that is easily measured. And because intuition isn't measurable, it doesn't carry much weight with people who want a quantitative assessment of the "value." As the social theorist Sir Geoffrey Vickers observed: "Because our culture has somehow generated the unsupported and improbable belief that everything real must be fully describable, it is unwilling to acknowledge the existence of intuition."

So Frank resisted formal evaluation as much as he resisted well-laid paths and attempts to control people's experiences. Exhibits at the Exploratorium were evaluated, of course—constantly. Frank used to say that exhibits would stay around as long as the staff didn't get bored with them. It was part of everyone's job (especially that of the exhibit developers themselves) to "play" with the props along with the visitors, to find out what worked and what didn't. But the obsession with formal evaluation (which had only just begun at the time of his death) struck him as a kind of poison. "How does one determine the social benefits of sightseeing?" Frank wrote. "Should one devise tests, follow-up interviews or questionnaires? Should one judge by the average daily attendance, the length of stay of the visitors, the admission price that people are willing to pay?"

We value many things that can't be measured, so the lack of adequate benchmarks in no way implies that a practice that seems good should be abandoned. (We can't readily measure kindness or honesty or wisdom either, which may account in part for the decreasing respect we seem to accord them in modern society.)

As Frank pointed out, people used fire to keep themselves warm long before Galileo invented the thermometer. "Why do we insist that there must always be a measure for the quality of learning?" he asked. "By thus insisting, we have limited our teaching to only those aspects of learning for which we have devised a ready measure . . . If we prematurely insist on a quantitative measure for the effectiveness of museums, we will have to abandon the possibility of making them important." Or, as he put it more emphatically elsewhere, "More than one enlightened attempt at educational improvement during the past half century has been plunged into darkness because it could not meet the demands for certification of the students."

Tested and focus-grouped to near extinction, public offerings in popular science (like fashion or movies or anything else) become narrowly channeled toward a well-defined (and probably fictitious) audience with single-minded interests, not much imagination, and little opportunity for choice. The process of formal evaluation easily becomes a filter that designs out aspects and options (a side trip to see the fairy slippers?) that could be just the experience that changes someone's life. You never know what people are going to be interested in; the thing you leave out of your design might be just the thing that grabs them.

What Frank wanted, above all, was to turn people on by whatever means he could, switch on those gold-rimmed glasses, get them taking apart lamps and wondering at seagull feet. Since people are so wonderfully diverse in their interests and attention spans, and since the tug of curiosity is so often unexpected, and since even a single individual may respond to very different types of seductions depending on the day or year or present company, there was nothing to do but set out a cornucopia of options and simply allow things to happen.

It's true, he acknowledged, that people can get frustrated if they aren't sure what exactly they're supposed to be seeing or learning. But how much spelling out would anesthetize them to the delicious process of addiction? Ultimately, Frank made his decision on this sole criterion: the number of learning addicts he could create.

"How many frustrated people is one addict worth?" he asked. "Since there is no going back if one gives away too much, we tend to lean toward the more radical answer to this arguable question. And we do have a large number of addicts who come back for more."

I happen to think that creating one addict is worth a lot of frustration — because I was lucky enough to become one.

Frank's Hippies

"The staff was no more organized than the place: probably less so."

On the surface at least, the staff was just as unstructured as the place — and also just as unruly. This was both deliberate and disconcerting. "It's one thing to have your friends say 'Break the rules,'" the writer Pat Murphy said. "But this is the guy in charge saying 'Break the rules.' Now it takes on a whole different meaning." Frank treated the staff (referred to by one observer as "Frank's hippies") much as he did his own children. "We were given unheard-of freedom to do as we liked and with little or no supervision," Mike Oppenheimer told me — so long as it was something his parents approved of.

Frank's refusal to establish rules of any sort was based on the strong belief that rules were excuses people made for failing to take the time to make thoughtful decisions. So he encouraged rule-breaking in the extreme — and often the risk-taking that went along with it.

One night he came in late to find a staffer driving a huge sky lift filled with teenage Explainers. When Frank asked him what he was doing, he answered, "I'm taking the Explainers for a joyride." Frank broke out in a big smile. "That's great!" On another occasion, the staff rigged up a swing that hung sixty-five feet from the Palace's ceiling. People would climb up into the sky lift, grab the seat, and fly — forty miles an hour, one person calculated, at the fastest point in the swing. After a while, the staff became worried about safety and talked about taking it down. It was one of the times Frank became really angry. He pounded his cane and said, "No, we're *not* taking the swing down." He felt that people weren't allowed to take enough risks in their lives. (This didn't mean the Exploratorium was a dangerous place. On the contrary, insurance company representatives were always stunned by the low accident rate.)

And of course, Frank broke rules himself, even when he was

pretending to be on his best behavior. When Jon Else was making his *Nova* documentary, the filmmaker put a small radio microphone on Frank to capture whatever he said throughout the day. "It took a lot of coaxing," Else said, "but he wore it." Even so, it wasn't long before Frank had figured out how to detune the wire antenna so that it would transmit practically anything *other* than his voice. So pretty soon Else began receiving the radio frequency from the fluorescent lights in the museum instead of Frank. One day, he was picking up Mozart on a local radio station. Another day, as Else was looking through his camera lens at Frank working in the shop, the soundman found he was tuned in to an Oakland A's baseball game.

The artist Gerald Marks delights in telling the story of the time Frank came to see a set design that Marks created for the theater company Mabou Mines. The play was based on Samuel Beckett's *Company;* the music was by Philip Glass; Marks's set used three giant satellite dishes to produce strange acoustic effects.

The huge dishes moved during the play on worm-gear tracks, cranked by three workers underneath. It was a fascinating piece of equipment, so Marks wasn't surprised when a friend's seven-year-old son sneaked under the satellite dishes when no one was looking and started cranking. The child didn't understand the Beckett, but he had a fine time playing with the dishes.

"So now it's a week later," Marks recalled, "and Dr. Frank Oppenheimer comes to see it at the Public Theater in New York. He is already old and sick and uses a cane. Physically, he is not a well old man. And he is wearing a suit. And he watches the play, and of course he grasps what Samuel Beckett is all about and the music of course. But right after the play, he's on the ground, underneath the satellite dish, cranking it! Just like the seven-year-old!"

Among my favorite Frank moments were those serendipitous adventures that came about when he ditched whatever it was he was supposed to be doing to do what he wanted to be doing. One morning, as Frank and I were on our way to the museum for a meeting, the rugged, sweeping headlands beyond the Golden Gate beckoned irresistibly, and we decided to play hooky. Frank chased

Orestes down the smooth, pebbled beach and up steep, craggy cliffs. I chased Frank. (Even with his lame leg and cane, he was a hard man to catch.) We talked about physics, art, the meaning of life, the danger of war. Back in the car, driving over the bridge, we joked that we'd wasted the morning. Frank used to get furious when people said his years on the ranch were "such a waste" — in fact, the idea that he'd squandered his time there (or anywhere else) struck him as "incomprehensible."

Even when Frank set out policies himself (like the one about not having rules), he expected people to do whatever they thought needed to be done, though he might scream about it later. "It's easier to get forgiveness than permission" was the modus operandi. Trying to guess what you were supposed to be doing was pointless, and sure to piss off Frank. He wanted *you* to figure that out; he wanted you to *invent* your job.

His tolerance for so much individual freedom came in part from his enormous faith in the abilities of people and his respect for individual minds. "That's the thing that comes back to me most vividly about Frank," one staff member said. "He had this idea that if you gave [people] the tools, they could figure things out; they could transform themselves. And he proved that over and over again."

If Frank liked you, the sky was the limit; you could take on almost any challenge that appealed to you. Leni Isaacs was just finishing college and playing bluegrass flute when she was hired as an Explainer. She wanted to stay around when that job was over, so she was hired to build circuit boards and solder. Leni and Frank bonded over flute playing, and soon she was put in charge of sound and music exhibits. "I didn't know anything about science or music theory," she said. She spent most of the first year reading, then started building exhibits. "That was one of the wonderful things about the Exploratorium," she said. "You had an idea, and you'd just start working on it."

Frank's vision of democracy was plainly at work in the way he "managed" his staff. He firmly believed that the Brownian motion of freely acting particles, connecting and colliding, would, in aggregate, always produce the best results. "The only reliable method of deciding the complex problems before our society is that of relying on the opinion of the majority," he said. "I dread any suggestion of government by experts."

Strictly speaking, the Exploratorium wasn't democratic, and Frank knew it. But it did capture democracy's best features: ingenuity, independence, and stability. The stability came from the inherent ability of a democracy to fix its own mistakes. An absolute ruler might, in some cases, be able to do a lot of good, Frank said. But inevitably, such a person would make mistakes — perhaps fatal ones. Democracy had a built-in mechanism to prevent that, because there were always (at least in principle) voices that would point out when the emperor (or the anarch, as the case may be) had no clothes — voices that, in a real democracy, would always be heard. So in that sense, at least, the Exploratorium *was* a democracy. "The Exploratorium is democratic in the sense that anything that's really terrible," Frank said, "the staff is strong enough to prevent from happening."

Ad-minister

"People ask: 'Why are you people at the Exploratorium like that? You're very idealistic. You're very philosophical.' And it's true. It's because of Frank."

Virtually every one of Frank's administrative decisions (if such a term even applies) was based on idealistic, not practical, considerations. This came as a shock to new staffers who had left more traditional workplaces, but sooner or later they got used to the idea that what might otherwise seem like business decisions would often be made on moral grounds. "He made a lot of impractical decisions that were beautiful decisions," as one staff member put it.

Those "beautiful" decisions involved everything from his choice of doorknobs to the many kindnesses built into the exhibits — from his willingness to "waste time" to his insistence on letting the high school Explainers be exactly who they were.

Other museum directors wondered how Frank pulled it off—how he managed to maintain such a pure, consistent vision seemingly independent of politics and money and "all the things that beset most of us," said Alan Freedman, director of the New York Hall of Science. "Frank is the only one I know who was able to keep that intellectual temperament while still being a successful museum director . . . If you look at the results—the quality of the exhibits, the influence of the institution, the people who grew up there, the financial survival of the institution—all these are measures of success." And yet, Freedman said, he never saw Frank make compromises. "I compromise all the time," he said. "I never heard him struggle with that. His attitude was 'We'll do it the right way, and we'll get by.'"

This idealism was all the more impressive to people who knew about Frank's past. The fact that he'd been so beaten up during the red scare and watched his brother destroyed (not to mention his role in bringing the first weapon of mass destruction into the world) didn't in the least tarnish his faith in people, his warmth, his hope. And that took a special kind of courage that naturally attracted people.

What really surprised a lot of people was how well Frank's insistent idealism served him as an administrative tool. He spent a lot of time persuading people of the rightness and import of what he was doing, and it paid off. They knew that even if he wasn't doing things in what they thought was the right way, he was doing them for the right reasons. That's often why they stuck around despite frequent arguments. "It wasn't about him," one staffer who'd had more than his share of disputes said. It was always about the principle.

Frank infused idealism even into budgets. In one of the earliest, he reflected philosophically: "Admittedly, it requires, at the moment, a certain amount of brazenness to suggest the expenditure of such a sum, and the corresponding effort, for an institution which is appropriate to an era of gentleness and prosperity; but if such an era does not materialize, a lot more will be at stake than a museum of science and technology."

I never saw Frank so angry as when someone dismissed idealistic considerations as irrelevant or impractical. The angriest (and, uncharacteristically, drunkest) I ever saw him was when the BBC aired a dramatization of his brother's downfall. Frank felt they trivialized the whole matter by turning it into a soap opera. But what really bothered him, as he later wrote to the producers, was that they'd focused on personality when in fact it was a story of ideas and ideals — and in that, they'd entirely missed the point. "The tragedy you intimate was not just a tragedy of a life," he wrote, "but the tragedy that the good and the noble things that so many people surrounding my brother had done were somehow subverted or negated by other forces that were at work in history and society."

Some people thought the Exploratorium was a sort of karmic reparation for Frank's role in the creation of the atom bomb, a way of giving something back to the world. Others thought it was his way of making up a world where no one (himself included) would be hurt by the kind of betrayal that wound up ruining his career and those of so many others.

Singular

"Frank would not have understood the phrase 'It's not personal.' There is always some personal aspect. He didn't try to hide from that."

In an era when "don't take it personally" seems to be the guiding principle of relationships, Frank's polar-opposite outlook took people aback in a way that stayed with them for a long time. Shortly after Frank died, the *San Francisco Chronicle* science writer David Perlman had a long talk with his wife about why they missed Frank so profoundly, though their interactions with him were relatively short and sparse. "We talked about his idealism and his optimism and his faith in the people and also the warmth that he exuded all the time," Dave told me. But mostly, he said, it was the intimacy. "You spent five minutes with the guy and you suddenly felt you were with someone you were very close to, even though it was a very brief kind of contact. You couldn't help but feel that he must love you and you must love him. Every contact I had was like that."

To Frank, running the museum was always a one-on-one affair. When some Exploratorium board members later accused him of being a poor administrator, he agonized for a long time over what the word "administer" meant. He decided that it meant literally to minister to people, an inherently individual process.

Since that was already his management style, he decided he had to do more of it. He'd come up to someone, lean over her shoulder, and ask, almost inaudibly, "Whatcha doin'?" Then he'd have a little discussion with the person and make suggestions. It was up to the staffer to decide whether to incorporate his suggestions or ignore them. (He let certain people ignore him a lot more than others.)

Frank would often troll the shop with Bob Miller, looking at whatever there was to see and stopping to chat with whomever they happened to meet. "He was like a happy cloud that wandered through the shop," photographer Nancy Rodger said. "He wanted to know what everyone was doing, and you felt really lucky that he wanted to talk with you."

Frank almost never met with people in his office (he was never *in* his office). Instead, you walked the floor with him, sometimes several times over, while he poked at things and adjusted what needed to be adjusted and talked to visitors. He was like a thoughtful host at a party, making sure everyone was having a good time. Occasionally other staff members or visitors would join you, and soon there'd be a group trailing behind him like the smoke from his cigarettes. "He was like the mother hen with all the little chicks," Nancy said.

For Frank, the personal trumped everything else. One day he was deeply involved in explaining something to a young student on the floor when he was supposed to be in a meeting with some dignitaries. Frustrated staff members became increasingly frantic that Frank was missing the meeting. Finally they sent a messenger out to retrieve him: "Dr. Oppenheimer, isn't there something you're supposed to be doing?" And Frank answered angrily, "No, *this* is what I'm supposed to be doing."

Everyone who knew Frank seemed to have a somewhat similar tale. A physics graduate student who'd heard of the Exploratorium drove from Texas to see it, only to learn that it was closed. After the long drive, he was understandably devastated. So he started banging on the door. To his astonishment, Frank answered. When the student explained why he was there, Frank invited him to come in and spend as long as he liked. For a whole day he had the place to himself. "You can imagine what it meant to me," he said. "And with Dr. Oppenheimer nearby!" The graduate student, Jorge Lopez, is now the chairman of the physics department at the University of Texas at El Paso, and produces a Spanish-language version of NPR's *StarDate,* which is broadcast by hundreds of stations in the United States and Mexico.

And it was always the personal things — even trivial ones — that made Frank really happy. Thomas Humphrey was having a conversation with Frank when they both happened to say the same thing at the same time. Frank was completely taken by the moment, Thomas remembered, and got that goofy smile on his face and said, "Isn't that lovely when two people say the same thing at the same time?"

When Frank couldn't connect with someone any other way, music or art could bridge the gap in an instant. There was a woodworker in the shop whom Frank didn't get long with and wanted to get rid of because he was too meticulous, too slow. But during the time the woodworker was there, Frank would sometimes hear someone playing the piano beautifully. He asked who it was, and someone told him it was the woodworker. So the next time he heard the music, Frank got his flute and went over to the piano and started playing.

Connecting personally meant that emotions ran high, deep, and often loud. Open conflict was fine, so long as the point wasn't to win but rather to get at the truth. Frank never argued to pull the rug out from under you; he argued to push ideas and test their resilience. He'd challenge you to back up what you said, but if you were right, he'd eventually agree. "When we would talk with Frank, we got to get upset," one staffer remembered. "We got to storm out if we wanted. We got to show how much we cared about stuff. There was fire in the room. There was passion. There was unrest."

In fact, nothing made Frank madder than someone who wouldn't or couldn't or didn't think it was appropriate to argue with him. When the mezzanine was being designed, the architects wanted to build offices with low partitions — the kind you can look over if you stand on your toes. They took Frank to see what such offices looked like, and he saw workers quietly going about their business in little cubbyholes. Suddenly Frank shouted something — just to see what would happen. Workers in every cubicle popped up like prairie dogs. Frank told the architect he wouldn't be able to shout in a place like that. "Such a place teaches people to talk softly. Well, I don't want to be taught — or teach anyone — to talk softly."

Frank's favorite sparring partner was Jackie. Michael remembers nights in Blanco Basin when he was awakened by an argument about something or other — the ranch, politics, personal things. One day some months after Jackie died, Liz Keim walked into his office and started yelling at Frank about how certain matters were being handled, and though she was only a weekend receptionist at the time she felt she had as much right as anyone to voice her opinion. "Wait a minute! Wait a minute!" Frank yelled back at her. He reached into his desk and got out a bottle of whiskey and poured two shots, one for each of them. "I want to hear about this!" he said, and they stood there with their glasses, shouting at each other for a good long time. Then Frank started laughing. "Why are you laughing?" Liz asked. "Nobody's come in here and wanted to yell at me since Jackie died," he said. "And I have to say, I thoroughly enjoy it."

On another occasion, Liz broke down in tears at a meeting, frustrated and furious at the way Frank had treated her. She had come up with what she thought was a workable solution to an emotionally charged problem; Frank uncharacteristically dismissed it as a dumb idea; she uncharacteristically started to cry and walked out. Later, he found her and tried to make amends. She apologized for acting so immaturely. And Frank said: "You don't need to apologize to me for crying in the meeting because that was demonstrative of your feelings, and how much you care about this situation. It's when you don't get upset, and you don't cry, and you've lost your feelings for things, *then* I want you to apologize to me."

This atmosphere tended to make anyone who worked with Frank pretty argumentative, a trait not all bosses appreciate. In truth, it made most of us pains in the butt. Not a few former staffers found it difficult to adjust to more traditional settings where superiors actually expected their orders to be followed.

It also made people blunt. In the spirit of both honesty and tolerance for emotional upset, everything was discussed openly, sometimes with chaotic (and not infrequently hurtful) effect. "The only way to cultivate trust is to be consistently honest and open in one's dealings," Frank said. Yet he also admitted that he was "completely baffled" as to how to go about eliminating secrecy as a way of doing business. So when there was enough money in the bank to give people raises, for example, the entire staff had to undergo a salary review where, one by one, staff members would be discussed, evaluated, and judged by their co-workers.

On the one hand, this insistence on total transparency was an indication of how much Frank respected the staff and trusted their wisdom and common sense. It was something the staff loved about him, and that kept them working for truly pitiful wages — on occasion, no wages at all. And it wasn't hard to see the political connection with his past — holed up on a ranch in Colorado because he refused to name names for the FBI. As one staff member put it, "He insisted on a level of openness which I suppose was a reaction to that closed, controlled world that he had lived in during those years."

At the same time, the salary review could be brutally painful, especially when funds were in short supply, which was normally the case. "They weren't kind dynamics," one staff member said. "People were saying good and bad things so openly. When one person was under discussion, I remember thinking, If I were that person, at this point I'd be under the table bleeding." Coming from a corporate atmosphere where nothing was discussed openly, the staffer had a hard time getting used to this assuredly eccentric system. "It was like being in a commune or something," she said.

Colliding Worlds

"His roots were deep. He was too subtle by half for the rest of those guys."

Because Frank's way of running things was so peculiar to Frank, we all worried endlessly about what would happen to the place (and to us) when Frank died. But nothing much was done to prepare beyond hiring a string of unfortunate "deputies" at the insistence of the board. None stayed around for long. Few felt really comfortable with the idea that the Exploratorium was more about developing intuition and aesthetics than conveying hard facts about science, or with an "administration" that was so ad hoc. "It was not a business model," one short-lived deputy complained. And a business model was clearly what the board wanted.

Frank did agree to create an organizational chart in 1983, and formed an executive council. And at some point he tried to work out an arrangement with the California Academy of Sciences, and then with Stanford University, to assure the survival of the Exploratorium, but both institutions turned him down.

The poor track record of the short-lived deputies was probably inevitable given the fact that Frank had no real interest in appointing a successor. He was confident that the staff could carry on just fine without him. "The staffing of the Exploratorium is really excellent," he reported to the board in 1982. "There is a sizeable group of people with talent, enthusiasm, and experience that is quite capable of administering the Exploratorium if the present director should expire."

There's no question that the chaotic and complex blend that made the Exploratorium so rich, and attracted so many talented people to Frank's side, also put off many people who didn't agree with (or didn't understand) his way of thinking. They thought he was muddled, flailing about in some idealistic neverland that could not survive in the real, hard world. Much as Frank tried — and despite the clear and overwhelming success of the institution he created — there were a fair number of people Frank never managed to convince that he knew what he was doing.

Ironically, this very complexity was the same quality that brought down Robert. Edward Teller's damning testimony boiled down to the argument that Robert Oppenheimer was "complex" and therefore couldn't be trusted. But as Alice Kimball Smith and Charles Weiner point out in their introduction to a collection of Robert's letters, "Complexity is not (in itself) a trait of personality; it indicates rather that the observer is puzzled." A lot of people were puzzled by Robert, and also by Frank.

To Frank's admirers, however, his complexity reflected his depth. "His behavior, his values were a mile deeper than anyone else's," said Mike Templeton, the former director of the Association of Science-Technology Centers. "The other museum directors knew he was smarter than they were, that he was taking risks that they wouldn't take. Even those who disagreed with him loved him for that."

In addition to urging Frank to appoint a deputy/successor, the board wanted him to make a financial plan and stick to it. This was not the way he was used to doing things. It isn't that he didn't care about a balanced budget; it was that other priorities got in the way. Shop director Tom Tompkins remembers telling Frank there was a kid he wanted to hire. Frank looked distraught, and said, "God, the bookkeeper is four months behind. I don't know that I've got any money. Go ahead and hire him."

No matter how much Frank needed money for the museum, though (and he was always fairly desperate for it), he could be stubborn when it came to key principles. In one incident, captured in Jon Else's *Nova* documentary, the staff was arguing about whether Explainers should have to dress up in black pants and white shirts in order to please some bankers who had rented the Exploratorium for a black-tie event. The staff had recently taken a pay cut, so the situation was pretty dire.

The development office staff, including Ginny, didn't understand why asking the Explainers to dress a certain way was such a big deal. But Frank did. "Just think how much nicer it would have been," Frank said, "if somebody had said to the Explainers, 'We want you to look nice, now figure out how to do it.'" He thought for a moment and then said, "Let's do it the nicer way."

The fact that funds might be jeopardized by his stubbornness was something he was fully prepared to accept. "Just because somebody is paying for it, we don't do things which we don't think are the right things to do," he said. "Christ, we've jeopardized a lot of funds because we're so funny."

Even after a $1 million four-year grant from the MacArthur Foundation* — a godsend — financial troubles persisted. Particularly devastating was the decision of the newly elected president Ronald Reagan to eliminate the National Science Foundation budget for science education. "Just 'Poof,' and they killed the entire Education Directorate," said George Tressel, who was then the director of the NSF's Public Understanding of Science department. At the 1981 Exploratorium fundraising dinner, Frank described the money situation as "outrageously hair-raising." With the Exploratorium's annual budget of nearly $2 million, the MacArthur grant provided only one-eighth of what was needed, and federal money was drying up. Efforts to raise money for an endowment were only minimally successful.

"I am disappointed that things look shaky at the moment," Frank said in a talk he gave around this time. "But I am confident that in the future . . . the support will come back."

Unfortunately, support didn't appear fast enough or in sufficient quantity to prevent the one thing Frank most hoped to avoid: an

* This was made possible largely by the efforts of Nobel laureate Murray Gell-Mann, then on the board of the foundation and a longtime supporter of the Exploratorium.

admission charge. The April/May issue of the *Exploratorium* magazine announced the new policy, explaining that the museum had already cut staff, was still in debt, and saw no alternative. Henceforth adults would pay $2.50 to get in, senior citizens $1.25. What made this palatable to Frank was the stipulation that anyone under eighteen could enter free, and that an admission ticket bought a six-month pass to the museum, with no limit on the number of visits.

Frank wanted to make sure that teenagers, in particular, would still have access to the place whenever they wanted — especially since it was one of the few cultural institutions in the area that regularly attracted young people. "We hope that this new policy does not exclude anyone who wants to come here and does add as much to our coffers as we estimate it will," the announcement read.

If the Exploratorium had to charge admission, Frank said, this was definitely "the nicer way" to do it.

As the Exploratorium board of directors became increasingly nervous over his management, Frank tried to win them over the same way he did everyone else. He made impassioned presentations about new and ongoing projects, trying to persuade the board that everything was on track with the same commitment and idealism that had served him so well in the past. The meetings were "very intense," Pief Panofsky remembers.

But the board was no longer made up mostly of other scientists, or of people who necessarily shared Frank's experiences and philosophies. Increasingly it included people who had a head for business, and they were harder to convince that everything would turn out fine if Frank ran his anarchy any way he pleased. They wanted the Exploratorium to have some financial stability, and worried that Frank's way wasn't going to carry his creation into the future.

And so Frank told the board only as much as he wanted to, and when he wanted to. The projects he described were often three-quarters finished by the time he got around to mentioning them.

"The board was probably running as fast as the staff was to figure out what Frank was doing," said F. Van Kasper, who became chairman of the board as the crisis was coming to a head.

Van wanted Frank to think more like a businessman, but Frank really didn't understand the need. By every conceivable measure, the Exploratorium was a roaring success. Frank's "crisis aversion" fundraising strategy was hair-raising, but so far it had worked, and he was trying to amass an endowment. His handling of the staff was unconventional at best, but again, the results spoke for themselves. Van wanted Frank to think about things like "employee productivity" and "value for money spent."

Their conversations were frustrating for both of them. Van found Frank complicated and hard to pin down. When Van pushed, Frank deployed his usual tactics. "He could ignore you with more style than anyone I ever met," Van said.

In the spring of 1983, the board told Frank that while he should remain the "conceptual and inspirational" leader of the Exploratorium, its day-to-day management should be handed over to somebody else. Frank was deeply hurt, and he agonized over what he should do. Finally he wrote Van a long letter saying that he'd suffered many sleepless nights trying to figure out—in effect—what the board's problem was. Frank's budget projections were better than on target, and the millions of dollars that had poured in from government and private funders testified to its stellar reputation. Respected scientists from fields ranging from neuroscience to physics had nothing but high praise for the exhibits. Detailed road maps assessing the status of the museum and its future direction ("Where Are We and Where Do We Go from Here?" parts I and II) were drawn up, in 1976 and 1981, and followed reasonably closely. The museum had gone into debt briefly during only three years of its operation—years in which Frank was being treated for lymphoma and Jackie was dying of cancer. In every other year, he'd ended with a small surplus of cash.

"There is an impression on the Board that there is a reverence for funkiness and sloppiness," Frank wrote Van Kasper. "I don't know where that impression comes from. I think it is merely that the staff has a different set of symbols for high performance than those that are more familiar to the Board." The staff, in fact, had a "meticulous respect" for high quality, he said. And while Frank admitted his managerial style did come from his "personal convictions," it also came from his experience working in successful research laboratories such as Rutherford's and Lawrence's. The idea of becoming an "inspirational leader" who would leave the management of the Exploratorium to someone else was unthinkable to Frank. After all, he'd built many of the exhibits himself. "That is not my nature," he wrote.

Frank felt betrayed and disrespected. Just as heartbreaking, he felt his staff was not valued by the board. He was scared about what might happen — that having created his world, he now might lose it because his methods were unconventional.

Nobody but Frank could have built the Exploratorium. But Frank couldn't get hired to run the Exploratorium today, Van told me. "Frank couldn't get hired to run *any* museum today." And because the job of director was now seen as mainly a matter of fundraising, "he wouldn't want to be hired either."

In truth, Frank was, in many ways, both behind and ahead of his time. The openness to experimentation in the 1960s and 1970s had surely slipped away under the onslaught of Reaganesque ideology. However, according to Warren Bennis,* a renowned expert on leadership, the way Frank ran the Exploratorium has once again become the model for creative leadership, including the whole idea of a flat hierarchy. In the business community, Bennis told me, it's increasingly recognized that the genius for change comes from people who, like Frank, have a rather "skewed vision of reality," who are not just open to the new, but "open to the unbidden," who "change the temperature of the room." Van's assessment that a Frank-like leader couldn't hack it in the so-called real world "is simply not true," Bennis said. "It's very not true."

In 1983, however, Frank couldn't convince his own board of

* Bennis is both University Professor and Distinguished Professor of Business Administration at the University of Southern California and the author of *On Becoming a Leader* and coauthor of *Judgment: How Winning Leaders Make Great Calls.*

directors that he could manage the world he created. The situation remained essentially unresolved, and within a year Frank would learn that he was suffering from terminal cancer.

In the midst of this meltdown, Frank went to have dinner at the home of his old friend Lou Goldblatt of the Longshoremen's Union, who'd been so helpful to Frank in getting the Exploratorium started. While he was inside eating and drinking, he left Orestes, as always, waiting in his Dodge Dart. But this time, somehow, the car caught fire—probably because of an electrical short caused by Orestes' chewing or an overheated muffler that ignited dry leaves. By the time Frank realized what had happened, Orestes was dead.

Frank was, of course, devastated. When he phoned to tell Mike, he was sobbing. It was the only time in his life, Mike said, that he'd heard his father express true "horror."

12

THE WORLD HE MADE UP

Dear Frank,

I can't call you because Bob tells me you still have tubes down your throat. What an awful indignity. (But he does tell me your handwriting is improving, which is one consolation!) You must be miserable. We are all miserable worrying about you. I know I told you to go into the hospital for a while so you could work quietly on your book, but this is too much. I take it all back. What can I do? My column on Electricity is appearing in the next issue of *Discover*. It is truly terrible. That's what happens when I don't have you to work with. You've got me playing my flute now every day because in some silly way it reminds me of you and makes me think it will make you get better faster. (It also reminds me of your superstitious rituals relating to the small bronze dachshund.) I'm sure Millie *[sic]* and Bob and Judy are taking good care of you, but I feel frustrated that I can't help. Please let us know what we can do, and when you are ready for company.

All our love,

K.C.

DESPITE THE TROUBLE with the board, continuing financial crises, and the death of Orestes, Frank remained his playful, cantankerous, sentimental old self—all the more remarkable given that, for much of this time, he was also gravely ill.

In 1977, Frank was diagnosed with lymphoma and over the next two years underwent several courses of chemotherapy. He eventually took himself off chemo because he thought it was making him tired, and the lymphoma did not, in fact, return. Then, following a routine chest x-ray in preparation for a hernia repair in late 1983, a tumor was found in his lung. The doctors performed a lobectomy, but this time the cancer was incurable.

Although it does not appear among Frank's letters at the Bancroft Library with the rest of his collection, I distinctly remember writing him what amounted to a thank-you note around this time. Among the "gifts" I know I mentioned were my new appreciation for chaos and the uses of bribery, for permission to "play" in museums, and a new delight in hot August days. Also for my flute playing, which he'd given me the incentive to try and the confidence to stick with through the rough spots.

Several weeks after Frank left the hospital, I flew to San Francisco for a visit, and he astonished me (I don't know why) by insisting on playing the flute. Lost lung be damned, Frank herded Milly and me toward the piano and we played our way through bits of Purcell and Bach as if nothing had changed.

As the 1984 Exploratorium awards dinner approached, Frank called to say he wanted to give me that year's Public Understanding of Science Award. Of course, I was thrilled. Previous recipients included such luminaries as Walter Sullivan of the *New York Times,* Philip Morrison, and Lewis Thomas, the author of *Lives of a Cell* and president of the Memorial Sloan-Kettering Institute. It was a minor miracle that the dinner went on at all that year. The roof was under repair, so there was scaffolding everywhere and the rain poured in; the staff was busy with squeegees, struggling to get the water out up to the last possible moment.

But it all came off beautifully, as always. William R. Hewlett, co-founder of Hewlett-Packard, was the main honoree. Frank spoke about "The Sentimental Fruits of Science." He presented me with a gold pocket watch from the Smithsonian that was almost entirely transparent, every gear and spring and jewel in full view.

Not long after this, perhaps in early January 1985, Frank phoned me in New York to ask if I'd come out and see him. I said I was very busy just then and probably couldn't make it for a month or so. He said in a weak voice, "I think you'd better come right away." That's when I knew he was dying.

I wish I could remember more about that last visit, but beyond Frank's uncharacteristic frailty, everything seemed pretty much the same. We still went to the Exploratorium every day.

Although Frank didn't want people at the museum to know how sick he was, everyone had begun to notice changes. He was always going off for medical treatments and slowly losing his strength. He was starting to let go of the reins a bit, which was clearly difficult for him. His old fight seemed to be draining away; he said he was tired of "having the same old arguments." Some people thought he was softer, not so much of a curmudgeon. Others thought he looked more and more afraid.

Soon everyone knew Frank was dying, and staff meetings were increasingly well attended and frequently tearful. Like Frank himself, the staff was worried about the future of the place. Everyone was looking for guidance, but Frank didn't offer much in the way of specifics. He said it wasn't so important to get a physicist to run the place, or an artist, or anyone from a particular field, as it was to find a person who was open and curious and had a vision for what the world and society could be.

Mainly, he spoke about ideals. "What he hoped to instill in us was the confidence to go after that passion," Liz Keim said. "Those were the last words I remember him saying to us as a group before he went home to die."

He also took care, as much as he could, to protect the legacy of the exhibits by making a series of videotapes on their proper function and maintenance, much as he had with his library of experiments at the University of Colorado. The tapes are still a pleasure to watch, despite the fact that Frank is clearly in pain. There he stands at his "Wave Machine" — a long spine of springy steel wire to which horizontal bars are soldered like ribs. He demonstrates for the camera how, as you send a pulse down the spine, the wave moves faster where the rods are shorter, because they have less inertia. It's "kind of a wave-guide transformer," he says. He shows how you can fix the rod at one end to reflect the wave, dissect the wave into parts, make the crests and troughs interfere with each other, or not, and make a dozen subtle adjustments.

He moves awkwardly, almost like a puppet, leaning heavily on his cane. He coughs a lot and has bags under his eyes. He talks about the importance of using the right kind of solder; he loosens a screw and removes the perforated paddles that dampen the wave, shows how the holes create turbulence in the oil, which increases the friction. He takes delight in all the tiny effects, the simple, clever technical fixes, the ingenious gadgetry.

At the end, he turns to the camera and says, "I believe that shows what all the adjustments can be made on this and what the performance can be." And then he giggles.

Frank took a lot of joy in these kinds of activities, and letting go of administrative battles freed him, in a way, to do what he'd always liked best. "He had all these meetings he was supposed to go to," Ned Kahn remembered, "but he was much happier trying to explain what electricity was. You could just see that in his body language and his tone of voice, and that's what he loved doing." Occasionally he began to use a wheelchair, especially after tiring medical treatments. Returning from a radiation session, he wheeled himself back and forth in front of an exhibit with a Geiger counter, enthralled to see how radioactive he was.

And just as Frank had called me in New York to ask that I come visit, he got in touch with many other old friends and supporters. He called George Tressel at the National Science Foundation in Washington, said he was coming to town, could they go out for a drink? "It was clear this was to say goodbye," said Tressel. He got the impression Frank was going all around the country doing the same thing.

Back at the Exploratorium, Frank had long philosophical conversations with staff members, telling more than one that he wasn't ready to die. There were too many things about the world he still wanted to find out, he said. He wanted to know if there would be a nuclear war. He worried that the younger generation felt they couldn't make a difference in the world. "I've never seen anybody fight so hard against dying," one staff member said.

All the while, he came to the Exploratorium every day he could, often with Mike pushing him around in his wheelchair. "Every moment was a moment to be alive in," Darlene Librero remembered. "And when he didn't have the energy to do this anymore, I knew he was near death."

Then one day, he didn't think he could drive himself home, and had to ask a staff member for a lift. After that, he stopped coming in altogether.

A Gathering of Friends

"Even with forty hours I would not find words enough to tell all that Frank meant to me."

Frank spent the last few weeks of his life at home, with Milly and Judy and Mike and Mike's wife, Jean. Visitors streamed in to see him as he lay on the couch in the living room. One by one, they went to say goodbye. Sometimes he was fully conscious, but at other times it was clear he was "already crossing over," one staff member said. "He wasn't really there anymore."

These final conversations ranged from the ordinary to the profound to the bizarre. "We got into these amazing discussions about art and science," Ned Kahn said. "It was weird because he was gasping for breath and kind of going in and out of consciousness." Frank and Ginny Rubin whispered together about the wonderful times they'd had together. He told Bob Miller he didn't think that seventy-two "was such a bad age to die." The last thing he said to Thomas Humphrey was a number.

Frank approached death like a scientist — fully conscious of the process, always focused on "noticing"; he faced his end as forthrightly as he had anything else. "He talked about the intensity of the physical sensations he was experiencing," Ned said. "Dying was just another phenomenon. And there was something really profound about that to me. I remember after hanging out with him I was in kind of a soaring mental state. Something about his death was incredibly inspiring to me, and I remember feeling really embarrassed about that feeling, because everyone else was all sad and coming unglued emotionally, and I was trying to suppress this incredible feeling of joy, or elation almost.

"I just thought, 'I want to have a death like that. Or a life like that.'"

Perhaps predictably, what marred this picture was a New York woman, an actress, whom Frank was having a relationship with, and who—to the hurt and horror of Mike and Judy and Milly—showed up at the house intent on professing her love and anguish. In retrospect, I did remember noticing that Frank spent a lot of time on the phone with a mysterious someone during my final visit, running into his office to catch the calls like an anxious teenager. I think I even went to the theater with the actress and Frank (and her husband, I believe) in New York, but never suspected the nature of the relationship. (I wouldn't be surprised if Frank's ability to act in such contradictory ways—that is, deeply love his family yet hurt them at the same time—had something to do with his embrace of the complementarity principle in quantum mechanics, where accepting mutually exclusive truths comes naturally out of the physics.)

It was in the midst of this odd stew of activity and emotion that Frank died, at home, on February 3, 1985. Writing in the *San Francisco Chronicle,* David Perlman described him as "a gentle, courtly, humorous and profoundly humanistic scientist . . . dedicated to the idea that understanding the phenomena of human perception as well as the physical laws of the universe involve as much art as science."

Three weeks later, there was a "gathering of friends" at the Exploratorium. On the invitation was one of my favorite Frank quotes — a very fitting one, I thought, as it was another kind of invitation in itself, an invitation to go exploring with Frank, to share his world and his endless wonder of it:

> If in the course of some wandering I come onto something delightful or exhilarating or beautiful or insightful, I want to tell someone else about what I have found. More than that, I want to bring them along with me to share what I have discovered: a view, a feeling, a person or a book or a new way of looking at physics or at justice, or a new way of teaching relativity.

We were all beneficiaries of his compulsion to bring everyone along whenever he could.

I flew out for the evening, and Bob Miller picked me up at the airport. It was just the kind of party Frank would have loved: everyone milling about the exhibits, telling Frank stories, catching up with old friends. One person said she felt as if she were part of a cult. A plant arrived from a porn theater on Chestnut Street where Frank was apparently a regular customer. In memory of Frank, the museum didn't charge admission that weekend — exactly what he would have wanted.

For the more formal part of the program, the staff put together a series of clips of people telling frequently funny and always moving stories about Frank. Van Kasper, the chairman of the board, read some of the letters that had come in from as far away as Bombay and New Zealand, and as close as a local third-grade class. So many of them said that meeting Frank had made them want to become "a better person."

Edward Lofgren recalled the frantic race to make uranium isotopes at Berkeley. Frank "gave a sense of purpose to many of the people at the lab," he said. Lofgren also reminisced about the cosmic ray experiments, the exhilarating chases after lost balloons. Showing pictures of their improvised equipment, he said, "The resemblance to an Exploratorium exhibit is not entirely superficial."

Ruth Newhall described how Frank had charmed the City of San Francisco into giving him the Palace of Fine Arts for his "marvelous monster." He was, she said, "a joyous spirit," and the Exploratorium was "an expression of his view that the world is a series of wonders."

George Tressel talked about how the National Science Foundation had become so inundated with proposals from museums promising to become "just like the Exploratorium" that the word had become a cliché.

Bob Wilson was the most moving speaker, and the only one to cry. The two physicists had been friends for forty years — through the war, the atom bomb, the postwar pursuit of peace, McCarthy, the ranch, the Exploratorium. They shared a deep love of art, music, physics, philosophy; they rode horses all over the New Mexico and Colorado mountains and wandered the streets of Frank's "beloved" New York. Like the other speakers, Wilson was given four minutes, but said, "Even with forty hours I would not find words enough to tell all that Frank meant to me."

The program notes included the closing lines of Frank's address to the Pagosa Springs High School graduating class of 1960, and in some sense, he captured his own life in those words better than anyone else did:

> I recommend that you be willing to become deeply involved in lots and lots of things and that you let yourself, perhaps even force yourself, to do things that you think are important and that you can take seriously.
>
> I make this recommendation to you because I believe that if you do, then even in the face of considerable adversity you will feel, as I do now, grateful for having lived.

Perhaps the most fitting tribute of all popped up a few months later on the posh corner of Madison Avenue and Fifty-sixth Street in Manhattan, where the IBM Gallery of Science and Art brought a piece of the Exploratorium into the heart of New York, through the efforts of its director, David Hupert. IBM wanted the exhibits to be shined up a bit, and we were all a bit concerned that the spit and polish would put people off from getting right in there poking around and playing. We needn't have been concerned. At the reception, even the waitresses in their crisp uniforms, juggling their trays of wine and canapés, couldn't keep their hands off the stuff.

A few weeks later, I sat in the glass-walled atrium and watched a scene that I knew would have gotten a giggle out of Frank — perhaps even a "Whoopee!" Women in mink coats blew enormous, shimmering, pink and green bubbles alongside construction workers on break; execs in Armani suits jostled with shop girls for a chance to listen to the "Pipes of Pan"; doormen and schoolkids tried to figure out how to get a four-hundred-pound pendulum swinging by tugging at it with tiny magnets "in tune" with its natural frequency (Bob Wilson's ingenious version of a particle accelerator).

Frank didn't live to see the reincarnation of the Exploratorium in his hometown, but he lived within it as surely as if he had been there in the flesh. It was as if the little boy who'd been taking apart coaster brakes and melting wires with house current on the Upper West Side had simply moved downtown, taken over a public square, spread out his toys, and invited the neighbors in to play.

Waves and Splashes

"Everywhere you go in the world, if you mention the Exploratorium it's like you're part of this huge fraternal organization, and people light up."

The Edsel made a splash. Elvis made waves. A splash is a onetime event, a flash in the pan; it comes and goes like a shooting star, its energy contained within. A wave, on the other hand, spreads its influence, sometimes forcefully, carrying energy and information far from the source, like a rumor passing through a crowd. A wave keeps right on going, ringing out long after whatever started it has gone quiet. An electron jiggles in the sun and the shake travels by means of a light wave ninety-three million miles through empty space to reappear as a gleam in your eye.

The wave that Frank made continues to shake things up in ways both global and intimate. Most of this book has addressed the intimate things — the increased neural activity and imaginings he set stirring inside the heads of those he touched; if you count all the people who have by this time encountered some incarnation or other of his ideas, the number adds up to millions.

The Exploratorium, according to the museum's current literature, has influenced 90 percent of U.S. science museums and 70 percent of museums worldwide. Today you can find Exploratorium spinoffs in Beijing (a few blocks from Tiananmen Square), Paris, Rio de Janeiro, Tokyo, Budapest, Moscow, Bombay, Canberra, and Seoul; in Renovat, Israel; Taskim, Turkey; and Riyadh, Saudi Arabia. "I've often said it's the most influential museum in the history of the world," Alan Friedman, the director of the New York Hall of Science, told me.

The late Harvard biologist Stephen J. Gould described Frank's work as "authentic genius." The former president of Caltech Marvin Goldberger described the Exploratorium as "a masterpiece of science, humanism, art, and technology." It was, he said, "the best and most exciting innovation in the country, and perhaps the world, for exciting everyone about science — real science and not Mickey Mouse stuff."

The social invention that Frank created from the stockpile of ideas that had rattled around in his head through all those years of experimenting and ranching and teaching and making art and music now has a life of its own — just like any other useful invention. It's an institution that lives dozens of different lives in dozens of different contexts, the kind of "big idea," like the veto power or a court of law, that Frank liked best.

And just as the idea of democracy is stronger than its various manifestations, so Frank's ideas will probably persist both in spite of and because of whatever people might want to make of them. Like many a parent, he built something bigger and stronger than he was, and while that's satisfying in one sense, it's also a little scary, because you never know how such offspring will turn out once they're all grown up.

Frank thought a lot about the natural human tendency to want to make waves rather than splashes—to create something that outlasts one's mortal life.* In fact, he thought the urge to have such "cultural children" was as central as the urge to have biological children. It was the same urge, he said, that compelled people to cultivate an immortal soul—and such an essential feature of human experience that in an ideal society the "freedom to create" ought to be considered as basic a right as the "freedom to procreate."

Frank felt he'd been extraordinarily lucky in the number of chances he'd had to be creative in his lifetime—whether through teaching or doing physics or building a museum or ranching or being part of some other creative enterprise that was larger than himself. "And, my God," he said, "there is an indelible sense of identification and a broadly relevant connection with immortality as a result of my four years of work on the damn atomic bomb."

Democracy, at least in theory, offered the freedom to create to large numbers of people, Frank said, because everyone—if only through voting—can contribute to the eventual character of the country.

The danger of identifying with one's creations was that one person's (or group's) creative efforts could threaten (or seem to threaten) those of others. Soldiers probably felt creative enough when they sacrificed themselves to protect certain ideals, Frank pointed out, but clearly, armies threatened each other. In Frank's view, armies, nations, and similar institutions should never be viewed as "avenues toward immortality" but rather as "tools for providing what people want and need . . . They should be regarded as useful devices, like highways, postal services or telephone networks."

But there were endless other ways people could engage in "cultural procreativity" that did not generate conflict—including schools, museums, sculptures, films, stories, buildings, music, new

* These ideas are hashed over in series of unfinished essays under titles such as "The Freedom to Create" and "Cultural ProCreativity."

ways of doing things, entertaining people, or "creating joy." Societies, he thought, should put a lot more effort into making such opportunities available to everyone, and perhaps the inventions of science could help. Certainly, Frank created a bounty of such opportunities at the Exploratorium.

And yet the Exploratorium itself only mirrors Frank's uncanny ability to make people see their *own* worlds anew — a creation, in a sense, of the self. Jamie Bell, a science teacher who is now helping create a new science museum in Kuala Lumpur, first came to the Exploratorium after Frank died and remembers how amazed he was at how everyone he met was always looking, asking, poking, prodding, noticing. His first thought was "These guys can't be serious." But then he himself started to view the world more closely. What he ultimately absorbed, as he put it, was "a culture of really engaging, of not being afraid to talk about everyday stuff, to keep uncovering, even if you don't really understand something."

He began to realize that people could take the feeling of confidence that came from understanding the world and transfer it to realms like politics — that feeling of "Yes, I can understand it. I can do it." In tough spots, Jamie even finds himself thinking "What would Frank do?" If he has trouble with a colleague or an employee, he takes that person out for a drink to talk with them. "Because it was so easy to imagine Frank doing that," he told me from Malaysia. "There were things like that about Frank that always kept me going."

This was, of course, just the legacy Frank wanted. "I forget exactly how he put it," Ned Kahn said, "but the gist of it was that the museum will eventually be full of stuff, and [I should] go out and do this stuff in the world, that's where it's really needed . . . He had that amazing connection with the bigger picture . . . that incredible grasp of something larger, and just kind of the whole progression of human knowledge or awareness."

In an age of cynicism and sound bites, Frank incited those who knew him to take things "seriously, but not personally," as he put it in his graduation address at Pagosa Springs High School.

Countless people who never knew Frank, but were transformed nonetheless, continue to push along his ideas and ideals in a hundred big and small and significant ways. Micah Garb built biology exhibits with Charlie Carlson in the early 1990s and is now an ob-gyn in Chicago. Today he feels that he uses what he absorbed indirectly from Frank with his patients every day. "I learned that anybody anywhere could understand anything, that nothing is beyond anyone's grasp; no matter how complicated it is, I can explain it to them."

Micah doubts whether he could have become a surgeon at all had it not been for his time in the machine shop – experiencing firsthand Frank's philosophy that separating doing from thinking was as artificial as separating art and science. "I was not particularly adept at working with my hands," he told me. But he soon learned that he could fabricate exhibits, and that made the manual aspects of doctoring a lot less intimidating.

Frank Millero became an intern at the Exploratorium in 1990, stayed ten years, and is now an industrial designer teaching at Pratt Institute. The museum opened his eyes, he said, to the fact that science can be intuitive, and art analytical. He noticed how strangers interacted with each other around exhibits, and how a toddler and an expert could get equally excited by some phenomenon, bringing together quite different perspectives. He realized that design is not so much about the object as it is about experiences people have around the object. He tells his students, "Go to the Exploratorium; see how the Exploratorium does it."

Karen Wilkinson was fusing plastic when I spoke with her. After a long career of developing innovative approaches to science teaching throughout the United States, she came to the Exploratorium just a few years ago. Working with people who knew Frank, she told me, allowed her to share what she called "incredibly uplifting moments." It was clear that the staff had a fundamental belief that whatever the problem, they'd puzzle through it together, she said. She began to appreciate that "even people with very little experience can notice really interesting things." She learned how to let herself get really stuck on an idea and "squeeze every bit of interesting juice out of it . . . I realized I had an ability I didn't think I had."

When Nicole Minor arrived at the Exploratorium in 1991, she was twenty-two years old, and felt Frank's presence constantly. "It felt like he was always around, keeping an eye on what was going on," she said. She left and came back to the museum several times since then, and finds it has affected the way she looks at everything around her. "I never get tired of having my mind blown, either by some new scientific concept or new windows into human perception. I stop all the time to look at shadows; I revel in little moments of light reflections as I wander the city."

Whether they knew Frank or not, many people who were influenced by him find themselves behaving in ways they realize are a homage to Frank — very "Frankesque," as one put it.

In the end, I suppose that's the real reason I'm writing this book — to make the world, if I can, a little bit more Frankesque. "If people read the book and feel that they'd like to be more like him than somebody else," David Perlman told me, "then that would be a great thing."

(It's worth noting that amid all this praise, not everyone appreciated what Frank was trying to accomplish. In December 2007, a request by House Speaker Nancy Pelosi for $300,000 for the Exploratorium showed up in "The Pork Report," compiled by Donald Lambro of the *Washington Times.* Stating that the Exploratorium was an obvious waste of taxpayers' money, he quoted from the museum's website that its mission was, absurdly enough, "to create a culture of learning through innovative environments, programs and tools that help people nurture their curiosity about the world around them.")

Postmortem

"I always thought we were a mobile army hospital for teachers."

Remarkably, nearly twenty-five years after Frank's death, roughly thirty staff members from the Frank era were still working at the Exploratorium when I visited in 2004 — quite a testimony to the family character that the place retains. Most feel deeply connected to his legacy. "They take care of it," one staff member said. "They nurture it." "I stayed because I was working for Frank's dream," said another.

By 2008, the Exploratorium had an annual budget of about $31 million, almost 350 employees (nearly 250 full time), and about 550,000 visitors a year — not to mention the 145 million people who visit Exploratorium exhibits worldwide (they are in 108 international and 113 U.S. science centers). As of 2008, some 270 artists had spent time in residence at the museum. More than 3,000 high school students had worked there as Explainers.

I think one of the things that would have pleased Frank most is the Exploratorium's reach as a center for teacher training and curriculum design. Some 450,000 teachers in dozens of states now benefit from Exploratorium educational programs. Hundreds of teachers participate in the on-site Teacher Institutes.

From what little I've seen firsthand, the high school teachers who come out of the Teacher Institutes are as inspired and refreshed by the experience as the teachers Frank taught during his years in the Colorado mountains. "I always thought we were the mobile army hospital for teachers, because they'd come in so shell-shocked," one staff member said. "They'd forgotten they were teachers."

Despite his dislike of computers, I think even Frank would have come around to liking the museum's award-winning website, which sends out some fifty webcasts a year — live from the inside of particle accelerators, from under the polar ice, from the edge of active volcanoes, and from laboratories all over the world. Webcasts of solar eclipses (Aruba, Turkey, Zambia, Greece) are of such high quality they are regularly picked up by major news media.

Some things Frank hoped would happen didn't pan out. He'd dearly wanted to have a "Minstrel Bus" that would roam the neighborhoods of San Francisco to bring families and children to the museum. "One can imagine the establishment of a tradition — somewhat like the Good Humor van — in which local people look forward to the gay and bell-ringing passage of the bus along outer Mission Street," he said. Neither did he manage to institute the thematically oriented programs he envisioned putting on with schools and libraries and other museums.

The Exploratorium has changed in some ways that, I think, would have upset Frank mightily (although predicting what Frank would have thought about anything is risky business — and for those who took part in creating the Exploratorium in even a small way, it's impossible to view the place today with anything approaching objectivity).

One of the changes that would have bothered him the most is the steep price of admission and the disappearance of the six-month pass. Inevitably, it meant that the character of the museum — now as expensive for a single visit as going to a movie — changed. It became more of a destination for tourists and less a community gathering place. Hardly a "public beach," the Exploratorium (in this sense at least) no longer really reflects Frank's love of openness, the lessons of his postwar exclusion, his all-around faith in the principles of democracy. In fact, it can be difficult to see the underlying idealism that was behind the museum's creation. When I tell people today about the Exploratorium's political and philosophical foundations, they are invariably surprised.

In other ways as well, the Exploratorium has (perhaps inevitably) come to look a lot more like other institutions. It has a large staff of professional evaluators. Meetings (including board meetings) are no longer open to all. Partly because the staff had felt left out of key decisions after Frank's death, the Exploratorium became a union shop in 1993 after a prolonged and bruising battle. The lines between workers and management have become much sharper, as have the lines between various departments. Salary differences are much wider, and no longer a matter for discussion. Workers don't spend much time in meetings making decisions, and managers don't spend time in the shop or on the floor. Partly because of the strict regulation of work hours, the shop tends to be deserted in the evenings. Certainly, it's no longer regarded as an extension of home.

Creations grow up; they change; to the extent that we don't accept this, we will always be disappointed. Yet ideals matter — and fighting hard to preserve some of Frank's original intent is a crucial mission that continues to drive many dedicated people to this day. The Exploratorium's current director, Dennis Bartels, is, fittingly enough, a person who seems to have absorbed Frank's essence by osmosis — though the two never met in person. "When you get right down to it, it's about democratizing science," he told me. "It's a humanist agenda. There's always another question to ask, and you always have a right to ask it. You never have to take anybody's claim at face value." Among the qualities he picked up from Frank, Bartels mentioned the confidence to rely on the intelligence of other people, to be at ease with the fact that things may not work, that the main thing is to make people comfortable. "If you think of this as a science museum, that's a huge mistake. It's about inquiry. It was never meant to teach what art is or what science is, but to teach us what learning is."

The World He Made Up

"We have refused to feel at home in the world as we found it, but have insisted upon finding a world in which we could feel at home."

Frank's ghost continues to haunt the Exploratorium in almost palpable ways — a fidgety spirit wearing a suit and tie and swinging a cane, trailing smoke, humming while he fixes up whatever needs adjusting after everyone else has gone home. His aesthetic is affixed to the wood and screws and concrete of the exhibits, but mostly to the ideas behind them. One new staffer who arrived after Frank died sensed that he was still firmly in charge. "Frank was still the electricity that ran the place," she said. "He was like an outlet." Others claim that every now and then they feel a certain presence, hear the tapping of a cane. "I felt that he actually inhabited some of the exhibits," another said.

He's a complex sort of specter — surely not one of those dour ghosts associated with séances or haunted houses. Rather, I see him sitting in the small theater, watching his favorite film, *Fiddle Dee Dee* — three and a half minutes of pure animated silliness, featuring hand-painted images on film dancing jerkily to an old-time fiddle. But he's hardly one of those kindly Casper-type ghosts either. "People talk about the golden era of Frank," one staff member said. "People who hadn't been working here!" They forget what a cantankerous old coot he could be.

Mike Templeton, who started calling Frank "Saint Frank" after his death, reminds us that "all saints have feet of clay. And they are all revered nonetheless." More than once, I felt almost desperate to escape Frank's often exhausting presence — but I miss him dearly, and he continues to haunt my life. I frequently find myself spouting Frankisms, which I've come to think of as "algorithms for living": "The worst thing a son-of-a-bitch can do is turn you into a son-of-a-bitch." "Artists and scientists are the official noticers of society." "It's not the real world; it's a world we made up."

In some ways, Frank took us over as completely as any horror-film body snatcher — leaving us nosy, principled, playful, argumentative, stubborn, addicted to learning, and taken to talking about Frank. When a friend found out he had been dead for twenty years, she couldn't believe it, because I tell Frank stories as if he were still around. In all the important ways, he is.

When Frank died, some obituaries mentioned that he'd spent life "in the shadow" of his older brother, but I don't believe that was true; on the contrary, he seemed to have done a better job, in the end, of escaping those shadows. He emerged from the dark side clear and strong, idealism intact, the world still an endless wealth of wonders. He created a place where playful exploration — together with science and art — was put in service of saving the world from self-destruction.

But given all the people and institutions he affected so deeply, it's striking to consider those things he had no effect on at all. By the 1960s, the government was already spending more on nuclear arms than on schools, the environment, law enforcement, scientific research, energy production, and social services combined. By the sixtieth anniversary of the Trinity test, in 2005, there were nine nuclear nations and tens of thousands of bombs — with thousands presumed by experts to be on the highest levels of alert. Our arsenals have done nothing to dissuade countries like North Korea and Iran from pursuing weapons programs of their own.

Frank never understood why he and the other atomic scientists had so little luck in persuading people that the bombs couldn't possibly create security in the long run. Efforts to scare people obviously didn't work — or rather, they succeeded in scaring people, but not in stopping the flood of nuclear bombs. "All of [the efforts] have failed," he said. "Every single one of them."

And as in the postwar era, the opinions of scientists closest to the issues today are being ignored. The George W. Bush administration effectively dismantled the scientific advisory apparatus for nuclear weapons; physicists who had been advising presidents through both Democratic and Republican administrations were effectively fired. For the first time, the control of our nuclear arsenal is in the hands of people who have never witnessed a nuclear explosion. Legislators and their aides — educated primarily as lawyers — are in general ignorant about the basic facts of science. "Many do not get that nature does not care about human politics," wrote Martin Hoffert, a New York University physicist with long experience testifying before Congress.

Instead of heeding the advice of physicists, our government has been spending billions on new weapons while the critical work of protecting nuclear weapons throughout the world so that they don't fall into the hands of terrorists is being ignored. According to the Natural Resources Defense Council, the Bush administration was "spending *more than 12 times* as much on nuclear weapons research and production than it [did] on urgent global nonproliferation efforts to retrieve, secure, and dispose of weapons materials worldwide." This despite the fact that Russia alone has some sixteen thousand nuclear weapons, a large portion of them unsecured and protected only by inadequately trained workers living close to poverty and taking second jobs simply to make ends meet — "blinky-eyed from lack of sleep."

As the great physicist Freeman Dyson told *Discover* magazine in June 2008, nuclear weapons remain the most serious threat to our survival. "People have more or less forgotten about them," he said. And yet there is still "a huge chance that some stupidity" could cause the destruction of civilization. Dyson suggested it's time for a new campaign to rid the world of them. "It's not hopeless."

In the face of this threat, we have been spending tens of billions of dollars a year on what can only be described as a "faith-based" missile defense system, which a broad consensus of physicists says can't work (and which has passed no practical tests).

It's not just in physics, of course, where evidence-based knowledge has been dismissed as irrelevant and scientists are treated as just another special interest group. The teaching of evolution — the bedrock of biology — is being debated anew in forty states.

Indeed, it's both striking and disturbing how much of Frank's story resonates with the politicization of science under the Bush administration. Once again, many scientists feared speaking out against the political powers that be — or suffered retribution for declining to support the prevailing political agenda. J. Robert Oppenheimer was deemed untrustworthy in large part because he argued that building a hydrogen bomb was pointless and dangerous — a view shared by many (if not most) of his Los Alamos colleagues. His public humiliation was his punishment for not playing along.

This time around, thousands of scientists and dozens of scientific societies signed petitions and took to the road to protest the administration's widespread manipulation of science. And the Obama administration, which began as this book was going to press, appears determined to restore science to its rightful place in culture and politics alike.

It's a very welcome sign. Given that we live in dangerous times, the inventions of scientists and social scientists are more important than ever. The perils we face cry out for the kinds of courageous and even outrageous new ideas Frank spent his life promoting.

To survive the nuclear era, Frank thought, people might have to reevaluate their ideas of what is ultimately worth fighting for. To say that a nuclear war is not worth fighting, he and others pointed out, is not a pacifist statement. It's simply a statement of fact. "It is a specific disclaimer that a war between major powers could settle nothing worthwhile."

Perhaps the things we think are worth fighting for — political systems, for example — are mere veneers, surrogates for more important values, symbols of a kind of life that *is* worth living and fighting for. Perhaps if we focused instead on providing ways to satisfy human needs — emotional and aesthetic needs as well as material ones — then we would come up with the means to protect these values in ways that are not so prone to generate conflict. Nicer ways.

As Frank kept reminding us, we live in a world we made up, and we have the power to change it. Or as Frank's favorite philosopher, Max Otto, put it, quoting Justice Felix Frankfurter: "There is no inevitability in history . . . except as men make it."

Frank's was, above all else, a philosophy of optimism, rooted in the belief that if people are prepared to think the best of each other, listen to each other, and not accept a world less sane or kind than the best artists and philosophers have imagined, then they will, in the end, make up a better one.

In the early 1980s, Liz Keim went with Frank to an exhibition at the Museum of Modern Art in New York that she remembers was called "The Human Condition." "It was very, very dark," she said. "Lots of slashing and burning." When they left the museum, Frank said to her angrily, "This is not the human condition! This has nothing to do with what humans are in their hearts."

In the summer of 2004, thirty-two years after I discovered Frank's "marvelous monster," I returned for a day of aimless sightseeing in his still fertile "woods." The first thing I noticed as I walked in the door was a concert going on in the shop. A performance artist named Annie Gosfield was conducting a symphony that merged music and industrial sounds — using saws, lathes, metal sheeting, dustbins, and compressed-gas cylinders as instruments with an end to destroying the boundaries between music and noise.

I thought of the early "musician" who shattered thousands of mirrors as part of his performance, and I knew Frank would be pleased.

The Exploratorium was crowded, so I had a hard time getting to exhibits. Instead, I mostly watched. In the math section, people were absorbed in balancing dangerously off-center logarithmic blocks and building a precarious catenary arch so stable you could push it with your finger and it would wiggle and settle like Jell-O. I watched people pass a straight baton through a curved slot and roll square wheels, building intuition in ways equations can't.

I wandered over to the "Water Spinner" and considered how the surface always knows to form a perfect parabola — a convincing demonstration, if there ever was one, that math is nature's native language (equations *do* grow on trees). I remembered how Frank put his head in the way of the spinning frame to prove that it really couldn't hurt anyone (much).

In a darkened corner, an Explainer blew shimmering pink-and-aqua-tinged soap bubbles that floated above an invisible layer of carbon dioxide. When visitors approached and puzzled at the effect, she explained that the vapor from the evaporating dry ice beneath the bubbles is heavier than air. As the vapor enters the bubbles and makes them heavier, they start to sink. She told them to watch carefully: some bubbles expand perceptibly before popping. She explained that unlike water, carbon dioxide "evaporates" directly into gas without going through a liquid phase.

Up on the mezzanine, things were a little quieter, the mood more contemplative. A father and his two towheaded boys spent thirty minutes playing with a bunch of gears, spools, motors, and rubber bands; if you connected everything just so, you could get wheels spinning, fans twirling, flags fluttering. A young girl in carpenter pants joined in. Another little boy, no more than four — no family in sight — ran frenetically from one side of the table to the other, lickety split, determined to try everything every which way. "I did it!"

Even family groups spin apart in such a minefield of distraction. I saw a pair of sweethearts struggling hard to hold hands while being torn apart by conflicting curiosities. Trying to keep people together at the Exploratorium is like trying to put an amoeba on a leash.

There was a break in the crowd, so I took the opportunity to play with a tableful of endlessly adjustable pendulums, and wondered once again at what a clever thing a clock escapement is. A young boy asked me, "Yo! What's up with that? What are we supposed to do?" It was the "we" that struck me. Strangers a second ago, now we were lab partners.

Nearby, I joined a white-haired black man who was scratching his head, puzzling over what was still my favorite exhibit, the "Relative Motion Swing." Two pendulums swung at right angles, in straight lines, together carving out circles and ellipses. It seemed impossible, but it's all a matter of frames of reference — and explains why ancient peoples saw the planets move in such bizarre (to us) ways. By extension, it explains why so many things other people "see" seem bizarre to us as well.

Then I came upon a really nice surprise, an exhibit I hadn't seen before, every bit as lovely and amazing as anything Frank himself could have come up with. Shiny brass balls hung like heavy fruit from a branch of a pendulum tree, each a different length, and therefore a different frequency. When I set them all going at the same time, a wave snaked through them — palpably real even though it didn't exist save as a relationship among moving parts. But then, I thought, neither do I exist except as a relationship among moving parts: since the atoms in my body are replaced, on average, every seven years, what makes me "me" is pattern, not matter.

Moving away from the pendulums, I wandered around to find old friends among the exhibits and see what was new. Some of my old favorites had gotten face-lifts — some fabulous, some unfortunate, some funny. Among the latter was the old shoe tester, which was now wearing sneakers instead of Mary Janes. Among the fabulous was a new version of "Bells," the exhibit in which you can run a bow over the edge of metal plates to make sand fly off except at the resonant nodes, resulting in beautiful patterns. It taught me why the rings of Saturn have gaps. In this new version, the frequencies were controllable, and it was easier to make the crystals collect in all the quiet places, sculpting delicate filigrees of sound.

Among the unfortunate was Bob Miller's "Sun Painting," which no longer welcomed visitors with splendorous shifting pure-colored shapes. Now it was way in the back, and smaller. The large circular zone plate that focused sound to a point by diffraction looked lonely without its optical counterpart — taken off the floor, I was told, because it was deemed too difficult for the public to appreciate.

Nevertheless, it was clear that Frank's legacy is very much intact. A new exhibit builder (a protégé of Ned Kahn) had turned a bit of plastic garbage bag into a flower that bloomed as it spun in front of a flickering computer screen. He'd made dry-ice "comets" that shot out into a liquid sky, spinning and spitting gas just like the real thing. A couple of young girls watched intently as tiny dry-ice bergs made their way along a little conveyor belt and down a small chute. "Whoa! Look at this!" said the one in braids. "Check it out! It's like little crystals. Ooh, it's cold. Hurry, here it comes. Wow! Comets! Shooting stars! Are they attracted to each other? This is incredible!" (Shawn Lani, who created these works, also never met Frank, but insists he feels his presence keenly: "Make no mistake . . . I smell the smoke, I hear the cane.")

A janitor paused on his rounds to watch the comets too, a rubber-gloved hand resting on the enormous wheeled gray trash bin hung with containers of cleaning fluid. He seemed mesmerized by all the motion — tiny comets spouting fountains, meeting, merging, splitting; wakes of curly smoke rising up.

I hoped Frank was watching, listening: "I figured it out." "I can do it." "I know what it is." "Look at that!" "Look it! Look it!"

What a wonderful playground this was — one thing after another you could fiddle with for no apparent reason, a sandbox for the mind. I came upon the "Adjustable Plaything," which is nothing but an electromagnet with some knobs, weights, screws — and here was that janitor again! The sign on it read: "Mess around with the brass controls to see if you can make this thing do something." Didn't that just sum up this whole crazy place?

Frank's original handmade exhibits were still scattered here and there. His coupled pendulums (built on Liz Keim's kitchen table) had acquired little yellow plastic feet sometime during the filming of the *Nova* documentary, and looked suitably ridiculous (though no less instructive). I found his old reverse-distance depth perception experiment, his shadow kaleidoscope, his exhibit on reaction time with its dangerously sharp-cornered meter stick. And, of course, there was his cane, hanging on a wall, inviting anyone who passed by to discover its center of gravity, just as Frank would have done in real life.

At closing time, I missed the brassy resonant ring of the old ranch school bell — now back in the family, on its way to Mike's new barn. Instead, a foghorn bellowed long and low — such a forlorn sound. People reluctantly got in their last licks and looks, and slowly made their way toward the door.

I stopped at the long "Echo Tube" that extended far out and up into the Palace, and clapped my hands. The sound came back at me, altered and multiplied, echoing from all along the tube, carrying news of the places it'd been. Rat-a-tat-tat. Like the fast, impatient tapping of a cane. Time to go home. Fiddle Dee Dee.

CODA: LIVING A FRUITFUL LIFE

ACKNOWLEDGMENTS • NOTES

BIBLIOGRAPHY • INDEX

CODA: LIVING A FRUITFUL LIFE

Speech to the 1960 graduating class of Pagosa Springs High School

By Frank Oppenheimer

I am grateful for the life I have lived. It has certainly not been as full as the lives of some people, and yet it has probably been richer in experience and in a sense of accomplishment than the lives of many.

I think that part of the sense of having lived a full and a rich life comes from an ability to continually take things seriously — but not too personally. Of a willingness, even a determination, to become deeply involved in what you are doing, but not obsessed by it.

What have you taken seriously? What has involved a lot of your attention, your time and worry: I can mention a tremendous variety of things: your school work, ball games, county fairs, science fairs, plays, concerts, talent shows, to name some of the obvious ones. But also some of you have been involved with a job or with the putting up of hay or doctoring sick animals. You have been concerned with events in your family, your relations with your friends and with things that you have made or bought. You have had to make decisions about what to do in the summer — and about what to do next. These form merely a suggestion of the kind of things that most of you have had some occasion to take seriously. What do you do when you take them seriously? You learn, you work, you worry and plan. It makes a difference to you whether one thing happens or another. Whether you get an A or a D, a kiss or a sneer, a victory or a loss.

I want to put a little more meaning into the phrase "taking things seriously." Perhaps I can best explain what I mean by talking about myself. I would say, for example, that I took my teaching in this school seriously. First of all, I thought it was an important job. I felt that if you learned some science, you would be able to lead better lives and that by trying to do a good job of teaching, I might have some effect not only on you individually but also on the school and the community. The teaching involved a lot of work and planning and I had to learn new things, not only about the subject matter, such as the names of the various geologic epochs, but also about how to present ideas that I was, at first, not able to get across. I stopped thinking of myself as a rancher or a nuclear physicist and thought of myself primarily as a high school teacher and wanted to be thought of as a good teacher. I wanted you to understand the things I enjoyed understanding, such as why a star got hot and stayed hot. I wanted you to get satisfaction from being able to do some of the things I found pleasure in doing, whether blowing glass or solving a problem. I felt an enthusiasm for the whole process of teaching.

Now let me give you another example, in retrospect a quite trivial one. I remember that at about the time I graduated from college, I took *coffee* seriously. I read about coffee and found out where and how it was grown and roasted. I wandered about New York City looking for coffee import houses, bought my own grinder, and learned to tell the difference in taste between Mocha and Java and Guatemalan and Brazilian and Costa Rican coffees. I drank my own mixtures and occasionally served them to my friends, each type of coffee for the proper time of day. Undoubtedly my friends thought I was nuts, but I thought of myself as a connoisseur, an expert. Now, twenty-five years later, I can chuckle at my former self; but obviously at the time it was not a trivial interest, or I would not now recall it so vividly.

I do not want to relive my life for you, but I would like to mention for the purpose of example a few more of the things that have absorbed me. During the War, it seemed enormously important to me that America develop an atomic bomb as quickly as possible and before anybody else did. Now the making of atomic bombs seems repugnant and evil to me. Before the War, I worked hard and long to help support the Spanish Loyalists . . . After the War, I gave speech after speech on the need for nuclear disarmament. During my years here in the Basin, I put my heart into my ranch, trying to make it a better one. I derived pleasure from the crops and animals that flourished, and learned the sick feeling that comes when one fails in helping a heifer to deliver a live calf or sees four or five cows dying on the range from having eaten larkspur.

Before coming here, I was in Minnesota for a couple of years. I remember how exciting it was when, with our high-altitude-balloon experiments, we discovered that not only hydrogen but the atoms of all the elements were in the cosmic rays coming from outer space.

In thinking about my life, I arrived at some ideas about what was necessary for a fruitful life. First, you become involved in projects that you can put your heart into. They seem important. What happens, the outcome of your efforts, must make a difference to you.

Second, the outcome must have, directly or indirectly, a wanted effect not only on you but on something outside you, on other people or on science or on a ranch or on a business.

Third, your project must involve some effort in doing, and especially in learning and experimenting.

Fourth, you have to really commit yourself by being willing to stand for something and to represent the kind of person to yourself and to others that is not inconsistent with what you are involved in . . .

It is not easy to explain why people take things seriously. If one thinks deeply and objectively about anything, even life itself, it can appear trivial and one can argue that it makes a negligible difference to a universe that is billions of years old and a billion light years in diameter. But such thinking is somehow irrelevant to the way humans act. I am aware of the immensity of the cosmos and yet I can take things seriously. So can you.

I do not really want to imply that I have no sense of values and that everything is of equal importance to me. Some kinds of pursuits and exploits that I could have at one time put my heart into now seem unimportant to me; but other, perhaps somewhat more channeled interests have appeared in their place. I find it increasingly hard to think that I am an expert in anything and yet I have taken very seriously the opportunity to talk and give advice to you tonight. I do not know what will capture my devotion in the future, but from past experience I have some confidence that it will be caught.

For you, I hope that there is another domain that will attract you. Throughout one's life one sees the perpetration of innumerable injustices and inhuman acts both at home and abroad. Usually one feels powerless to do anything about them, but I recommend to you that when an opportunity to intervene for justice arises, either for you alone or in concert with others, you take these opportunities seriously, and consider them important.

I have gone a little astray from my main purpose tonight, which was merely to remind you that you have a long life ahead of you and to say that I hope it will be a good one. I have talked about just one small aspect of how you live your life, but I think it is an aspect over which you have some measure of control and also one that you might not have been aware of. I recommend that you be willing to become deeply involved in lots and lots of things and that you let yourself, perhaps even force yourself, to do the things that you think are important and that you can take seriously.

I make this recommendation to you because I believe that if you do, then even in the face of considerable adversity you will feel, as I do now, grateful for having lived and always looking forward with eagerness to more of the same.

ACKNOWLEDGMENTS

Where to begin? I've been working on this book in one way or another since 1980, and it's almost impossible to remember all the people who helped along the way. Early encouragement was essential, so for that I thank especially Charles Weiner at MIT and Loretta Barrett, then at Doubleday, who took Frank's ideas seriously and thought they should be shared with the general public. Thanks to the *New York Times,* which let me write about "my friend the physicist" ad nauseam, and also *Newsday* and the *Washington Post* for the same. *Discover* magazine put up with this for five years or more, so many thanks to then editor Leon Jarroff for liking what I (and through me, Frank) had to say. More recently, Michael Parks, my editor in chief at the *Los Angeles Times,* and especially Joel Greenberg, my science editor for years, encouraged my rather sideways approach to science journalism. Thanks to Paul Glickman and the folks at KPCC for letting me try out some of these ideas in my radio commentaries.

Many of the people who helped me along the way are no longer with us, but I want to thank them nonetheless. In particular: physicists Philip Morrison and Victor Weisskopf, Robert Wilson and Pief Panofsky. Educator David Hawkins. *Scientific American* editor Dennis Flanagan.

Countless scientists have shared their insights and interests with me over the years, and I'm grateful for all the time they spent helping me understand physics (that's the least of it). Special thanks to Murray Gell-Mann for writing the foreword, and to David Kaiser — physicist and science historian at MIT — who read the entire manuscript (more than once) and made invaluable comments and corrections, as did Charles Weiner. Also special thanks to physicist Bob Park, who announced on his weekly update, "What's New," that I was looking for people who may have known Frank. Through Bob, I met several former students of Frank from the exile years, when he was teaching high school in Colorado. Among them were James Heckman of the University of Chicago, who went on to win a Nobel Prize in economics, and Stanley Fowler, now at the University of South Carolina School of Medicine, who discovered life in hot springs under Frank's tutelage.

Another special thanks to physicist Al Bartlett at the University of Colorado, who sent me Frank's lab books and introduced me to the people who worked with him when he was setting up the Library of Experiments there, and Jerry Leigh, for showing me around. Also to Richard Gregory, whose work on perception research so influenced Frank, for hosting me in Bristol and telling me Frank stories. Thanks to Tilla Durr (daughter of Clifford and Virginia Durr) for many conversations and e-mails, filling me in on Frank's worst (and best) years at the ranch and during the HUAC hearings.

A number of librarians and archivists helped along the way, including the ever-accommodating people at the Bancroft Library in Berkeley, who made endless copies of papers for me and let me sit and peruse boxes of letters and documents over many weeks. Thanks also to the Cornell University Archives, which produced some of the most touching letters between Frank and Robert Wilson, and revealed sides of Frank I hadn't known. Thanks to the University of Minnesota, which allowed me to take an uncensored look into Frank's tragic past there, when both he and the university were caught up in the rampant red scare of the day. I should also thank the FBI for making it relatively easy to get Frank's papers and providing the kinds of behind-the-scenes footage I couldn't have gotten anywhere else (although it's astounding how much is still marked out). Thanks to Adrienne Kolb at the Fermilab Archives, who helped me and my assistant, Jenny Lauren Lee, find yet more Bob Wilson letters bearing on Frank's situation.

In addition, Charles Weiner of MIT allowed me to make use of the many hours of tape-recorded conversations he had with Frank detailing most aspects of his life (for this I also thank the Bohr Library, where the tapes reside). Also Judith Goodstein of Caltech, for interviews with Frank that led to insights about his years at Caltech. And Jon Else, who kindly sent me all the interviews connected with his two astonishing films involving Frank: *The Day After Trinity* and *Palace of Delights*. And, of course, thanks to all the many other people who allowed me to interview them about Frank over these past many years.

I have the most trouble trying to adequately thank people at the Exploratorium, because this entire book, in some ways, is the result of their continuing support, and many of them have read and commented on the entire manuscript (more than once). Thanks to Ron Hipschman, for keeping Frank alive on the Internet while I struggled with the book, creating ingenious indexes and posting Frank's talks, papers, and random thoughts. Thanks to Thomas Humphrey, the best explainer of physics ever, for deep insights about Frank and what it was like to work with him. Thanks to Rachel Meyer, Ned Kahn, Darlene Librero, Bob Miller, Dave Barker, Larry Shaw, Charlie Carlson, Ester Kutnick, Liz Keim, Pete Richards, Brenda Hutchinson, Kurt Feichtmeir, Michael Pearce, Rob Semper, Lynn Rankin, Ginny Rubin, Tom Tompkins, Pat Murphy, Linda Dackman, Joe Ansel, Leni Isaacs, Jon Boorstin, Ruth Brown, Nancy Rodger, Jamie Bell, and so many others. Thanks to the former chairman of the Exploratorium board of directors F. Van Kasper for sharing his thoughts with me. Thanks to Dennis Bartels, the Exploratorium's new director, for "getting it."

The Exploratorium helped in many institutional ways as well

—for example, letting me host an impromptu (and entirely unserious) "séance" for Frank. Thanks to Mary Miller for the summer Osher fellowship, which resulted in my being able to transcribe fifteen-year-old tapes of conversations between Frank and myself. Thanks to then director Goéry Delacôte for making it possible.

Special thanks to Rose Falanga and Meg Bury for research assistance, especially in finding the wonderful photographs that accompany the text.

I could never have completed this book without my writer/editor friends, who may ultimately have labored over this book more than I did (just hearing me talk about it for nearly thirty years qualifies them for sainthood). Ginger Barber, the original agent on this book, somehow sold it to Harcourt. The marvelous historian and teacher Patricia O'Toole taught me to use archives and write biography—and led me to the fat and endlessly rich "KC Cole" folders in the Frank Oppenheimer Collection at the Bancroft Library. Claudette Sutherland, Evelyn Renold, and Mary Lou Weisman read various drafts and tried ever so diplomatically to nudge me in the right direction, although I resisted every step of the way. I never stood a chance against Jane Isay, then at Harcourt, my most brilliant editor ever, who saw the vision for the book, then kindly tore apart what I'd written, cut it in half, and made me put it back together in a way I initially thought was just short of insane. Of course, she was right, and the structure and tightness of the book belong to her. A wonderful surprise entered the picture when Andrea Schulz took over the reins at Harcourt (now Houghton Mifflin Harcourt) and immediately understood the book better than I did. Heartfelt thanks to all of them, especially to the lovely and amazing Mary Kay Blakely, who took the time to expertly line edit the manuscript. And to Tom Siegfried, now editor of *Science News,* who always points out my mistakes (I always argue anyway), let me just say: OK. You're almost always right.

Many other friends and colleagues helped enormously with their continuing interest in and advice on the project, often reading and commenting on drafts, sometimes cheering from the sidelines, and

I'd like to mention at least a few of them: Roald Hoffmann, Dava Sobel, Elsa Feher, Alan Alda, Nancy Linehan Charles, Gwen Roberts, David Perlman, George Tressel, Jack Miles, Kip Thorne, Carolee Winstein, Victor Navasky, Rosie Mestel, Adam Frank, Bonnie Garvin, Jim Lafferty, Alice Powell, John and Nancy Romano, Eric Lax, Jonathan and Ann Kirsch, Candace Barrett, Carolyn See, Carol Tavris, Malcolm Gordon, Richard Larson, Stan Wojcicki, Paul Preuss, Marcelo Gleiser, Douglas Hofstadter, David Hupert, Susan Diamond, Susan Chase, Whitney Green, and many others. Thanks to Sherry Frumkin and Yossi Govrin for letting me keep Frank's dream alive at Santa Monica Art Studios. Thanks to Kay Mills for inviting me to her biography club. Thanks to all the women at JAWS (Journalism and Women Symposium) for letting me try out some of my ideas on them. Thanks to Don Byrd for editorial and artistic advice, a constant supply of soul food, and so much more.

My father, Bob Cole, was on the board of the Exploratorium early on, and both of my children, Pete and Liz, were Explainers, as were my stepdaughters, Katie and Kristen, so in a sense, the journey has been a family affair. Thanks to all of you.

Most recently, I am pleased by the support I've received from colleagues at the University of Southern California, particularly from Warren Bennis, University Professor and Distinguished Professor of Business Administration, who read the entire manuscript and taught me about leadership; also Amy Parish, Doe Mayer, Jed Dannenbaum, Clifford Johnson, and Richard Reeves, who told me the book was a "love story." Thanks to Ed Cray for reading parts of an early draft. Thanks to Bob Sheer, for inviting me to talk about Frank in his class. Thanks to Steve Ross and the Los Angeles Institute for the Humanities for letting me try out an early version of the book on this fascinating group. Thanks to everyone at the Annenberg School who tried to help me find a researcher to help with the endnotes and bibliography. That turned out to be Lilly Fowler, who did a heroic job in a very short time; thank you, Lilly. Many thanks also to all the students who have helped along the way, including, in particular, Jenny Lauren Lee (who worked on the project for the past five years) and Joshua Rodriguez. Thanks to Ashley Carter, Liz Janssen, and Phil and Carolyn Canterbury, for typing up Frank's papers.

I suppose in the end the biggest thanks should go to Michael and Judy Oppenheimer, who not only helped me through it all, but also remained gracious, warm, and wise. I love you both.

NOTES

1. Palace of Delights

4 *reopened in 1968: Palace of Fine Arts: A Brief History*, Exploratorium publication.

belly of a whale: This image is sometimes attributed to Frank, but it comes from a paper he and I wrote together, probably the first. "The Exploratorium: A Participatory Museum." *Prospects* 4, no. 1, Spring 1974.

"all the charm of a blimp hangar": Thomas Albright, "From Electric Music Boxes to Solar Energy Art: Chaotic Funkiness and Fun at Exploratorium," *San Francisco Chronicle,* October 22, 1970.

5 *cards spelled out how to play:* K. C. Cole, "San Francisco's Scientific Fun House," *New York Times,* July 9, 1978.

8 *"this is heaven!":* Hutchinson, Brenda, in discussion with the author.

9 *"having a good time":* Friedman, Alan, in discussion with the author.

"point of emphasis": Brown, Ruth, in discussion with the author.

too big to go through the door: Hipschman, Ron, in discussion with the author.

10 *saw through things, beyond things:* Lynn Rankin, Exploratorium staff reminiscences at Frank Oppenheimer Memorial, 1985.

"half of what he was saying": Dackman, Linda, in discussion with the author.

"without the cane!": Oppenheimer, Judy, in discussion with the author, October 2005.

11 *you had to watch out:* Ester Kutnick, Frank Oppenheimer Memorial.

gently on the ankle: Brown, Ruth, in discussion with the author.

"nature as in people": Frank Oppenheimer quoted in Ruth Reichl, "Dreams of a Mad Scientist," *New West,* July 17, 1978.

12 *toward its next destination:* Frank Oppenheimer and K. C. Cole, "The Exploratorium: A Participatory Museum."

13 *"nooks along the way":* Frank Oppenheimer, "Everyone Is You . . . or Me." *Technology Review* 78, no. 7 (June 1976).

"things one sees are so beautiful": Frank Oppenheimer, "The Pleasures and Per-

sonal Satisfactions of Science," lecture delivered to alumni groups at Grand Junction and Glenwood Springs, 1964.

14 *"against being fooled or misled"*: Early fundraising brochure for the Exploratorium, undated.

15 *"'American way of life?'"*: Exploratorium, "Frank Oppenheimer: Founder of the Exploratorium," collection of favorite quotes, put together by staff members, http://www.exploratorium.edu/frank.

"like neighbors": Oppenheimer, Frank, interview by Jon Else, *Palace of Delights*, *Nova,* October 10, 1981.

"same with scientific phenomena": House of Representatives, Subcommittee on Elementary, Secondary, and Vocational Education, Testimony by Frank Oppenheimer and Ian Chabay, September 30, 1982.

16 *"three-dimensional neurophysiological world"*: *Exploratorium,* directed by Jon Boorstin, Jon Boorstin Productions, 1974.

"there are no trees around": Oppenheimer, interview by Else, *Palace of Delights*.

make an exhibit "misbehave": Oppenheimer and Cole, "Exploratorium."

"for adults as well as children": Frank Oppenheimer, "The Unique Educational Role of Museums," lecture, Belmont Conference on the Opportunities for Extending Museum Contributions to Pre-College Science Education, January 1969.

"'We trust you'": Hipschman, Ron, in discussion with the author.

17 *"Can you break rules?"*: Liz Keim to Frank Oppenheimer, February 1982.

"everything are part of that": Meyer, Rachel, in discussion with the author.

"between propriety and wildness": Janssen, Katie, in discussion with the author.

"or just enjoy themselves": Oppenheimer, "Everyone Is You . . . or Me."

18 *"unbearable and ineffective in school"*: Oppenheimer, "The Unique Educational Role of Museums."

"don't mind losing some control": Oppenheimer, Frank, interview by Linda Dackman, "Invisible Aesthetic: A Somewhat Humorous, Slightly Profound Interview with Frank Oppenheimer," *Museum International* 38, no. 2 (1986).

"the people who built them": Frank Oppenheimer, "The Exploratorium: A Playful Museum Combines Perception and Art in Science Education," *American Journal of Physics* 40 (July 1972).

"something incredibly wonderful happens": Frank Oppenheimer, "Adult Play," *Exploratorium* 3, no. 6 (February/March 1980).

19 *"remain playful in their work"*: Ibid.

tub in a single swoosh: Oppenheimer, Frank, in discussion with the author.

"to see what was happening": Pearce, Michael, in discussion with the author.

20 *"like a champagne cork"*: Semper, Robert, in discussion with the author.

21 *"how much of life works"*: John Barth, *The Floating Opera* (New York: Doubleday, 1956).

"sitting all by itself": Meyer, Rachel, in discussion with the author.

22 *for someone else to deal with:* Liz Keim to Frank Oppenheimer, February 1982.

"profoundly humanistic": David Perlman, "Scientist Frank Oppenheimer Dies at 72," *San Francisco Chronicle*, February 4, 1985.

never raised his voice: Zurin, Hal, in discussion with the author.

understanding and perceptive: Heckman, James, in discussion with the author.

knock his block off: Kutnick, Ester, in discussion with the author.

came out crying: Tompkins, Tom, in discussion with the author.

"I met Frank the Flawed": Humphrey, Thomas, in discussion with the author.

in Hiroshima and Nagasaki: Oppenheimer, interview by Else, *Day After Trinity.*
"yes" to everything: Miller, Bob, in discussion with the author.

23 *enjoying the whole thing immensely:* Hipschman, Ron, in discussion with the author.

25 *"He Did, He Did":* Robert Wilson, Frank Oppenheimer Memorial.

2. A Little Royal Family

26 *"out of an Italian":* Melba Phillips, Frank Oppenheimer Memorial.

27 *and no command of English:* Sam Maddox, "The Other Oppenheimer: A Life in the Shadows," *Sunday Camera,* March 24, 1985.
jackets during World War I: Richard Rhodes, *The Making of the Atomic Bomb* (New York: Touchstone, 1988).
his eye for fine fabrics: Oppenheimer, interview by Else, *Day After Trinity.*
New York's Museum of Modern Art: Maddox, "The Other Oppenheimer."
Society for Ethical Culture: Rhodes, *The Making of the Atomic Bomb.*
"love of the right": Philip M. Stern, *The Oppenheimer Case: Security on Trial* (New York: Harper and Row, 1969).
"the educated classes": S. S. Schweber, *In the Shadow of the Bomb: Bethe, Oppenheimer, and the Moral Responsibility of the Scientist* (Princeton, N.J.: Princeton University Press, 2000).
"still life or a landscape": Frank Oppenheimer, interview by Charles Weiner, February 1973, Bohr Library.
"long black lashes": Kai Bird and Martin J. Sherwin, *American Prometheus: The Triumph and Tragedy of J. Robert Oppenheimer* (New York: Alfred A. Knopf, 2005).
and apparently never mentioned: Rhodes, *The Making of the Atomic Bomb.*
who died in infancy: Bird and Sherwin, *American Prometheus.*

28 *"out of an Italian painting":* Melba Phillips, Frank Oppenheimer Memorial.
and art dealers: Oppenheimer, interview by Weiner, Bohr Library.
"old friends": Alice Kimball Smith and Charles Weiner, *Robert Oppenheimer: Letters and Recollections* (Cambridge, Mass.: Harvard University Press, 1980).
trees much more carefully: Frank Oppenheimer, "Growing Up in the Arts," *National Elementary School Principal* 56, no. 1 (September/October 1976).
with oils on canvas: Ibid.
marvelous colors and textures: Frank Oppenheimer, unpublished reminiscence about Robert.
galleries on Fifty-seventh Street: Frank Oppenheimer, "Growing Up in the Arts."

29 *popular at the time:* Frank Oppenheimer, "Museums for the Love of Learning – A Personal Perspective," unpublished manuscript.
the needs of society: Frank Oppenheimer, lecture, Association of School Administrators 110th Annual Convention, Atlanta, February 1978.
live-in maids, a chauffeur: Oppenheimer, interview by Weiner, Bohr Library.
Frank's grandmother: David Cassidy, *J. Robert Oppenheimer and the American Century* (New York: Pi Press, 2005).
Julius loved to argue: Bird and Sherwin, *American Prometheus.*
"a lot of conversation": Oppenheimer, interview by Else, *Palace of Delights.*
"pompous congratulatory speech": Oppenheimer, reminiscence about Robert.
tonsils out in his bedroom: Oppenheimer, interview by Weiner, Bohr Library.

and her captain awaited: Oppenheimer, interview by Else, *Palace of Delights.*
30 *would be strewn with petals:* Smith and Weiner, *Robert Oppenheimer.*
"goings on all the time": Ibid.
edition of Chaucer's works: Oppenheimer, interview by Weiner, Bohr Library.
"playing for a little royal family": Ibid.
New York Flute Club: Oppenheimer, "Growing Up in the Arts."
Baltimore Bach Club: Ibid.
tube stops in London: Boorstin, Leni Isaacs, in discussion with the author.
living, breathing people: Oppenheimer, Michael, in discussion with the author.
grandfather in Hanau, Germany: Rhodes, *The Making of the Atomic Bomb.*
"Germans at that time": Oppenheimer, interview by Else, *Palace of Delights.*
"shoved into another country": Oppenheimer, reminiscence about Robert.
31 *French occupying army:* Oppenheimer, interview by Else, *Palace of Delights.*
second (and frequently first) home: Oppenheimer, reminiscence about Robert.
receptacles for holy water: Ibid.
responsible, caring citizens: Ethical Culture Fieldston School, http://www.ecfs.org.
he got a tutor: Oppenheimer, interview by Weiner, Bohr Library.
32 *until at least 1982:* Algernon D. Black to Frank Oppenheimer, December 23, 1982, New York Society for Ethical Culture.
science teacher Augustus Klock: Oppenheimer, interview by Weiner, Bohr Library.
replicas of the switchboard: Oppenheimer, Frank, in discussion with the author.
on a regular basis: Ibid.
"the last 25 years": Augustus Klock to Frank Oppenheimer, March 19, 1956, Bancroft Collection, University of California, Berkeley.
"it shall run noiselessly": Smith and Weiner, *Robert Oppenheimer.*
when he was only twelve: Rhodes, *The Making of the Atomic Bomb.*
entranced by their beauty: Oppenheimer, Frank, in discussion with the author.
so well it dissolved: Oppenheimer, interview by Else, *Palace of Delights.*
33 *"junk he gave me":* Oppenheimer, Frank, in discussion with the author.
hauled it off to school: Ibid.
"so full of emotion": Ibid.
"repulsively good little boy": Bird and Sherwin, *American Prometheus.*
putting them back together: Oppenheimer, Frank, in discussion with the author.
instructor for his playmates: Frank Oppenheimer, "Teaching and Learning," speech, January 1982.
"for the fifteen-year-olds": Oppenheimer, Frank, in discussion with the author.
on real *train tracks:* Morrison, Philip, in discussion with the author.
"these things I was doing": Oppenheimer, Frank, in discussion with the author.
34 *"the permission to do so:"* Frank Oppenheimer, "Curiosity," excerpt published in *San Francisco Sunday Examiner*, September 25, 1983.
became a regular customer: Oppenheimer, interview by Weiner, Bohr Library.
"tear me off the ladder": Letter to or from Frank Oppenheimer, not clear.
"place had enchanted me": Ibid.
"think it ever stopped": Oppenheimer, reminiscence about Robert.
35 *"This is to get a new lock":* Ibid.
the Trimethy: Oppenheimer, interview by Weiner, Bohr Library.
to go look for them: Ibid.
had to be towed back: Oppenheimer, reminiscence about Robert.
"write your own speech": Smith and Weiner, *Robert Oppenheimer.*
"things like relativity": Ibid.

into art, music, philosophy: Ibid.

36 "AS A DUTY": Ibid.

"it shall run noiselessly": Ibid.

"jealous or deep-down angry": Oppenheimer, reminiscence about Robert.

"irreducible 'Fragestellung'": Smith and Weiner, *Robert Oppenheimer.*

"tearful time for me": Oppenheimer, reminiscence about Robert.

"your eyes and unaltered": Smith and Weiner, *Robert Oppenheimer.*

37 *"for one evening there with you":* Ibid.

"a Perry Mason *show on TV":* Oppenheimer, reminiscence about Robert.

"very sad when we departed": Oppenheimer, interview by Else, *Palace of Delights.*

met on his previous excursions: Frank Oppenheimer to M. Eugene Sundt, president of Albuquerque Gravel Products, February 18, 1970, Bancroft Collection, University of California, Berkeley.

"Hot Dog!": Frank Oppenheimer to Robert D. Purrington, chairman and professor of physics, Tulane University, June 15, 1982, Bancroft Collection, University of California, Berkeley.

Robert finally bought it: Oppenheimer to Sundt.

known for his crêpes: Oppenheimer, interview by Else, *Day After Trinity.*

for his cakes: Oppenheimer, reminiscence about Robert.

Robert Serber: Oppenheimer, interview by Else, *Day After Trinity.*

38 *in front of a roaring fire:* Melba Phillips, Frank Oppenheimer Memorial.

"act kind of silly, I guess": Oppenheimer, interview by Else, *Day After Trinity.*

1,000 miles a summer: Ibid.

"but he was all right": Ibid.

killed his own brother: Oppenheimer, reminiscence about Robert.

of fallen timbers: Oppenheimer, interview by Else, *Day After Trinity.*

11,000-foot pass: Oppenheimer, reminiscence about Robert.

39 *having broken his brother's arm:* Oppenheimer, interview by Else, *Day After Trinity.*

"we never got to Pasadena": Oppenheimer, reminiscence about Robert.

when Frank was nineteen: Oppenheimer to Sundt.

"to cheer and comfort them": Smith and Weiner, *Robert Oppenheimer.*

"gentle beyond all telling": Ibid.

when Frank was twenty-four: Bird and Sherwin, *American Prometheus.*

40 *"and really felt part of it":* Oppenheimer, Frank, in discussion with the author.

knew a lot of the faculty: Ibid.

three years later, in 1933: Official biography of Frank Oppenheimer, author unknown.

profound and delicious sort: Oppenheimer, Frank, in discussion with the author.

"than you can possibly imagine": Oppenheimer, interview by Else, *Day After Trinity.*

41 *his eyes so clearly saw:* Oppenheimer, Frank, in discussion with the author.

"for being so dumb": Ibid.

"gloomier and gloomier": Ibid.

42 *spent from 1933 to 1935:* Bird and Sherwin, *American Prometheus.*

"big names" were there: Judith R. Goodstein, "Interview with Frank Oppenheimer," California Institute of Technology Oral History Project, 1985.

being part of it all: Oppenheimer, Frank, in discussion with the author.

to talk about his research: Ibid.

might be interesting to work on: Ibid.

"exciting place to be": Goodstein, "Interview with Frank Oppenheimer."
worked for eighteen months: Linda Dackman, official Exploratorium biography/obituary of Frank Oppenheimer, 1985.
results of these experiments: Hilda Hein, *The Exploratorium: The Museum as a Laboratory* (Washington, D.C.: Smithsonian Institution Press, 1990).
no reason to come back: Goodstein, "Interview with Frank Oppenheimer."
He got a pilot's license: Paul Preuss, "On the Blacklist," *Science 83*, June 1983.
Gipsy Moth biplane: Louis Goldblatt to William Coblenz, September 26, 1968.
hike in the snow: Hawkins, David, in discussion with the author.

43 *local ski factory:* Frank Oppenheimer, "Mountain People," unpublished essay.
seen some sharks: Frank Oppenheimer, "Desert People," unpublished essay.
at the Osservatorio di Arcetri: Goodstein, "Interview with Frank Oppenheimer."
late into the night: Oppenheimer, Frank, in discussion with the author.
at the Uffizi Gallery: Dackman, official biography/obituary.
memorizing the paintings: Preuss, "On the Blacklist."
"three-star generals": Frank Oppenheimer, "Museums for the Love of Learning – A Personal Perspective," unpublished manuscript.
so normal on the surface: Oppenheimer, Frank, in discussion with the author.
"whole society seemed corrupt": Goodstein, "Interview with Frank Oppenheimer."
"could make a difference": Preuss, "On the Blacklist."
In 1935, Frank followed: Arlene Silk, "Dr. Frank Oppenheimer Receives Caltech Distinguished Alumni Award," California Institute of Technology press release, May 17, 1979.

44 *"from the laboratory"*: Stern, *The Oppenheimer Case.*
"smell a vacuum?": Goodstein, "Interview with Frank Oppenheimer."
"products of shellac": Oppenheimer, Frank, in discussion with the author.
nuclear physics facility: Goodstein, "Interview with Frank Oppenheimer."

45 *"It was beautiful to see that"*: Ibid.
"was not terribly good": Ibid.
what they were doing: Oppenheimer, Frank, in discussion with the author.
"as far away as the Athenaeum": Robert Serber, *Peace and War* (New York: Columbia University Press, 1998).
physicist Richard Tolman: Oppenheimer, interview by Weiner, Bohr Library.
"and being not of it": Smith and Weiner, *Robert Oppenheimer.*

46 *"'down with the bosses' movement"*: Oppenheimer, Frank, in discussion with the author.
Jacquenette Quann: Preuss, "On the Blacklist."
working-class origins: Bird and Sherwin, *American Prometheus.*
particular need to join: Ibid.
second year as a graduate student: Smith and Weiner, *Robert Oppenheimer.*
a radical newspaper: Goodstein, "Interview with Frank Oppenheimer."
"you could eat and charge it": Ibid.
"stranger would interrupt them": Ibid.
communism in the Soviet Union: Ibid.
have to go hungry: Ibid.

47 *"good music"*: Ibid.
"could make a difference": Preuss, "On the Blacklist."
Jackie joined the Communist Party: Ted Morgan, *Reds* (New York: Random House, 2003).

unemployment on the home front: Victor S. Navasky, *Naming Names* (New York: Hill and Wang, 1980).
frightening people: Morgan, *Reds.*
the Pink Decade: Ibid.
like a perfect society: Ibid.
"the Roosevelt New Deal": Hawkins, David, in discussion with the author.
suspicions to themselves: Morgan, *Reds.*
48 *the person of Klaus Fuchs:* Ibid.
fascism abroad: Navasky, *Naming Names.*
back on Thursday: Oppenheimer, Frank, in discussion with the author.
And so in 1937: House Committee on Un-American Activities, Hearings Regarding Communist Infiltration of Radiation Laboratory and Atomic Bomb Project, University of California, Berkeley, June 14, 1949.
application for membership: Oppenheimer, interview by Weiner, Bohr Library.
"became very active": Goodstein, "Interview with Frank Oppenheimer."
he never used it again: Bird and Sherwin, *American Prometheus.*
to be his alias: Papers from Frank Oppenheimer's FBI files.
"quite open about it": Goodstein, "Interview with Frank Oppenheimer."
49 *called the marriage "infantile":* Bird and Sherwin, *American Prometheus.*
would talk with anyone: Hawkins, David, in discussion with the author.
with his FBI tail: Leigh, Jerry, in discussion with the author.
"fairly upset by that": Goodstein, "Interview with Frank Oppenheimer."
50 *"difference in physics":* Oppenheimer, interview by Else, *Day After Trinity.*
"just not a Marxist": Morgan, *Reds.*
left the party in 1940: Frank Oppenheimer to Peter R. E. Goodchild, British Broadcasting Corporation, May 21, 1980, Bancroft Collection, University of California, Berkeley.
United States evaporated: Morgan, *Reds.*
the arts or aesthetics: Oppenheimer, interview by Else, *Day After Trinity.*
"started the Exploratorium": Ibid.

3. The Uncle of the Atom Bomb

51 *"like is a cuckoo clock":* Alan Alda, *Things I Overheard While Talking to Myself* (New York: Random House, 2007).
52 *the United States as well:* Oppenheimer, interview by Else, *Day After Trinity.*
his political activities: Oppenheimer, interview by Weiner, Bohr Library.
"Stanford was awful": Oppenheimer, in discussion with the author.
Palo Alto Civic Center: Oppenheimer, interview by Weiner, Bohr Library.
the not so distant future: Bird and Sherwin, *American Prometheus.*
53 *"some of the surrounding territory":* Kenneth M. Deitch, *The Manhattan Project: A Secret Wartime Mission,* Perspectives on History Series (Discovery Enterprises, Inc., 1995).
Army Corps of Engineers: Ibid.
54 *extremely difficult to extract:* Ibid.
mix of uranium isotopes: Exploratorium staff, Frank Oppenheimer Chronology, undated.
separated and collected: Richard Rhodes, *The Making of the Atomic Bomb.*
brand-new 184-inch cyclotron as well: Ibid.

"overtaxed electrical equipment": Frank Oppenheimer, talk delivered to Berkeley Democratic Club, November 27, 1945.
"people at the lab": Ed Lofgren, Frank Oppenheimer Memorial.
overseeing their operation: Exploratorium staff, Frank Oppenheimer Chronology.

55 *Oak Ridge into operation:* Louis Goldblatt to William Coblenz, September 26, 1968.
slacked off on the job: Oppenheimer, interview by Weiner, Bohr Library.
"failed in our jobs": Philip Morrison, quoted in Bird and Sherwin, *American Prometheus.*
a "magic place": Smith and Weiner, *Robert Oppenheimer.*
well-known physicists: Bird and Sherwin, *American Prometheus.*
virtually overnight: Deitch, *The Manhattan Project.*

56 *bathtubs in town:* Ibid.
babies were being born: Rhodes, *The Making of the Atomic Bomb.*
interested in their secrets: Jane S. Wilson and Charlotte Serber, *Standing By and Making Do: Women of Wartime Los Alamos* (Los Alamos, N.M.: Los Alamos Historical Society, 1997).
West Point graduate: Deitch, *The Manhattan Project.*
objected strongly to this policy: Rhodes, *The Making of the Atomic Bomb.*
"very good judgments": Oppenheimer, interview by Else, *Day After Trinity.*
first atomic bomb test: Oppenheimer, interview by Weiner, Bohr Library.
"a very tiny Mike": Robert Wilson, Frank Oppenheimer Memorial.

57 *and their friendship "blossomed":* Ibid.
"recognized the irony": Oppenheimer, interview by Weiner, Bohr Library.
"met that Nazi challenge": Robert Wilson, Frank Oppenheimer Memorial.
"lost it!": Robert Wilson, questionnaire regarding various aspects of the development of the atomic bomb, Fermi National Accelerator Laboratory, undated.
"caught us in its trap": Oppenheimer, interview by Else, *Day After Trinity.*

58 *"we completed our job":* Frank Oppenheimer, talk delivered to Berkeley Democratic Club.
"illusion of illimitable power": Freeman Dyson, *Disturbing the Universe* (New York: Harper and Row, 1979).
"its lights and values": Jeremy Bernstein, *Oppenheimer: Portrait of an Enigma* (Chicago: Ivan R. Dee, 2004).
"three person'd God": Rhodes, *The Making of the Atomic Bomb.*
chart the winds: Oppenheimer, interview by Weiner, Bohr Library.
"very long stem": Oppenheimer, Frank, in discussion with the author.
evacuating his own shelter: Frank Oppenheimer, talk delivered to Berkeley Democratic Club.

59 *the atmosphere on fire:* Joe Verrengia, "20th Century Forever Altered by Atomic Bomb," *Daily Herald,* December 28, 1999.
"and we became human again": Oppenheimer, interview by Else, *Day After Trinity.*
driven by Philip Morrison: Rhodes, *The Making of the Atomic Bomb.*
"this nuclear power": Trinity site brochure.
"coming together": Oppenheimer, interview by Weiner, Bohr Library.
flashed all around him: Jon Else, *The Day After Trinity,* documentary script.
Groves watched from base camp: Trinity site brochure.
in control bunker S-10000: Rhodes, *The Making of the Atomic Bomb.*
about five miles away: Deitch, *The Manhattan Project.*

60 *"that actually happened . . . echoing went on and on"*: Oppenheimer, interview by Weiner, Bohr Library.
"seemed to hang there forever": Oppenheimer, interview by Else, *Day After Trinity.*
"but that's all I remember": Ibid.
dump had exploded: Trinity site brochure.
oven door opening: Ibid.
"had just been born": Rhodes, *The Making of the Atomic Bomb.*
"sons of bitches": Ibid.

61 *"what we have seen"*: Deitch, *The Manhattan Project.*
"one way or another": Rhodes, *The Making of the Atomic Bomb.*
"let it get away from them": J. Robert Oppenheimer, *Atom and Void: Essays on Science and Community* (Princeton, N.J.: Princeton University Press, 1989).
"future war unendurable": Deitch, *The Manhattan Project.*
"No more wars": Oppenheimer, *Atom and Void.*
"pot of boiling black oil": Rhodes, *The Making of the Atomic Bomb.*
million more Allied casualties: Oppenheimer, *Atom and Void.*
to the Soviet Union: Barton J. Bernstein, "The Atomic Bombings Reconsidered," *Foreign Affairs,* January 1995.

62 *no legitimate military purpose:* Gar Alperovitz, "A Dubious Advantage," *Bulletin of the Atomic Scientists,* July/August 2005.
"use of any new weapon": Rhodes, *The Making of the Atomic Bomb.*
stay out of politics: Bird and Sherwin, *American Prometheus.*
dropped on a population center: Oppenheimer, interview by Weiner, Bohr Library.
a nuclear arms race: Rhodes, *The Making of the Atomic Bomb.*
to President Harry Truman: Nina Byers, "Physicists and the 1945 Decision to Drop the Bomb," lecture, UCLA Department of Physics and Astronomy, October 13, 2002.
and allowed to surrender: David Cassidy, *J. Robert Oppenheimer and the American Century* (New York: Pi Press, 2005).
150 Manhattan Project scientists: Bird and Sherwin, *American Prometheus.*
Japanese officials could be invited: Byers, "Physicists and the 1945 Decision."

63 *never saw these documents either:* Bird and Sherwin, *American Prometheus.*
until Japan surrendered: Kaiser, David, in conversation with the author.
easy to find and destroy: Bird and Sherwin, *American Prometheus.*
to assess the damage: Jon Else, *The Day After Trinity,* documentary script.
"was gone after Hiroshima": Alexander Rabinowitch, "Founder and Father," *Bulletin of the Atomic Scientists,* January/February 2005.
from front or back: Rhodes, *The Making of the Atomic Bomb.*
"a pot of boiling black oil": Ibid.
"no longer see the city": Ibid.

64 *"about sixty thousand feet"*: Deitch, *The Manhattan Project.*
dead within five years: Rhodes, *The Making of the Atomic Bomb.*
"all those flattened people": Oppenheimer, interview by Else, *Day After Trinity.*
50 percent of the population: Rhodes, *The Making of the Atomic Bomb.*
and without discussion: Bird and Sherwin, *American Prometheus.*
"Woe is me": Einstein exhibit at the Museum of Natural History, New York.

65 *"to their moral strength"*: Bird and Sherwin, *American Prometheus.*
"they cannot lose": S. S. Schweber, *In the Shadow of the Bomb.*
"we have used it": Mark Strauss, "Essence of a Decision," *Bulletin of the Atomic Scientists,* July/August 2005.

called Robert a "cry baby": Bird and Sherwin, *American Prometheus.*
"be so in the future": Frank Oppenheimer to Peter K. Hawley, United Office and Professional Workers of America, February 14, 1946.
"not for lack for trying": Robert Wilson, Frank Oppenheimer Memorial.

66 *their part in securing the peace:* Frank Oppenheimer, untitled talk.
"It wasn't enough": Oppenheimer, interview by Weiner, Bohr Library.
"fledgling political movement": Jessica Wang, *American Scientists in an Age of Anxiety: Scientists, Anticommunism, and the Cold War* (Chapel Hill: University of North Carolina Press, 1999).

67 *to sit tight:* Stern, *The Oppenheimer Case.*
"outrage with each other": Oppenheimer, reminiscence about Robert.
"Nobody was explaining anything": Oppenheimer, Frank, in discussion with the author.
"bodily thrown from the bus": Frank Oppenheimer, unpublished essay, around 1948.
"too likely to be wrong": Ibid.

68 *when they see them and elect them:* Ibid.
"the ability to walk": Ibid.
"may have some decisive effect": Ibid.
"you were waiting": Ibid.

69 *"deluxe and king-sized war":* Ibid.
naval officers and labor unions: Oppenheimer, Frank, in discussion with the author.
and so he did: Ibid.
"'into the radiation lab!'": Ibid.
using the same equipment: Frank Oppenheimer, "A Factor of a Thousand," undated essay, version published in the *Saturday Review* as "The Mathematics of Destruction."

70 *new approaches were needed:* Frank Oppenheimer to Peter K. Hawley, February 14, 1946.
"board of consultants": Oppenheimer, Frank, in discussion with the author.
most of the board's report himself: Cassidy, *J. Robert Oppenheimer and the American Century.*
transparent to everyone: Bird and Sherwin, *American Prometheus.*
world government, for example: Cassidy, *J. Robert Oppenheimer and the American Century.*
with no vested interests: Frank Oppenheimer, untitled talk.
"where we must begin": Frank Oppenheimer, "Science and Peace," lecture, Boulder, Colorado, 1946.
"then we are surely sunk": Frank Oppenheimer, untitled talk.
"forced to practice": Frank Oppenheimer, speech, American Association of University Women, 1945 or 1946.

71 *"build their own bombs":* David Kaiser, "The Atomic Secret in Red Hands? American Suspicions of Theoretical Physicists During the Early Cold War," *Representations* 90 (Spring 2005): 28–60.
"the handle of a hunting knife": Ibid.
it should not be internationalized: Robert Wilson to the editor, *San Francisco Chronicle,* September 28, 1945. Emphasis in original. Fermi National Accelerator Laboratory.
to the armed services: Oppenheimer, speech, American Association of University Women.

"utter disregard for the truth": Ibid.

72 *"at the same time make progress":* Ibid.

spill into research and teaching: Ibid.

"enemies of scientific progress": Frank Oppenheimer, "Public Welfare in the Atomic Age: The Support of Science," lecture, Minnesota Bankers Association, June 1948.

basis for public discussion: Oppenheimer, "Science and Peace."

veto power in the U.N.: Cassidy, *J. Robert Oppenheimer and the American Century.*

it fell flat: Oppenheimer, Frank, in discussion with the author.

"familiar with its nature": Oppenheimer, "Science and Peace."

73 *"however remote":* Frank Oppenheimer and Charles L. Critchfield to editor, *New York Times,* October 6, 1948.

"solution of these grave problems": Byers, "Physicists and the 1945 Decision."

"more seriously than they should be": Rhodes, *The Making of the Atomic Bomb.*

"equally authoritative": Frank Oppenheimer to Mr. Binder, editor, *Minneapolis Morning Tribune.*

74 *"we made in the 18th century":* Frank Oppenheimer quoted by K. C. Cole, "The Rewards of a Most Unusual Friendship," *Newsday,* June 2, 1985.

4. Un-American

75 *"Sincerely, Frank":* Frank Oppenheimer to Ernest Lawrence, undated.

76 *"the Grand Cayman Island":* Frank Oppenheimer, undated essay, probably 1947 or 1948.

"dense nuclei of atoms": Oppenheimer, "Public Welfare in the Atomic Age."

first experiments on it: Oppenheimer, interview by Weiner, Bohr Library.

first proton accelerator: Preuss, "On the Blacklist."

"a lot of really nice work": Panofsky, Wolfgang, in discussion with the author.

split wood or drive a car: Oppenheimer, Frank, in discussion with the author.

he accepted: Oppenheimer, interview by Weiner, Bohr Library.

77 *through Earth's atmosphere:* Frank Oppenheimer, "Practical and Sentimental Fruits of Science," Fifteenth Anniversary Award Dinner speech, *Exploratorium Quarterly,* January 25, 1985.

"investigations and anything practical": Oppenheimer, "Public Welfare in the Atomic Age."

"We had made a discovery!": Oppenheimer, "Practical and Sentimental Fruits of Science."

78 *with cosmology than nuclear physics:* Ibid.

"landmark research": Quoted by Dackman, official biography/obituary.

"capacity for hard work": Hans Bethe, quoted in Frank Oppenheimer Chronology.

"cursing and cussing": Oppenheimer, Michael, in discussion with the author.

79 *"looked pretty inexperienced":* Frank Oppenheimer, "Recovery," unpublished essay, 1947 or 1948.

for their balloons: Frank Oppenheimer, "Air Craft Carrier Caper," unpublished essay, 1947 or 1948.

beans, bananas, and pineapples: Oppenheimer, "Recovery."

"still amazes me": Ibid.

80 *immediately offered his machete:* Ibid.

had swelled to thirty-three: Ibid.
"fought off the sun with our arms": Ibid.
"energy to our legs": Ibid.
the course in any event: Ibid.
pull themselves up with both hands: Ibid.
81 *"walked up the tree":* Ibid.
from flammable vines: Preuss, "On the Blacklist."
"intense and disagreeable effort": Oppenheimer, "Air Craft Carrier Caper."
"Why don't you guys give up?": Ibid.
"sickening hard luck": Frank Oppenheimer to Robert Wilson, probably late 1948 or 1949, Carl A. Kroch Library, Cornell University.
"completion of our experiments": Oppenheimer, "Air Craft Carrier Caper."
82 *"stares me in the face":* Frank Oppenheimer to Robert Wilson, Carl A. Kroch Library, Cornell University.
"physics before": Letter from "John" to Frank Oppenheimer, February 3, 1984, Bancroft Collection, University of California, Berkeley.
"Frank in another way": Else, Jon, in discussion with the author.
83 *"Communist songs were sung":* FBI files.
every postwar talk Frank gave: Ibid.
"A Practical Spanish Grammar": Ibid.
84 *Practices Commission:* Ibid.
in contact with Soviet agents: Ibid.
"confidential informants": Jessica Wang, *American Scientists in an Age of Anxiety: Scientists, Anticommunism, and the Cold War* (Chapel Hill: University of North Carolina Press, 1999).
as a real and present danger: Morgan, *Reds.*
gravest threat: Ellen Schrecker, *No Ivory Tower: McCarthyism and the Universities* (New York: Oxford University Press, 1986).
"ideological police force": Navasky, *Naming Names.*
"Hotel Curtis": Ibid.
85 *"Oh, yes, he's left":* Oppenheimer, interview by Weiner, Bohr Library.
a series of twelve lectures: R. Woxen to Frank Oppenheimer, September 9, 1947, Bancroft Collection, University of California, Berkeley.
Royal Institute of Technology: Robert R. Wilson, director, Laboratory of Nuclear Studies, Cornell University, to Frank Oppenheimer, September 9, 1959.
"issued to you at this time": R. B. Shipley to Frank Oppenheimer, August 6, 1947.
"subject to considerable criticism": FBI files.
"no longer welcome in this laboratory": Oppenheimer, Frank, in discussion with the author.
86 *"I had taught there":* Ibid.
was a precursor: Wang, *American Scientists in an Age of Anxiety.*
"and he was fond of me": Oppenheimer, Frank, in discussion with the author.
"should have said 'No comment'": Oppenheimer, Frank, in discussion with the author.
"I don't know why": Oppenheimer, Frank, in discussion with the author.
a member of the Communist Party: Oppenheimer, Frank, in discussion with the author.
87 *"A-Bomb":* Peter J. Goodchild, *Robert Oppenheimer: Shatterer of Worlds* (New York: Fromm International, 1985, 1980).
Frank's party membership numbers: Stern, *The Oppenheimer Case.*

draw up a statement: Oppenheimer, Frank, in discussion with the author.

"the Communist Party": Frank Oppenheimer to T. R. McConnell, August 16, 1947, Bancroft Collection, University of California, Berkeley.

"I just should have said 'No comment'": Oppenheimer, Frank, in discussion with the author.

very dangerous situation: Wang, *American Scientists in an Age of Anxiety.*

lied to protect Robert: Victor Cohn, "Noted Physicist Must Plead to Teach," *Minneapolis Morning Tribune,* June 26, 1958, Bancroft Collection, University of California, Berkeley.

attack his brother: Malcolm Willey, memo, July 22, 1947, University of Minnesota Archives.

88 *of avoiding a nuclear arms race:* Robert Wilson to President Deane W. Malott, July 6, 1959, Fermi National Accelerator Laboratory Archives.

"investigations of the Cold War era": David Kaiser, "The Atomic Secret in Red Hands?"

"in an atomic war": Dean McConnell to President J. L. Morrill, July 25, 1947, University of Minnesota Archives.

"involved in atomic energy": Ibid.

"know all kind of things": Oppenheimer, Frank, in discussion with the author.

files to the Times-Herald *reporter:* FBI files.

89 *information about Frank:* Ibid.

"in a position to give us": Malcolm Willey to J. Edgar Hoover, July 23, 1947, University of Minnesota Archives.

"hold its files confidential": J. Edgar Hoover to Malcolm Willey, July 28, 1947, University of Minnesota Archives.

protested his actions: Unsigned memo, July 14, 1947, University of Minnesota Archives.

"Mr. Hoover for 'advice'": T. R. McConnell to M. M. Willey, July 25, 1947, University of Minnesota Archives.

offended by the letter: Ibid.

"in my mind of his loyalty": K. T. Bainbridge to Malcolm Willey, July 28, 1947, University of Minnesota Archives.

"Oppenheimer in our hands": Malcolm Willey to "Dear Lew," July 17, 1947, University of Minnesota Archives.

"University had done nothing": Malcolm Willey to "Dear Lew," July 23, 1947, University of Minnesota Archives.

"have dropped the matter": Malcolm Willey to "Dear Lew," July 18, 1947, University of Minnesota Archives.

"subject to incipient *criticism":* Malcolm Willey to J. Edgar Hoover, July 23, 1947, University of Minnesota Archives.

90 *"handling of the atomic bomb":* T. R. McConnell to Malcolm Willey, July 18, 1947, University of Minnesota Archives.

"nonviolent and consensual": Schrecker, *No Ivory Tower.*

"do not question his denial": Statement of the associates of Dr. Frank Oppenheimer relating to accusations that he was at one time a member of the Communist Party, undated, University of Minnesota Archives.

"any issue outside the university": University of Minnesota Archives.

91 *"myself into a real pickle":* Oppenheimer, Frank, in discussion with the author.

"experiments involving nuclear energy": FBI files.

"wholeheartedly into his new work": Ibid.

"the least to be feared": Kenneth Ford, introduction of Frank Oppenheimer, Millikan Lecture, American Association of Physics Teachers, June 1973.

92 *"relating to academic freedom"*: Frank Oppenheimer to J. W. Buchta, June 13, 1949, University of Minnesota Archives.

wouldn't be acted upon: John A. Salmond, *The Conscience of a Lawyer: Clifford J. Durr and American Civil Liberties, 1899–1975* (Tuscaloosa: University of Alabama Press, 1990).

"Oppenheimer had written was untrue": FBI files.

but to let Frank go: Buchta, in Frank Oppenheimer Chronology.

left the office in tears: Oppenheimer, Frank, in discussion with the author.

"and accepting a truth": E. B. Veen to "President, University of Minnesota," June 15, 1949, University of Minnesota Archives.

"feeling of shame": Roy Tollefson to "President Morrell [sic]," June 15, 1949, University of Minnesota Archives.

"on the part of the university": Luther N. Johnson to J. W. Morrill, June 15, 1949, University of Minnesota.

"he had achieved was in jeopardy": Charles J. LaVine to J. L. Morrill, July 20, 1949, University of Minnesota Archives.

93 *"the freedom to make mistakes"*: Edward Teller to J. W. Buchta, June 27, 1949, University of Minnesota Archives.

"think independently and critically": Phyllis Frier to J. W. Morrill, June 22, 1949, University of Minnesota Archives.

students sent a petition: Letter to J. L. Morrill, June 14, 1949, University of Minnesota Archives.

"mistake made in the past": Sophie Oleksa to J. M. Morrill, June 29, 1949, University of Minnesota Archives.

"Congratulations Oppenheimer matter": John Bricker to J. L. Morrill, date unclear, University of Minnesota Archives.

"queer as a witch": J. L. Morrill to James E. Pollard, June 28, 1949, University of Minnesota Archives.

"isn't worth a thing": Hokah Chief, letter or op-ed piece, undated, University of Minnesota Archives.

way to bring Frank back: J. W. Buchta to J. L. Morrill, October 9, 1949, University of Minnesota Archives.

94 *"along with Paul Robeson"*: Schrecker, *No Ivory Tower.*

"joining the ACLU": Johnson, Clark, in discussion with the author.

"gentlest man I had ever met": Emerson, Ed, in discussion with the author.

"needlessly, pointlessly, broken": Joseph Alsop, "Matter of Fact: The Case of Dr. Oppenheimer," Bancroft Collection, University of California, Berkeley.

"because one wants them to": Ibid.

"from gas on the stomach": Ibid.

95 *"very disturbed young people"*: Salmond, *The Conscience of a Lawyer.*

the civil rights movement: Durr, Tilla, in discussion with the author.

keeping the dialogue active and open: Ibid.

"who was supporting who and how": Ibid.

"how much Frank loved his brother": Ibid.

96 *"seemed to go on forever"*: Ibid.

"individuals and their families": Ibid.

warned that it could be risky: Salmond, *The Conscience of a Lawyer.*

could be cited for contempt: Oppenheimer, Frank, in discussion with the author.

witnesses who came after: Schrecker, *No Ivory Tower.*
during Jackie's testimony: House Committee on Un-American Activities, Hearings Regarding Communist Infiltration of Radiation Laboratory and Atomic Bomb Project.
and #1001 in 1939: Ibid.
97 *"of any of my friends":* Ibid.
"I know of no such cases": Ibid.
to clear themselves: Schrecker, *No Ivory Tower.*
he continued to refuse: House Committee on Un-American Activities, Hearings Regarding Communist Infiltration of Radiation Laboratory and Atomic Bomb Project.
"laws of the United States": Ibid.
"in our success": Ibid.
"a Communist in Labor": Frank Oppenheimer, letter reporting on Jackie's thoughts at HUAC hearing, no addressee, undated.
98 *"about being Un American":* Ibid.
"laws should be made": House Committee on Un-American Activities, Hearings Regarding Communist Infiltration of Radiation Laboratory and Atomic Bomb Project.
went into executive session: Ibid.
"and he lost everything": Kutnick, Ester, in discussion with the author.
toss ideas around openly: Frank Oppenheimer to Peter R. E. Goodchild.
99 *"survive throughout adversity":* Phyllis Freier, collected testimony on Frank Oppenheimer's impact, undated.
"to cooperate with us": Oppenheimer, Frank, in discussion with the author.
going to work out: Ibid.
"I realized what the wall was": Ibid.
cosmic rays in India: Frank Oppenheimer to Homi Bhabba, undated.
inclined to accept: Oppenheimer, Frank, in discussion with the author.
"a hearing on this question": Frank Oppenheimer to Dean Acheson, undated.
100 *"he had mended his ways":* FBI files.
"present a few difficulties": Arthur H. Snell to Frank Oppenheimer, December 16, 1958, Bancroft Collection, University of California, Berkeley.
"has evaporated": Wolfgang K. H. Panofsky to Frank Oppenheimer, January 13, 1959, Bancroft Collection, University of California, Berkeley.
"They didn't": Schrecker, *No Ivory Tower.*
"You're still a Communist!": Oppenheimer, Frank, in discussion with the author.

5. Exile

101 *"from the artificial values":* Else, Jon, in discussion with the author.
Pagosa Springs, Colorado: Oppenheimer, Judy, in discussion with the author.
"September 8, for $24,000": FBI files.
place to live "someday": Oppenheimer, interview by Weiner, Bohr Library.
the McCarthy storm there: Robert Wilson, Frank Oppenheimer Memorial.
102 *their new surroundings:* Oppenheimer, interview by Weiner, Bohr Library.
"parts of the nervous system": Frank Oppenheimer to Bob Wilson, undated, roughly fall 1949, Carl A. Kroch Library, Cornell University.
"people sweat it out for me": Ibid.

"in there pitching and working": Frank Oppenheimer to Robert Wilson, Carl A. Kroch Library, Cornell University.
with evident satisfaction: Ibid.
"doing it with them": Frank Oppenheimer to Robert Wilson, roughly winter 1949, Carl A. Kroch Library, Cornell University.
letters and petitions of support: Ibid.
"nincompoop": Most of Frank's letters are typed, but this one is in Frank's almost illegible longhand. I'm fairly confident the word is "nincompoop" (it sounds like Frank), but not 100 percent sure.

103 *"either impotent or monsters"*: Frank Oppenheimer to Robert Wilson, Carl A. Kroch Library, Cornell University.
"ceases to believe in anything": Ibid.
"my attention during the summer": Frank Oppenheimer to "Sam," undated.
"during the past season": Frank Oppenheimer to Bruno, November 3, no year.
"I want to be in it": Frank Oppenheimer to Robert Wilson, roughly fall 1949, Carl A. Kroch Library, Cornell University.

104 "was *a rancher"*: Oppenheimer, interview by Weiner, Bohr Library.
"damn it": Frank Oppenheimer to Robert Wilson, roughly 1948, Carl A. Kroch Library, Cornell University.
only about 3,000: Oppenheimer, interview by Weiner, Bohr Library.
the entire Basin: Durr, Tilla, in discussion with the author.

105 *"And they were atheists!"*: Richards, Peter, in discussion with the author.
"could wear with a tie": Oppenheimer, Judy, in discussion with the author.
"little grassy parks": Frank Oppenheimer to "Phil," Bancroft Collection, University of California, Berkeley.
"like the Swiss Alps": Richards, Peter, in discussion with the author.
still sit outside: Oppenheimer, Judy, in discussion with the author.
reading and long rides: Oppenheimer, interview by Weiner, Bohr Library.
Durrs and Linus Pauling: Oppenheimer, Judy, in discussion with the author.
secure the trailer: Ibid.
"*What hay?"*: Frank Oppenheimer to "Phil," Bancroft Collection, University of California, Berkeley.
"not just any old grass": Oppenheimer, interview by Weiner, Bohr Library.

106 *"knew nothing about a cow"*: Ibid.
nearly ten years: Oppenheimer, interview by Else, *Day After Trinity.*
one of Frank's van Goghs: Jeremy Bernstein, *Oppenheimer: Portrait of an Enigma* (Chicago: Ivan R. Dee, 2004).
"come in mighty handy": Frank Oppenheimer to Robert Wilson, roughly fall 1949, Carl A. Kroch Library, Cornell University.
"confused about money": Oppenheimer, Michael, in discussion with the author.
tackle to pull it out: Oppenheimer, Judy, in discussion with the author.

107 *newborns from freezing:* Preuss, "On the Blacklist."
"survive the cold spring": Frank Oppenheimer to "Phil," Bancroft Collection, University of California, Berkeley.
"with the scours": Frank Oppenheimer, "The Freedom to Create," draft II, September 1982, unpublished.
"good obstetricians": Preuss, "On the Blacklist."
"rotate the calf": Oppenheimer, Judy, in discussion with the author.
"unintelligible to anyone else": Frank Oppenheimer to "Phil," Bancroft Collection, University of California, Berkeley.

burlap and bales of hay: Ibid.
ditches to clean: Frank Oppenheimer to Robert Wilson, roughly fall 1949, Carl A. Kroch Library, Cornell University.
caring for the animals: Oppenheimer, interview by Weiner, Bohr Library.
"entropy and disorder": Oppenheimer, interview by Else, *Palace of Delights.*

108 *"organize the place together":* Frank Oppenheimer to Robert Wilson, Carl A. Kroch Library, Cornell University.
to think and reflect: Oppenheimer, interview by Else, *Palace of Delights.*
October through April: Frank Oppenheimer to "Phil," Bancroft Collection, University of California, Berkeley.
just to feed cattle: Ibid.
"empty of people": Ibid.
"grinding away in his pocket": Richards, Peter, in discussion with the author.
get it up and going: Miller, Bob, in discussion with the author.

109 *"a really nasty mood":* Richards, Peter, in discussion with the author.
sparing people's feelings: Ibid.
"Don't perjure yourself": Oppenheimer, Michael, in discussion with the author.
"a wonderful time": Oppenheimer, interview by Else, *Day After Trinity.*
living as a rancher: Oppenheimer, Judy, in discussion with the author.
lonesome and hard for her: Oppenheimer, interview by Weiner, Bohr Library.
"marry my mother": Oppenheimer, Judy, in discussion with the author.
"did interesting things": Richards, Peter, in discussion with the author.

110 *"I have* two *nice girls":* Hawkins, Francis, in discussion with the author.
"we all forgave him": Ibid.
suitcases down the stairs: Miller, Bob, in discussion with the author.

111 *"was worth many years":* Oppenheimer, Michael, in discussion with the author.
"the oil and the feel of it": Ibid.
"family hi-fi amplifier": Ibid.
only six years old: Ibid.
"I don't know what I answered": Ibid.
a half-dozen students: Durr, Tilla, in discussion with the author.
the only students: Oppenheimer, Judy, in discussion with the author.
named Old Snorty: Frank Oppenheimer to Robert Wilson, roughly fall 1949, Carl A. Kroch Library, Cornell University.

112 *her husband was fired:* Durr, Tilla, in discussion with the author.
"in the physical world": Ibid.
the sound of Frank's flute: Ibid.
"told the FBI to go to hell": Oppenheimer, interview by Weiner, Bohr Library.
"I was thrilled!": Oppenheimer, Michael, in discussion with the author.
"of any other person": Frank Oppenheimer to Hon. John S. Wood, chairman, House Committee on Un-American Activities.

113 *had grilled Frank about:* FBI files.
he'd better cooperate: Oppenheimer, Frank, in discussion with the author.
"Here come the hard-boiled eggs!": Oppenheimer, Michael, in discussion with the author.
"odd in his manner": FBI files.
"make me an outcast": Frank Oppenheimer, "The Tail That Wags the Dog," 1949.

114 *eccentrics and alcoholics:* Richards, Peter, in discussion with the author.
"didn't touch the people": Oppenheimer, interview by Else, *Palace of Delights.*
agricultural subcommittee: Louis Goldblatt to William Coblenz, September 26, 1968.

"what people's troubles were": Oppenheimer, interview by Weiner, Bohr Library.
"can't get away from it": Ibid.
"unless he stopped": FBI files.

115 *"that people could get to":* Oppenheimer, Frank, in discussion with the author.
"might have come from Mars": Fowler, Stanley, in discussion with the author.
arts as well as physics: Morley Ballantine, "Smiling Through the Witchhunt."
make electric motors: Oppenheimer, Michael, in discussion with the author.
"I thought you were": Oppenheimer, interview by Weiner, Bohr Library.
license to teach: Fowler, Stanley, in discussion with the author.

116 *motions of planets:* Frank Oppenheimer, article about mathematics for teaching credentials, Pagosa Springs, 1958.
around with it mathematically: Ibid.
over a hundred students: Personal History and Qualifications for Frank Oppenheimer, undated.
the school ever had: Victor Cohn, "Noted Physicist Must Plead to Teach," *Minneapolis Morning Tribune,* June 26, 1958.
School of Medicine: Biography of Stanley Fowler, University of South Carolina School of Medicine, http://www.med.sc.edu.
dropped in from Mars: Fowler, Stanley, in discussion with the author.
Frank wore a tie: Ibid.
room for a smoke: Richards, Peter, in discussion with the author.
"I know that one!": Ibid.

117 *made a pet out of it:* Carlson, Charles, in discussion with the author.
"totally into it": Richards, Peter, in discussion with the author.
"was really delightful": Cohn, "Noted Physicist Must Plead to Teach."
"fun and gaiety of it": Frank Oppenheimer to "Mr. Fleming," undated.
"don't need a football team": Oppenheimer, interview by Weiner, Bohr Library.
"Pagosa Springs? Who?": Bartlett, Albert, in discussion with the author.

118 *from Pagosa Springs:* Harold Zirin, in discussion with the author.
"I read articles about physics": Fowler, Stanley, in discussion with the author.
"I turned down Harvard and everything!": Ibid.
science curriculum: Louis Goldblatt to William Coblenz, September 26, 1968.
science and technology: http://www.hq.nasa.gov/office/pao/History/sputnik.
"inspiring contributions": Testimony on Frank Oppenheimer's impact, Exploratorium Archives.

119 *"in the hands of students":* Hawkins, David, in discussion with the author.
fired up about science: Heckman, James, in discussion with the author.
"from another planet": Ibid.

120 *"gave me an A-plus":* Ibid.
Pagosa Springs High School: Oppenheimer, Michael, in discussion with the author.
evenings throughout the year: Oppenheimer, interview by Weiner, Bohr Library.
"that his brother didn't have": Hawkins, Frances, in discussion with the author.
"robots for their students": Oppenheimer, interview by Weiner, Bohr Library.

121 *made all the difference:* Frank Oppenheimer to David Rockefeller, June 23, 1976.
"as much as it might have": Frank Oppenheimer, "Teaching and Learning," Millikan Lecture, American Association of Physics Teachers, June 1973.
"wrong with schooling": Oppenheimer, interview by Else, *Palace of Delights.*
"on top of the job": Frank Oppenheimer to David Rockefeller.

"learning anything": Frank Oppenheimer to Robert Wilson, around winter 1957, Carl A. Kroch Library, Cornell University.
expensive over time: Frank Oppenheimer, letter, probably to *Denver Post.*
"almost killed him": I. I. Rabi, interview by Else, *Day After Trinity.*

122 *"You never do that"*: Else, Jon, in discussion with the author.
secretary of state: Stern, *The Oppenheimer Case.*
a Hiroshima-scale atomic bomb: Bernstein, *Oppenheimer: Portrait of an Enigma.*
"keep his shirt on": Ibid.

123 *gallons of water into steam:* Rhodes, *The Making of the Atomic Bomb.*
a mile in diameter: Else, Jon, in discussion with the author.
"which is intolerable": Bernstein, *Oppenheimer: Portrait of an Enigma.*
"what it was like": Ibid.

124 *detailing his activities:* Stern, *The Oppenheimer Case.*
enough to start a stampede: Ibid.
"the common defense and security": Bernstein, *Oppenheimer: Portrait of an Enigma.*
held in March 1954: Avalon Project, "Findings and Recommendations."
went on for four weeks: Stern, *The Oppenheimer Case.*
seeing much of the evidence: Jon Else, *Day After Trinity,* narration.
it succeeded brilliantly: Stern, *The Oppenheimer Case.*
Spanish Civil War: Avalon Project, "Findings and Recommendations."

125 *"falsified and fabricated"*: Stern, *The Oppenheimer Case.*
to protect Frank: Greg Herken, *The Brotherhood of the Bomb: The Tangled Lives and Loyalties of Robert Oppenheimer, Ernest Lawrence, and Edward Teller* (New York: Henry Holt, 2002), 214.
"more coercive forces": Oppenheimer, interview by Else, *Day After Trinity.*
"not to ask me these questions": Stern, *The Oppenheimer Case.*
"and therefore trust more": Bernstein, *Oppenheimer: Portrait of an Enigma.*
"defects in his 'character'": Stern, *The Oppenheimer Case.*

126 *"and he was gone too"*: Bird and Sherwin, *American Prometheus.*
"humanity at the same time": Freeman Dyson, *Disturbing the Universe* (New York: Harper and Row, 1979).
"longer than it should have": Smith and Weiner, *Robert Oppenheimer.*
"fulfilled of the two": Dennis Flanagan to K. C. Cole, May 11, 1990.

127 *"hard to not want it"*: Oppenheimer, interview by Else, *Day After Trinity.*
"I loved him very much": Ibid.
Some people nearly starved: Stern, *The Oppenheimer Case.*
just another weapon: Oppenheimer, interview by Else, *Day After Trinity.*
"nations would adopt": Frank Oppenheimer to Peter R. E. Goodchild.
"create a less dangerous world": Ibid.

6. An Intellectual Desert — and a Library of Experiments

128 *"to talk to each other"*: Oppenheimer, Frank, in discussion with the author.
gain him entry: Ibid.
"beauty and order of nature": Frank Oppenheimer to Robert Wilson, around spring 1958 or 1959, Carl A. Kroch Library, Cornell University.
as a research associate: Frank Oppenheimer, "The Sentimental Fruits of Science,"

speech given on receiving the Oersted Award by the American Association of Physics Teachers, San Antonio, Texas, January 31, 1984.

129 *"list it as for sale"*: Frank Oppenheimer to Robert Wilson, around spring of 1958 or 1959.

member of the Communist party: Frank Oppenheimer to Robert Wilson, roughly spring 1958 or fall 1959, Carl A. Kroch Library, Cornell University.

"it turns out slightly nasty": Frank Oppenheimer to Robert Wilson, probably fall 1959, Carl A. Kroch Library, Cornell University.

"back into the profession": Robert Wilson to President Deane W. Malott, Cornell University, July 6, 1959, Fermi National Accelerator Laboratory Archives.

"ways of expressing my affection": Ibid.

130 *needed constant attention:* Frank Oppenheimer, interview by Weiner, Bohr Library.

"silver belt buckle": Bartlett, Albert, in discussion with the author.

"ordinary, friendly guy": Finegold, Leonard, in discussion with the author.

anything but certain: Bartlett, Albert, in discussion with the author.

went to bat for him: Ballantine, "Smiling Through the Witchhunt."

wrote a glowing appraisal: Exploratorium staff, Frank Oppenheimer Chronology.

his considerable influence: Preuss, "On the Blacklist."

"very fortunate": Victor Weisskopf to Provost Oswald Tippo, University of Colorado, November 2, 1960.

131 *"loyal to the United States"*: Buchta, quoted in Frank Oppenheimer Chronology.

full professor in 1964: Personal History and Qualifications for Frank Oppenheimer.

very rare . . . events: Oppenheimer, interview by Weiner, Bohr Library.

before and during the war: Preuss, "On the Blacklist."

the Atomic Energy Commission: Louis Goldblatt to William Coblenz, September 26, 1968.

held much excitement for him: Oppenheimer, interview by Weiner, Bohr Library.

graduates in any other field: David Kaiser, "The Postwar Suburbanization of American Physics," *American Quarterly* 56 (December 2004).

"for material rewards": Ibid.

132 *"from the postwar Americans"*: Ibid.

considered a waste of time: Oppenheimer, Frank, in discussion with the author.

"the place to try and find it": Frank Oppenheimer, "Physics Competition," unpublished essay.

"the insights into interconnections": Ibid.

133 *"to understand their textbooks"*: Oppenheimer, "Curiosity."

"displays of aggressiveness": Frank Oppenheimer, "A War in the Shadow of the H-Bomb," *Bulletin of the Atomic Scientists,* May 1968.

alternatives to war: Letters to editors of newspapers including the *New York Times,* the *Washington Post,* the *Christian Science Monitor,* the *New York Herald Tribune,* the *St. Louis Post-Dispatch,* mostly around 1966.

"in Vietnam and elsewhere": David E. Rosenbaum, "In the Fulbright Mold, Without the Power," *New York Times,* May 3, 2004.

"in the shadow of the H-bomb": Oppenheimer, "A War in the Shadow of the H-Bomb."

134 *"its horror and immorality"*: Ibid.

"childlike displays of aggressiveness": Ibid.

"number of professors": Frank Oppenheimer, "A Factor of a Thousand."

"has bad effects": Frank Oppenheimer to Roy McVicker, House of Representatives, February 21, 1966.

"America changes": Frank Oppenheimer to editor, *New York Times,* April 28, 1965.

135 *"admiration for our right":* Frank Oppenheimer to Roy McVicker, February 21, 1966.

"act like one": Frank Oppenheimer to Norman Cousins, editor, *Saturday Review,* July 16, 1960.

136 *"underpasses and cloverleafs":* Frank Oppenheimer, unpublished essay.

"nobody could get angry": Bartlett, Albert, in discussion with the author.

"personally preoccupied": Frank Oppenheimer, "A Proposal for an Alternative Course Structure."

"flunks a museum": Oppenheimer, "The Unique Educational Role of Museums."

for speaking their minds: Oppenheimer, Frank, in discussion with the author.

137 *directed at these groups:* Frank Oppenheimer, "Stacked Deck," *Colorado Daily,* undated.

"to do it your way": Ibid.

"you couldn't hack it": Oppenheimer, Frank, in discussion with the author.

138 *credit for doing them:* Leigh, Jerry, in discussion with the author.

were pure drudgery: Albert A. Bartlett to K. C. Cole, with attachments on new physics building.

were always available: Oppenheimer, interview by Weiner, Bohr Library.

"you can hope to encounter": Bartlett, Albert, in discussion with the author.

$100,000 in total: Frank Oppenheimer and Malcolm Correll, "The Library of Experiments at the University of Colorado," *American Journal of Physics* 32, no. 3 (March 1964).

more than eighty experiments: Oppenheimer, interview by Weiner, Bohr Library.

found difficult to grasp: Ibid.

"could understand these things": Leigh, Jerry, in discussion with the author.

jury-rigged creations: Oppenheimer and Correll, "The Library of Experiments."

139 *"unfailing labor":* Leigh, Jerry, in discussion with the author.

"best of friends": Ibid.

unclassified work: FBI files.

"He just laughed": Else, Jon, in discussion with the author.

piquing their curiosity: Oppenheimer and Correll, "The Library of Experiments."

"the way Frank wanted": Bartlett, Albert, in discussion with the author.

resembled a small museum: Oppenheimer, Frank, in discussion with the author.

"a university library": Oppenheimer and Correll, "The Library of Experiments."

140 *"what he was doing":* Bartlett, Albert, in discussion with the author.

141 *to follow his lead:* Malcolm Correll to Oswald Tippo, May 17, 1963, Bancroft Collection, University of California, Berkeley.

project stagnated: Leigh, Jerry, in discussion with the author.

"Frank's equipment!": Ibid.

"I think it's tragic": Ibid.

"to reach ordinary people": Hawkins, David, in discussion with the author.

two Guggenheim fellowships: Exploratorium staff, Personal History and Qualifications for Frank Oppenheimer.

Palais de la Découverte: Louis Goldblatt to William Coblenz.

school-based learning needed: Frank Oppenheimer, "Partial Vitae," attached to letter from Louis Goldblatt to William Coblentz, September 26, 1968.

exhibits in working order: Frank Oppenheimer, "The Role of Science Museums," in *Museums and Education,* edited by Eric Larrabee (Washington, D.C.: Smithsonian Institution Press, 1968).

"part of their learning": Frank Oppenheimer, "Museums for the Love of Learning."

142 *the exhibits as props:* Oppenheimer, "The Role of Science Museums."
science museum experience: Oppenheimer, "Museums for the Love of Learning."
learn by making mistakes: Oppenheimer, "The Role of Science Museums."
in the state of Colorado: Ibid.

143 *"my life significantly"*: Oppenheimer, "Museums for the Love of Learning."
"stole the show": Hilda Hein, *The Exploratorium: The Museum as a Laboratory* (Washington, D.C.: Smithsonian Institution Press, 1990).

7. A Museum Dedicated to Awareness

147 *"I had thought"*: Frank Oppenheimer, interview by P. K. Kabir, May 30, 1974, Bancroft Collection, University of California, Berkeley.
returned a year later: Ibid.
one way or the other: Oppenheimer, interview by Weiner, Bohr Library.
"Get thee to a typewriter!": Oppenheimer, interview by P. K. Kabir, Bancroft Collection, University of California, Berkeley.
in Curator *magazine:* Hein, *The Exploratorium.*

148 *"woods of natural phenomena"*: Frank Oppenheimer, "Museums and Toys," *Exploratorium*, February 13, 1981.

149 *in the everyday world:* Oppenheimer, Frank, in discussion with the author.
images and experiences: Librero, Darlene, in discussion with the author.
could be made accessible: Oppenheimer, "The Unique Educational Role of Museums."

150 *should serve the people:* Oppenheimer and Cole, "Exploratorium."
"a decent society": Oppenheimer, "Museums and Toys."
help solve human problems: "The Exploratorium: A Synopsis," undated, but post-1981.
for Christmas tree sales: Ibid.
into the right circles: Ruth Newhall, Frank Oppenheimer Memorial.
the city's supervisors: Hein, *The Exploratorium.*

151 *interested in the building:* Oppenheimer, interview by P. K. Kabir, Bancroft Collection, University of California, Berkeley.
"never going to fly at all": Perlman, David, in discussion with the author.
what he wanted to do: Ruth Newhall, Frank Oppenheimer Memorial.
"a remarkable thing to watch": Ibid.
cultural/political scene: Hein, *The Exploratorium.*
signed on as volunteers: Ibid.

152 *"your fears to the surface"*: Oppenheimer, Frank, in discussion with the author.
"You improve them": Panofsky, Wolfgang, in discussion with the author.
wrote exhibit labels: Ruth Newhall, Frank Oppenheimer Memorial.
"what's happening in the world": Frank Oppenheimer in an early filmed interview.
"and out the door again": Oppenheimer, interview by Else, *Palace of Delights.*
"about the turn-downs": Frank Oppenheimer to Kevin Smith, Educational Development Corp., February 13, 1970.

153 *budget was "brazen"*: Oppenheimer, interview by P. K. Kabir, Bancroft Collection, University of California, Berkeley.
"it all came off": Ruth Newhall, Frank Oppenheimer Memorial.

"what one had to do": Oppenheimer, interview by Else, *Palace of Delights.*
and liked to do: Oppenheimer, interview by P. K. Kabir, Bancroft Collection, University of California, Berkeley.
"with wonderful people": Oppenheimer, interview by Else, *Palace of Delights.*
"apply their own values": Richards, Peter, in discussion with the author.
"said sure, why not?": Oppenheimer, Michael, in discussion with the author on Lummi Island.
"other than it was interesting": Ibid.

154 *a handy multipurpose tool:* Oppenheimer, interview by Else, *Palace of Delights.*
focus of many exhibits: Frank Oppenheimer, "Where Are We and Where Do We Go from Here – I," June 1, 1976.
and frames of reference: Oppenheimer, interview by Else, *Palace of Delights,* 24.
biological evolution: Hein, *The Exploratorium.*
effects were rigged: Hipschman, Ron, in discussion with the author.
chamber from NASA: Hein, *The Exploratorium.*
helped Frank to get it: Ibid.
was played continuously: Oppenheimer, Michael, in discussion with the author.
a glider, a forklift: Oppenheimer, interview by Else, *Palace of Delights.*

155 *effects of refraction:* Oppenheimer, Frank, in discussion with the author.
"whole museum this way": Friedman, Alan, in discussion with the author.
around in endless circles: Hein, *The Exploratorium.*
could study the optics: K. C. Cole, "New Exploratorium Catalogue. Facets of Light: Colors, Images and Things That Glow in the Dark," *Exploratorium* 4, no. 2 (June/July 1980).
he'd refuse it: Oppenheimer, interview by Else, *Palace of Delights.*
wonderfully eclectic character: Oppenheimer, Frank, in discussion with the author.
with duct tape: Kahn, Ned, in discussion with the author.
new one, and he did: Keim, Liz, in discussion with the author.
by Frank's lights: Templeton, Michael, in discussion with the author.
passive and addictive: Frank Oppenheimer, "Where Are We and Where Do We Go from Here – II," November 1982.

156 *"an unmitigated disaster":* Frank Oppenheimer, "Chips and Changes," *Exploratorium Quarterly,* Spring 1984.
"the oil from a lathe": Oppenheimer, interview by Else, *Palace of Delights.*

157 *"'They learn something from that'":* Bainum, Kurt, in discussion with the author.
of Big Bang fame: Grinell, Sheila, in discussion with the author.
Dedicated to Awareness: Ibid.
"plant torture exhibit": Carlson, Charles, in discussion with the author.
we use to extend our senses: Frank Oppenheimer, "The Exploratorium: A Playful Museum Combines Perception and Art in Science Education," paper presented at the joint APS-AAPT Symposium on New Directions in Science Education, San Francisco, February 21, 1972.
"humanistic atmosphere": Oppenheimer, "The Exploratorium: A Playful Museum."
no right or wrong answers: Frank Oppenheimer, "The Study of Perception as a Part of Teaching Physics," *American Journal of Physics* 42 (July 1974): 531–37.
"answer sensible questions": Ibid.

158 *Alice's magic mushrooms:* Oppenheimer, interview by Else, *Palace of Delights.*
"as long as they did": Carlson, Charles, in discussion with the author.

interpretation to accept?: Frank Oppenheimer, "The Content of the Museum," 1967.

"political and personal ones": K. C. Cole, "Vision: In the Eye of the Beholder," *Exploratorium,* 1978.

159 *to inhumanity or ugliness?:* Frank Oppenheimer, "Getting Used to the World," undated.

"the behavior of nature": Oppenheimer, "Where Are We and Where Do We Go from Here – I."

and its converse, logarithms: Oppenheimer, interview by Else, *Palace of Delights.*

160 *"perception of meaning":* Oppenheimer, "Where Are We and Where Do We Go from Here – II."

"something you didn't expect": Kay, Alan, in discussion with the author.

"experience everyday phenomena": Oppenheimer, "The Exploratorium: A Playful Museum."

"requires further investigation": Frank Oppenheimer, "The Importance of the Role of Science Pedagogy in the Developing Nations" (Budapest, 1965).

"to these surprises": Oppenheimer, interview by Else, *Palace of Delights.*

explored by anyone: Oppenheimer, "The Importance of the Role of Science Pedagogy."

161 *"it doesn't work well":* Oppenheimer, interview by Else, *Palace of Delights.*

the night before: Humphrey, Thomas, in discussion with the author.

they stayed for hours: Bainum, Kurt, in discussion with the author.

"and make it go": Tom Tompkins, Frank Oppenheimer Memorial.

could readily see: Ansel, Joe, in discussion with the author.

162 *exhibit he liked best:* Frank Oppenheimer, "History of Exhibits," transcription of tape recording.

"blacker than black": Ibid.

and so it did: Ibid.

multitude of ways: Robert Semper in Exploratorium film clip.

"we didn't know it": Ansel, Joe, in discussion with the author.

163 *"talking down to them":* Barker, David, in discussion with the author.

"American education for you": Oppenheimer, Michael, in discussion with the author.

"somebody else clever?": Oppenheimer, "Curiosity."

164 *even the power supply:* Humphrey, Thomas, in discussion with the author.

"it's gotta look cheap": Friedman, Alan, in discussion with the author, July 2002.

"the real phenomena out there": Ibid.

eventually be like: Carlson, Charles, in discussion with the author.

"deliberately fabricate data": Oppenheimer, "Practical and Sentimental Fruits of Science," Fifteenth Anniversary Award Dinner speech, Exploratorium, January 25, 1985.

165 *"people and democracy":* Frank Oppenheimer, AAUW speech.

"are connected underneath": Muriel Rukeyser, "Islands," *Exploratorium* 6, no. 1 (April/May 1982).

"and brothers and sisters": Oppenheimer interview by Else, *Palace of Delights.*

sure they existed: Frank Oppenheimer, "The Wave Equation," *Exploratorium Magazine on Mathematics* 5, no. 4 (December 1981/January 1982).

166 *bouncing balls:* Frank Oppenheimer, "Exhibit Design and Conception," paper presented to the joint Association of Science-Technology Centers/CIMUSET Conference, Monterrey, Mexico, October 24–25, 1980.

167 *"natural phenomenon as well":* Frank Oppenheimer in Exploratorium film clip.
"they are connected underneath": Rukeyser, "Islands."
a theme in a symphony: Oppenheimer, "Everyone Is You . . . or Me."
"the coherence of his composition": Oppenheimer, "Exhibit Design and Conception."
"called the optic nerve": Videotapes of various Explainer functions, provided by Darlene Librero.
"You mean tomorrow?": Carlson, Charles, in discussion with the author.
168 *trailer as an assistant:* Richards, Peter, in discussion with the author.
"Sure, why not?": Humphrey, Thomas, in discussion with the author.
"two acres of asphalt": Ibid.
"He really wanted to know!": Meyer, Rachel, in discussion with the author.
who had ideas: Templeton, Michael, in discussion with the author.
how to communicate it: Carlson, Charles, in discussion with the author.
on many different levels: Humphrey, Thomas, in discussion with the author.
"people who had ideas": Templeton, Michael, in discussion with the author.
169 *"as opposed to mathematical":* Humphrey, Thomas, in discussion with the author.
his proudest accomplishment: Ibid.
explain things to a jury: Videotape of thirtieth Explainer reunion, provided by Darlene Librero.
170 *could figure things out:* Ibid.
"to the Exploratorium": Videotape of twenty-fifth Explainer reunion, provided by Darlene Librero.
experience for visitors: Librero, Darlene, in discussion with the author.
the collective right answer: Oppenheimer, interview by Else, *Palace of Delights.*
"whole feeling of learning: Librero, Darlene, in discussion with the author.
how science really works: Oppenheimer, interview by Else, *Palace of Delights.*
sure he'd meet them: Kutnick, Ester, in discussion with the author.
"he sort of poked at": Exploratorium staff, Frank Oppenheimer Memorial.
171 *out of sight:* Kutnick, Ester, in discussion with the author.
took him for walks: Videotape of Explainer thirtieth reunion, provided by Darlene Librero.
excellent floor staff: Kutnick, Ester, in discussion with the author.
with their homework: Librero, Darlene, in discussion with the author.
"See what happens": Videotapes of various Explainer functions, provided by Darlene Librero.
172 *slimy bits around the table:* Ibid.
a teenager's life: Kidder, Rushworth M. "Museum Guides Turn On to Science," *Christian Science Monitor.*
now have analogous programs: Librero, Darlene, in discussion with the author.
"'there's a deer'": Frank Oppenheimer, "Persuasion," talk in honor of Muriel Rukeyser, Sarah Lawrence College, December 9, 1978.
to point it out to: Oppenheimer, interview by Else, *Palace of Delights.*
to talk to each other: Meyer, Rachel, in discussion with the author.
"one's hands in despair": Frank Oppenheimer, "Let the Teachers Teach and the Learners Learn," undated.
what they were studying: Frank Oppenheimer, Presentation on being awarded the American Association of Museums Distinguished Service Award, June 21, 1982.

173 *"subjected to education":* Oppenheimer, "The Role of Science Museums," in Larrabee, ed., *Museums and Education.*

some kind of product: Oppenheimer, "Let the Teachers Teach and the Learners Learn."

subverted for that purpose: Frank Oppenheimer, "The Exploratorium and Other Ways of Teaching Physics," paper presented at joint meeting of American Association of Physics Teachers and American Physical Society, Anaheim, California, January 31, 1975.

measured in schools: Oppenheimer, "Let the Teachers Teach and the Learners Learn."

"a grade point average": Ibid.

out of adults' way: Ibid.

more diverse than ever: Ibid.

"the schools into factories": Oppenheimer, "The Exploratorium and Other Ways of Teaching Physics."

174 *education from schools:* Oppenheimer, "Let the Teachers Teach and the Learners Learn."

"adjunctive organizations": National Committee on Excellence in Education, Testimony at Hearing on Science, Mathematics and Technology, Statement of Frank Oppenheimer, April 1983.

were much in demand: House of Representatives, Subcommittee on Elementary, Secondary, and Vocational Education, Testimony by Frank Oppenheimer and Ilan Chabay, September 30, 1982.

well, that was fine: Frank Oppenheimer, "A Photo Essay by Nancy Rodger with Notes from Visiting Students," *Exploratorium* 4, no. 4 (October/November 1980).

175 *"yes, they are":* Perlman, David, in discussion with the author.

"you just didn't care": Tressel, George, in discussion with the author.

remarkably accurate: Hein, *The Exploratorium.*

why it was worth supporting: Oppenheimer, interview by P. K. Kabir, Bancroft Collection, University of California, Berkeley.

it was hard going: Hein, *The Exploratorium.*

"so it wasn't a school": Frank Oppenheimer to Arlon Elser, program director, W. K. Kellogg Foundation, August 27, 1984.

176 *"incredibly badly treated":* Gregory, Richard, in discussion with the author.

"turning down a proposal": Tressel, George, in discussion with the author.

than he was making: Rubin, Virginia, in discussion with the author, summer 1999.

couldn't turn them down: Richards, Peter, in discussion with the author.

"having guards and docents": Rubin, Virginia, in discussion with the author.

177 *"walking into the office":* Ibid.

but never quite stable: Oppenheimer, interview by P. K. Kabir, Bancroft Collection, University of California, Berkeley.

"respect for Frank": "Séance: Interviews with Exploratorium Staff," summer 1999.

even for short visits: "The Exploratorium: A Synopsis."

"cost-effective form of learning": Frank Oppenheimer to Lou Branscomb, chairman, National Science Board, IBM, December 16, 1981.

finally installed: Oppenheimer and Cole, "The Exploratorium."

visitors a month: Frank Oppenheimer, "Teaching and Learning," Millikan Lecture, American Association of Physics Teachers, June 1973.

acquired 400 exhibits: Oppenheimer, "Where Are We and Where Do We Go from Here — I."
college Explainers: "The Exploratorium: Brief Description," May 11, 1976.
a great deal to him: Ford, Introduction of Frank Oppenheimer, Millikan Lecture.

8. A Decent Respect for Taste

179 *fine as the Sun Painting:* Oppenheimer, "Everyone Is You . . . or Me."
180 *broad concept of nature:* Frank Oppenheimer, "On Understanding."
"It's all around you": Pusina, Jan, in discussion with the author.
scientists make aesthetic ones: Oppenheimer and Cole, "The Exploratorium."
"anything goes": Hein, *The Exploratorium.*
181 *"like real life":* Muriel Rukeyser, "The Sun Painter."
such fantastical effects: K. C. Cole, "New Exploratorium Catalogue. Facets of Light: Colors, Images and Things that Glow in the Dark," *Exploratorium* 4, no. 2 (June/July 1980).
Corcoran Gallery: Oppenheimer, Michael, in discussion with the author.
so long he got hoarse: Oppenheimer, Frank, in discussion with the author.
then helped set it up: Presentation on being awarded the AAM Distinguished Service Award.
a major role at the Exploratorium: Richards, Peter, in discussion with the author.
182 *end of a mystery story:* Oppenheimer, interview by Else, *Palace of Delights.*
"artificial harmonics": Pusina, Jan, in discussion with the author.
brushed-metal plates: History of Exhibits, transcribed tape.
an overhead projector: Ibid.
in the Oakland Museum: Frank Oppenheimer, "The Common Bonds Between Art and Science," May 1977.
183 *sometimes wild experiments:* Richards, Peter, in discussion with the author.
remained but shattered glass: Oppenheimer, Michael, in discussion with the author.
"That just kept up": Frank Oppenheimer, interview by Milton Savage, *California Confederation of the Arts Newsletter* 1, no. 1 (Fall 1985).
"happening in the Exploratorium": Oppenheimer, interview by Else, *Palace of Delights.*
184 *"it would infuriate him":* Keim, Liz, in discussion with the author.
"along with the rest of us": Oppenheimer, "Persuasion."
newly discovered phenomena: Oppenheimer, interview by Savage.
"precision of meaning and imagery": Frank Oppenheimer, "The Language of Poetry and Science: Forefront Readings at the Exploratorium."
"starkly descriptive or intensely polemic": Oppenheimer, "Persuasion."
185 *"what it is makes a difference":* Ibid.
"made an aesthetic decision": Frank Oppenheimer, "The Arts: A Decent Respect for Taste," *National Elementary Principal,* October 1977.
186 *"It's not an imitation":* Franklin, Allen, in discussion with the author.
"what it is like to use one": Oppenheimer, quoted in Dackman, "Invisible Aesthetic: A Somewhat Humorous, Slightly Profound Interview with Frank Oppenheimer."
Reveling in it all: K. C. Cole, "The Rewards of a Most Unusual Friendship," *Newsday,* June 2, 1985.

187 *"and so from each other"*: Frank Oppenheimer, "Smells: An Introduction," *Exploratorium* 7, no. 1 (Spring 1983).

188 *"So we use them"*: Ibid.

"Some machinery feels nice": Ibid.

"I'd like to happen to me": Ibid.

paintings they did in school: Frank Oppenheimer, "Growing Up in the Arts," *National Elementary Principal* 56, no. 1 (September/October 1976).

189 *"order-disorder transition"*: Oppenheimer, "The Arts: A Decent Respect for Taste."

"shriek of delight": Ibid.

"they do not feel nice": Ibid.

at tin cans: Ibid.

used to feel and sound: Frank Oppenheimer, "Pleasant Play," *Exploratorium* 3, no. 5 (December 1979/January 1980).

190 *"nicer world in which to live"*: Ibid.

"ingenious gadgets": Gregory, Richard, in discussion with the author.

"make life stay human": Frank Oppenheimer, lecture, Association of School Administrators 110th Annual Convention, Atlanta, February 1978.

191 *"homes for the aged"*: Oppenheimer, "The Arts: A Decent Respect for Taste."

"would not be tolerated": Oppenheimer, "The Unique Educational Role of Museums."

then it probably was: Oppenheimer, "The Arts: A Decent Respect for Taste."

192 *their music:* Frank Oppenheimer, "Art and Science: Meaning, Tools, and Discipline," written for catalogue of the exhibition "The Expanding Visual World: A Museum of Fun."

"respect to other stars": Frank Oppenheimer, lecture, Smithsonian Institution Conference on Museums and Education, University of Vermont, August 1966.

between man and woman: Frank Oppenheimer, "Aesthetics and the Right Answer."

"relate to people well": Oppenheimer, interview by Else, *Palace of Delights.*

shapeless, amorphous, and emotionless?: Oppenheimer, "Growing Up in the Arts."

193 *with equations or charcoal:* Oppenheimer, interview by Savage.

"what they started with": Oppenheimer, lecture, Association of School Administrators.

reveal new insights: Oppenheimer, "Aesthetics and the Right Answer."

creating order out of confusion: Ibid.

know about human feelings: Oppenheimer, talk given at the Association of School Administrators.

194 *the role of artists:* Ibid.

"to make good decisions": Oppenheimer, "Art and Science."

"awareness and invention": Oppenheimer, "Aesthetics and the Right Answer."

"the other is electricity": Oppenheimer, interview by Savage.

195 *with our daily lives:* Frank Oppenheimer and K. C. Cole, "The Exploratorium: A Participatory Museum."

196 *"forefront of the field"*: Ibid.

"meaningless aesthetic experiences": Oppenheimer, "Growing Up in the Arts."

also through meaning: Oppenheimer, "The Common Bonds Between Art and Science."

197 *took over the class:* Humphrey, Thomas, in discussion with the author.

"fun I had with Frank": Semper, Rob, in discussion with the author.
198 *"more like a scientist":* Richards, Peter, in discussion with the author.
"the result of that relationship": Ibid.
200 *all this time with him:* Kahn, Ned, in discussion with the author.
201 *"they're physically unpredictable":* Ibid.
"done by an artist": Oppenheimer, interview by Else, *Palace of Delights.*
"no right or wrong in art": Oppenheimer, Frank, in discussion with the author.
202 *"provide any answers at all":* Ibid.
and not ruin it: Oppenheimer, interview by Savage.
had yet perceived: Oppenheimer, "Aesthetics and the Right Answer."
shadows of buildings or hills: Oppenheimer, Frank, in discussion with the author.
203 *"sensations to the brain":* Frank Oppenheimer, "Color," *Exploratorium* 1, no. 2 (June/July 1977).
"of the museum world": Oppenheimer, Presentation on being awarded the AAM Distinguished Service Award.

9. The Man with the Gold-Rimmed Glasses

204 *"like a nice bath?":* K. C. Cole to Frank Oppenheimer, undated.
205 *"so much stealth drive":* Kennedy, Donald, in discussion with the author.
"had an ambulance take her": Ansel, Joe, in discussion with the author.
206 *"that stupid retriever":* K. C. Cole to Frank Oppenheimer, June 12, 1978.
207 *and the dog's body:* Else, Jon, in discussion with the author.
208 *"in his trigonometry class":* Miller, Bob, in discussion with the author.
209 *"licking his manhood":* Ansel, Joe, in discussion with the author.
210 *leave the car running:* Carlson, Charles, in discussion with the author.
by his discovery: Rubin, Virginia, in discussion with the author.
"the wind was blowing": Sam Maddox, "The Other Oppenheimer: A Life in the Shadows," *Sunday Camera,* March 24, 1985.
211 *"watch it all happen":* Pearce, Michael, in discussion with the author.
"private world, is understandable": Oppenheimer, "On Understanding."
"artificially celibate": Oppenheimer, interview by Else, *Palace of Delights.*
the salt they need: Oppenheimer, "Curiosity."
212 *"'Curiosity killed the cat'":* Ibid.
"to dampen curiosity": Oppenheimer, "The Role of Science Museums."
encouraged such play: Oppenheimer, "Curiosity."
"keep our curiosity alive": Ibid.
"why you do it, normally": Oppenheimer, interview by Else, *Palace of Delights.*
213 *fear of being a nuisance:* Oppenheimer, "Curiosity."
214 *"to an interrelated family":* Oppenheimer, "On Understanding."
"to do very badly": Oppenheimer, "Museums and Toys."
"I have to take a leak": Oppenheimer, Frank, in discussion with the author.
"a very basic instinct": Frank Oppenheimer, talk given to Advisory Committee, Office of Science and Society, National Science Foundation, March 6, 1980.
play it over and over: Oppenheimer, "Teaching and Learning."
215 *"people dish out to us":* A scrap of paper of unknown origin.
"by alien events and forces": Frank Oppenheimer, "Practical and Sentimental Fruits of Science," Fifteenth Anniversary Award Dinner speech.

"they can make a difference": Ibid.

216 *"I don't believe that"*: Oppenheimer, interview by Weiner, Bohr Library.

"discover as a grownup": Else, Jon, in discussion with the author.

217 *"what they are doing"*: Oppenheimer, "Curiosity."

"'wiggle your ears'": Oppenheimer, "Teaching and Learning."

"I love this stuff": K. C. Cole, "The Rewards of a Most Unusual Friendship," *Newsday,* June 2, 1985.

"'That's really interesting'": Carlson, Charles, in discussion with the author.

a week or so before: Ibid.

218 *by little moth clouds:* Ibid.

"stopped on these projects": Ibid.

219 *"not do it himself"*: Oppenheimer, "Teaching and Learning."

"contagious and addictive": Oppenheimer, talk given to Advisory Committee, Office of Science and Society, National Science Foundation.

"useful and sympathetic person": Frank Oppenheimer, speech, Pagosa Springs PTA, 1957.

"when I was a child": Frank Oppenheimer, video clip, http://www.exploratorium.edu.

"I get really gloomy": Ibid.

220 *tweak someone's interest:* Wheaton, Bruce, in discussion with the author.

"were working just fine": Librero, Darlene, in discussion with the author.

"how fast it is going": Frank Oppenheimer, unidentified snippet of writing.

221 *understand and enjoy poetry:* Ibid.

"a triumph": Oppenheimer, Presentation on being awarded the AAM Distinguished Service Award.

"that made sense to me": Kutnick, Ester, in discussion with the author.

222 *"of other things as well"*: Oppenheimer, "Exhibit Design and Conception."

"Best wishes, Frank Oppenheimer": Frank Oppenheimer to Fred Duncan, May 7, 1984.

"seeing a particle accelerator": K. C. Cole to Frank Oppenheimer, undated.

"and notice things": Oppenheimer, interview by Else, *Day After Trinity.*

223 *"mechanisms can't detect"*: K. C. Cole to Frank Oppenheimer, undated.

"people take for science fiction": Ibid.

"without ever hinting at why": Ibid.

how could this be?: Ibid.

224 *"something he noticed"*: Oppenheimer, interview by Savage.

225 *from Earth at midnight:* Oppenheimer, Frank, in discussion with the author.

226 *"erupting sun spots"*: Frank Oppenheimer, "Introduction: Diaphanous," *Exploratorium* 8, no. 3 (Fall 1984).

228 *"Walking Beats"*: Flier on Founders' Day.

229 *"as well as piano strings"*: Oppenheimer, interview by Else, *Palace of Delights.*

"commercial and political life": Oppenheimer, "Everyone Is You . . . or Me."

230 *"with creating images"*: K. C. Cole, *Sympathetic Vibrations: Reflections on Physics as a Way of Life* (New York: William Morrow, 1985).

"narrow-minded scientists": Oppenheimer, interview by Savage.

"talent we can find": Oppenheimer, "Where Are We and Where Do We Go from Here — II."

no longer like to read: Oppenheimer, "Everyone Is You . . . or Me."

commercially profitable: Ibid.

231 *"in charge of graphics"*: Oppenheimer, "Where Are We and Where Do We Go from Here — II."

"divided by editor": Murphy, Pat, in discussion with the author.

232 *"to find that honesty":* Humphrey, Thomas, in discussion with the author.

"approaches the disingenuous": Philip Morrison and Kosta Tsipis, "Rightful names; and another thing . . . ," *Bulletin of the Atomic Scientists* 59, no. 3 (May 1, 2003): 77.

10. The Sentimental Fruits of Science

234 *"so does the Exploratorium":* Oppenheimer, interview by Else, *Day After Trinity.*

236 *universe and each other:* Frank Oppenheimer, "The Sentimental Fruits of Science," speech given on receiving the Oersted Award from the American Association of Physics Teachers, January 31, 1984.

"relationship to nature": Oppenheimer, Presentation on being awarded the AAM Distinguished Service Award.

as a vocational subject: Oppenheimer, "The Sentimental Fruits of Science."

"have a decent society": Oppenheimer, "On Understanding."

237 *"mutual attractions and repulsions":* Oppenheimer, foreword in Cole, *Sympathetic Vibrations.*

thinking into human realms: Oppenheimer, "The Pleasures and Personal Satisfactions of Science."

"part of culture and philosophy": Oppenheimer, foreword in Cole, *Sympathetic Vibrations.*

"more aware of their humanity": Frank Oppenheimer, "The Importance of the Role of Science Pedagogy in the Developing Nations," 1965.

238 *"human welfare are inseparable":* Max Otto, *Science and the Moral Life* (New York: New American Library, 1949).

"development of atomic energy": J. Robert Oppenheimer, *Atom and Void.*

"or its present borderlands": J. Robert Oppenheimer, *Science and the Common Understanding.*

239 *"promise for human values":* John Herman Randall, *Newton's Philosophy of Nature: Selections from His Writings* (Hafner Library of Classics, 1953).

"Bacon's dream come true": Richard Gregory, *Odd Perceptions* (New York: Routledge, 1989).

"interests of life are doomed": Max Otto, *Science and the Moral Life.*

240 *"profoundly conceived":* Ibid.

"from becoming probable": Oppenheimer, "Persuasion."

"to the will of God": Oppenheimer, "The Sentimental Fruits of Science."

he never forgot: Oppenheimer, Frank, in discussion with the author.

241 *dance to human sacrifice:* Frank Oppenheimer, "Science and Fear — A Discussion of Some Fruits of Scientific Understanding," *Centennial Review* 5, no. 4 (Fall 1961).

tall trees during thunderstorms: Ibid.

"of each other or of nature": Oppenheimer, "On Understanding."

behavior and economic disaster: Frank Oppenheimer and K. C. Cole, "Fear," *Exploratorium,* Special Issue: Frank Oppenheimer, 1912–1985 (March 1985).

242 *know what to do:* Oppenheimer, "Science and Fear."

response to the situation: Ibid.

"the quantity of this terror": Frank Oppenheimer, talk given to Advisory Committee, Office of Science and Society, National Science Foundation, March 6, 1980.
"which parts are effective": Oppenheimer, "Science and Fear."
on dark city streets: Ibid.

243 *"futile agony of a World War"*: Oppenheimer and Cole, "Fear."
"the possibility of progress open": Frank Oppenheimer, K. C. Cole's notes for "The Rewards of a Most Unusual Friendship," *Newsday,* June 2, 1985.
"disconnected to us": Oppenheimer, "The Sentimental Fruits of Science."

244 *"different from ourselves"*: Ibid.
more easily bridged: Frank Oppenheimer, "Museums: A Versatile Resource for Learning and Pleasure," unpublished talk, undated.
"separate kind of being": Oppenheimer, "The Sentimental Fruits of Science."
"makes science so wonderful": Ibid.

245 *"strange and frightening thing"*: Frank Oppenheimer, K. C. Cole's notes for "The Rewards of a Most Unusual Friendship."

246 *"present faults and limitations"*: Ibid.
"living at the moment": Oppenheimer, "The Pleasures and Personal Satisfactions of Science."
every single second: Philip Morrison, *The Ring of Truth.*

247 *religion or something else:* Oppenheimer, interview by Else, *Palace of Delights.*
"we should behave": Oppenheimer, Frank, in discussion with the author.

248 *"some complicated chemistry"*: Oppenheimer, "The Sentimental Fruits of Science."
"present and the past": Ibid.
"enjoy listening to music": Frank Oppenheimer, "Science and the Ethics of Coercion," unpublished essay.

249 *"often appears to be"*: Oppenheimer, "The Sentimental Fruits of Science."
"or very little": Oppenheimer, Frank, in discussion with the author.

250 *radio waves:* Frank Oppenheimer, "Science and Invention," date unclear, 1963 or 1968.
"not being oracles": Oppenheimer, "The Sentimental Fruits of Science."
"beavers build dams": Oppenheimer, "Science and Fear."

251 *"rather than repress it"*: Oppenheimer, "The Importance of the Role of Science Pedagogy in the Developing Nations."
"a pickle or a Picasso": Oppenheimer, "Science and the Ethics of Coercion."
"lump it all together": Oppenheimer, Frank, in discussion with the author.

252 *milling machines, and lasers:* Oppenheimer, "The Sentimental Fruits of Science."
"growing sociological understanding": Oppenheimer, "Science and Fear."
"to improve their lot": Oppenheimer, interview by Else, *Palace of Delights.*
the basis for invention: Oppenheimer, Presentation on being awarded the AAM Distinguished Service Award.

253 *"think somewhat outrageously"*: Oppenheimer, Frank, in discussion with the author.
all social inventions: Frank Oppenheimer, "Science and Immunity," *Scientific World,* 1969.
unacceptable to either party: Oppenheimer, Frank, in discussion with the author.

254 *"renaissance of creativity"*: Oppenheimer, "Persuasion."
to live on dry land: Oppenheimer, "Science and Immunity."
"in the academic world": Frank Oppenheimer, "The Character of a University," unpublished essay, 1964.

day of reflection: Oppenheimer, "Science and Immunity."
value of their work: Frank Oppenheimer, "Testimony with Regard to House Joint Resolution 639 Calling for White House Conference on the Humanities," January 4, 1978.

255 *most areas of human life:* Oppenheimer, "Science and Immunity."
"made science flourish": Oppenheimer, "The Sentimental Fruits of Science."
"undreamed-of problems": Oppenheimer, "The Character of a University."
autonomous and free: Ibid.

256 *"better relationship with people":* Oppenheimer, K. C. Cole's notes for "The Rewards of a Most Unusual Friendship."
Coercion, in fact: Oppenheimer, "Persuasion."
"cross it in a ship": Oppenheimer, "The Sentimental Fruits of Science."
on and off at will: Oppenheimer, "Science and the Ethics of Coercion."

257 *"made up our own mind":* Ibid.
"auto industry is in trouble": Oppenheimer, K. C. Cole's notes for "The Rewards of a Most Unusual Friendship."
"through art and science": Oppenheimer, "Persuasion."

258 *"make life miserable for you":* Ibid.
"from one another": Oppenheimer, "Science and the Ethics of Coercion."
"incredible delivery systems": Oppenheimer, "Persuasion."
"decide how to act": Oppenheimer, "Science and the Ethics of Coercion."

259 *"lied to all the time":* Oppenheimer, "Public Understanding of Science."
"their best interest": Semper, Robert, in discussion with the author.

260 *"much to be proud of":* Frank Oppenheimer to Norman Cousins, July 16, 1960.
"to act like one": Ibid.
mirror up to ourselves: Humphrey, Thomas, in discussion with the author.

11. The Anarch

261 *"we were getting famous":* Rubin, Virginia, in discussion with the author.
old spark had left him: Oppenheimer, Michael, in discussion with the author.

262 *disloyalty to Frank:* Hawkins, Francis, in discussion with the author.
"continuum": Oppenheimer, Michael, in discussion with the author.
thirty thousand square feet: Oppenheimer, "Where Are We and Where Do We Go from Here – I."
in May 1980: Hein, *The Exploratorium.*
"else in the museum": Oppenheimer, "Where Are We and Where Do We Go from Here – II."

263 *"exists anywhere else":* Victor Weisskopf, letter written on behalf of the Exploratorium, spring 1981.
museums around the country: Oppenheimer, "Where Are We and Where Do We Go from Here – II."
was going strong: Oppenheimer and Chabay, "Testimony Presented Before the U.S. House of Representatives Subcommittee on Elementary, Secondary, and Vocational Education."
Distinguished Service Award: Presented by Arnold R. Weber, president, H. H. Arnold, secretary of the university, and the Board of Regents, May 23, 1980.
Distinguished Alumni Award: Arlene Silk, "Dr. Frank Oppenheimer Receives Caltech Distinguished Alumni Award," May 17, 1979.

Distinguished Service: Frank Oppenheimer, recipient of the AAM Award for Distinguished Service, by Kenneth Starr, AAM Annual Meeting, Philadelphia, June 21, 1982.

Physics Teachers: Frank Oppenheimer, Millikan Lecture, American Association of Physics Teachers, June 1973.

"we were getting famous": Rubin, Virginia, in discussion with the author.

and make a fuss: Tompkins, Tom, in discussion with the author.

264 *"leader of a commune":* Bainum, Kurt, in discussion with the author.

very gentle "That way": Shaw, Larry, in discussion with the author.

"when you were around Frank": Exploratorium staff clips, Frank Oppenheimer Memorial.

265 *Everyone took their shot:* Meyer, Rachel, in discussion with the author.

he was in a good mood: Exploratorium staff clips, Frank Oppenheimer Memorial.

he was coming back: Tompkins, Tom, in discussion with the author.

a continual theme: Wheaton, Bruce, in discussion with the author.

they actually did: Exploratorium staff clips, Frank Oppenheimer Memorial.

"What's this about wax?": Brown, Ruth, in discussion with the author.

266 *"it doesn't work in art":* Frank Oppenheimer, quoted in memo to Exploratorium staff summarizing his papers for staff retreat, January 6, 1987.

entirely flat: Hipschman, Ron, in discussion with the author.

"the fish tanks": Tompkins, Tom, in discussion with the author.

the time of day: Kutnick, Ester, in discussion with the author.

to do the same: Joseph G. Ansel Jr., "The Unionization of the Exploratorium," in *Institutional Trauma: Major Change in Museums and Its Effect on Staff,* edited by Elaine Heumann Gurian (Washington, D.C.: American Association of Museums, 1995).

to start sweeping: Carlson, Charles, in discussion with the author.

or curriculum development: "The Exploratorium: A Synopsis."

267 *or written material:* Ansel, Joe, in discussion with the author.

"hand and the mind": Oppenheimer, "Where Are We and Where Do We Go from Here — I."

didn't have to be: Shaw, Larry, in discussion with the author.

was happy to help: Ansel, Joe, in discussion with the author.

"an academic kibbutz": Feher, Elsa, in discussion with the author.

and then do it: Semper, Robert, in discussion with the author.

268 *really their own:* Ansel, Joe, in discussion with the author.

exactly what Frank wanted: Ibid.

who built it: Carlson, Charles, in discussion with the author.

"'severely impact?'": Hipschman, Ron, in discussion with the author.

"He was respectful": Hutchinson, Brenda, in discussion with the author.

reveled in it: Carlson, Charles, in discussion with the author.

"That's science": Exploratorium, directed by Jon Boorstin.

269 *were their own:* Bainum, Kurt, in discussion with the author.

"catch with the ball": Oppenheimer, "Everyone Is You . . . or Me."

"with other exhibits": Oppenheimer, K. C. Cole's notes for "The Rewards of a Most Unusual Friendship."

270 *"told to you":* Oppenheimer, on being awarded the AAM Distinguished Service Award.

"have to destroy them": Oppenheimer, interview by P. K. Kabir, Bancroft Collection, University of California, Berkeley.

271 *"in science classes"*: Oppenheimer, "Learning with an Opportunity for Choice," presented at AAAS Annual Meeting, 1980.

"intensely preoccupied": Oppenheimer, on being awarded the AAM Distinguished Service Award.

"my mountain walks": Oppenheimer, "Everyone Is You . . . or Me."

272 *"was just their own"*: *Exploratorium*, directed by Jon Boorstin.

"could understand and accept": Perlman, David, in discussion with the author.

"the whole of it": Robert Wilson, letter written on behalf of the Exploratorium, spring 1981.

"experiences into ideas": Oppenheimer, interview by P. K. Kabir, Bancroft Collection, University of California, Berkeley.

"outcomes of play": Oppenheimer, "The Unique Educational Role of Museums."

273 *neurophysiology*: Frank Oppenheimer, "A Broader View," *Exploratorium*, 1977.

"and it was safe": Keim, Liz, in discussion with the author.

"attract teenagers": Oppenheimer, "Exhibit Design and Conception."

274 *"the existence of intuition"*: Judith Wechsler, ed., *On Aesthetics in Science* (Cambridge, Mass.: MIT Press, 1978).

"willing to pay?": Oppenheimer, "The Unique Educational Role of Museums."

"making them important": Ibid.

275 *"back for more"*: Oppenheimer, "Everyone Is You . . . or Me."

276 *"probably less so"*: Ansel, "The Unionization of the Exploratorium."

"different meaning": Murphy, Pat, in discussion with the author.

his parents approved of: Oppenheimer, Michael, in discussion with the author.

the low accident rate: Meyer, Rachel, in discussion with the author.

277 *baseball game*: Else, Jon, Frank Oppenheimer Memorial.

"the seven-year-old!": Marks, Gerald, in discussion with the author.

278 *"incomprehensible"*: Frank Oppenheimer, "The Freedom to Create," September 1982.

"forgiveness than permission": Hipschman, Ron, in discussion with the author.

"over and over again": Richards, Peter, in discussion with the author.

"start working on it": Oppenheimer, Frank, interview by P. K. Kabir, Bancroft Collection, University of California, Berkeley.

279 *"government by experts"*: Frank Oppenheimer, miscellaneous pages on political convictions.

would always be heard: Oppenheimer, Michael, in discussion with the author.

"to prevent from happening": Oppenheimer, Frank, in discussion with the author.

"It's because of Frank": Tompkins, Tom, in discussion with the author.

made on moral grounds: Murphy, Pat, in discussion with the author.

"were beautiful decisions": Dackman, "Invisible Aesthetic: A Somewhat Humorous, Slightly Profound Interview with Frank Oppenheimer."

280 *"beset most of us"*: Friedman, Alan, in discussion with the author, July 2002.

"are measures of success": Ibid.

"we'll get by": Ibid.

his hope: Perlman, David, in discussion with the author.

attracted people: Barker, Dave, in discussion with the author.

frequent arguments: Ansel, Joe, in discussion with the author.

"wasn't about him": Freichtmeir, Kurt, in discussion with the author.

281 *"science and technology"*: Frank Oppenheimer, "Proposal for a Palace of Art and Science," 1967.
into a soap opera: Rubin, Virginia, in discussion with the author.
"in history and society": Oppenheimer to Peter R. E. Goodchild.
of the atom bomb: Kahn, Ned, in discussion with the author.
of so many others: Kutnick, Ester, in discussion with the author.
"to hide from that": Murphy, Pat, in discussion with the author.

282 *"was like that"*: Perlman, David, in discussion with the author.
more than others: Hipschman, Ron, in discussion with the author.
"to talk with you": Rodger, Nancy, in discussion with the author.
"all the little chicks": Ibid.

283 *"supposed to be doing"*: Hewitt, Paul, in discussion with the author.
United States and Mexico: Jorge Lopez, e-mail message to author, January 10, 2005.
at the same time: Humphrey, Thomas, in discussion with the author.
started playing: Pearce, Michael, in discussion with the author.

284 *"There was unrest"*: Librero, Darlene, in discussion with the author.
"to talk softly": Oppenheimer, quoted in Dackman, "Invisible Aesthetic."
"thoroughly enjoy it": Keim, Liz, in discussion with the author.

285 *"apologize to me"*: Ibid.
orders to be followed: Murphy, Pat, in discussion with the author.
"completely baffled": Frank Oppenheimer to Mr. Binder, editor of the *Minneapolis Morning Tribune.*
"in during those years": Kutnick, Ester, Frank Oppenheimer Memorial.

286 *"commune or something"*: Brown, Ruth, in discussion with the author.
"the rest of those guys": Templeton, Michael, in discussion with the author.
"not a business model": Chabay, Ilan, in discussion with the author.
executive council: Ansel, "The Unionization of the Exploratorium."
turned him down: Semper, Robert, in discussion with the author.
"should expire": Oppenheimer, "Where Are We and Where Do We Go from Here – II."

287 *"Complexity is not"*: Alice Kimball Smith and Charles Weiner, *Robert Oppenheimer: Letters and Recollections* (Cambridge, Mass.: Harvard University Press, 1980).
"loved him for that": Templeton, Michael, in discussion with the author.
"Go ahead and hire him": Tompkins, Tom, in discussion with the author.

288 *"because we're so funny"*: Oppenheimer, *Palace of Delights.*
"entire Education Directorate": Rodney T. Ogawa, Molly Loomis, and Rhiannon Crain, "Social Construction of an Interactive Science Center: The Institutional History of the Exploratorium," presented at the biennial meeting of the European Science Educational Research Association, August 24, 2007, Malmö, Sweden.
was drying up: Frank Oppenheimer, Exploratorium Awards Dinner talk, November 7, 1981.
minimally successful: Ansel, Joe, in discussion with the author.
"support will come back": Statement of Frank Oppenheimer, Testimony at Hearing on Science, Mathematics, and Technology, National Committee on Excellence in Education, April 1983.

289 *"we estimate it will"*: Frank Oppenheimer, "New Admissions Policy," *Exploratorium* 5, no. 1 (April/May 1981).

"very intense": Panofsky, Wolfgang, in discussion with the author.
290 *"what Frank was doing":* Kasper, F. Van, in discussion with the author.
"for money spent": Ibid.
"anyone I ever met": Ibid.
surplus of cash: Frank Oppenheimer to F. Van Kasper, April 6, 1983.
291 *"to be hired either":* Kasper, F. Van, in discussion with the author.
"It's very not true": Bennis, Warren, in discussion with the author.
292 *Orestes was dead:* Oppenheimer, Michael, in discussion with the author.
true "horror": Ibid.

12. The World He Made Up

293 *"All our love, K.C.":* K. C. Cole to Frank Oppenheimer, undated, Bancroft Collection, University of California, Berkeley.
294 *in fact, return:* Oppenheimer, Judy, in discussion with the author, October 1, 2005.
cancer was incurable: Ibid.
the last possible moment: Robert Miller, "What I Did at the Exploratorium and Why I Left."
was the main honoree: Hein, *The Exploratorium.*
295 *begun to notice changes:* Librero, Darlene, in discussion with the author.
losing his strength: Kahn, Ned, in discussion with the author.
difficult for him: Freichtmeir, Kurt, in discussion with the author.
draining away: Barker, Dave, in discussion with the author.
"the same old arguments": Librero, Darlene, in discussion with the author.
a curmudgeon: Barker, Dave, in discussion with the author.
more and more afraid: Kahn, Ned, in discussion with the author.
and frequently tearful: Oppenheimer, Michael, in discussion with the author.
the way of specifics: Bainum, Kurt, in discussion with the author.
and society could be: Hutchinson, Brenda, in discussion with the author.
"went home to die": Keim, Liz, in discussion with the author.
296 *heavily on his cane:* Videotape of Frank Oppenheimer demonstrating exhibits, provided by the Exploratorium.
then he giggles: Ibid.
"what he loved doing": Kahn, Ned, in discussion with the author.
how radioactive he was: Hipschman, Ron, in discussion with the author.
doing the same thing: Tressel, George, in discussion with the author.
297 *nuclear war:* Exploratorium staff clips, Frank Oppenheimer Memorial.
"so hard against dying": Unidentified staff member, "Séance: Interviews with Exploratorium Staff."
in his wheelchair: Oppenheimer, Michael, in discussion with the author.
"he was near death": Librero, Darlene, in discussion with the author.
coming in altogether: Freichtmeir, Kurt, in discussion with the author.
"Frank meant to me": Robert Wilson, Frank Oppenheimer Memorial.
"really there anymore": Dackman, Linda, in discussion with the author.
"in and out of consciousness": Kahn, Ned, in discussion with the author.
they'd had together: Rubin, Virginia, in discussion with the author.
was a number: Humphrey, Thomas, in discussion with the author.

298 *"a life like that":* Kahn, Ned, in discussion with the author.
her love and anguish: Richards, Peter, in discussion with the author.
"as much art as science": David Perlman, "Scientist Frank Oppenheimer dies at 72," *San Francisco Chronicle,* February 4, 1985.

299 *"of teaching relativity":* Invitation for Frank Oppenheimer Memorial, February 26, 1985.
regular customer: Freichtmeir, Kurt, in discussion with the author.
museum didn't charge admission: Invitation for Frank Oppenheimer Memorial.
"a better person": F. Van Kasper, Frank Oppenheimer Memorial.
"not entirely superficial": Edward Lofgren, Frank Oppenheimer Memorial.

300 *"a series of wonders":* Ruth Newhall, Frank Oppenheimer Memorial.
become a cliché: George Tressel, Frank Oppenheimer Memorial.
"Frank meant to me": Robert Wilson, Frank Oppenheimer Memorial.
"grateful for having lived": Invitation for Frank Oppenheimer Memorial.

301 *particle accelerator:* K. C. Cole, "Things Your Teacher Never Told You About Science," *Newsday,* March 23, 1986.
"people light up": Explainer reunion video, Exploratorium Archives.

302 *museums worldwide:* Linda Dackman, "Thomas C. Rockwell Named New Director of Exhibitions and Programs," March 1, 2005.
from Tiananmen Square: Susan Ciccotti, "The Exploratorium's China Trade," *Museum News,* March/April 2001.
Saudi Arabia: "Exploratorium Exhibit Network: International Reach Map," November 2003.
"history of the world": Friedman, Alan, in discussion with the author.
"and technology": Arlene Silk, "Dr. Frank Oppenheimer Receives Caltech Distinguished Alumni Award," May 17, 1979.
"Mickey Mouse stuff": Marvin L. Golberger, letter written on behalf of the Exploratorium, spring 1981.
to make of them: Friedman, Alan, in discussion with the author.

303 *"freedom to procreate":* Oppenheimer, "The Freedom to Create," Bancroft Collection, University of California, Berkeley.
"damn atomic bomb": Ibid.
character of the country: Ibid.
"or telephone networks": Ibid.

304 *"science could help":* Ibid.
"always kept me going": Bell, Jamie, in discussion with the author.
"knowledge or awareness": Kahn, Ned, in discussion with the author.

305 *"explain it to them":* Garb, Micah, in discussion with the author.
"Exploratorium does it": Millero, Frank, in discussion with the author.

306 *"didn't think I had":* Wilkinson, Karen, in discussion with the author.
"I wander the city": Minor, Nicole, in discussion with the author.
as one put it: Isaacs, Leni, in discussion with the author.
"be a great thing": Perlman, David, in discussion with the author.
"world around them": Donald Lambro, "Hog-tied Spending Bills," *Washington Times,* November 12, 2007.
"hospital for teachers": Falanga, Rose, in discussion with the author.

307 *to this day:* List of current Exploratorium staff since 1985, May 18, 2004.
"They nurture it": Barker, Dave, in discussion with the author.
"for Frank's dream": Ansel, Joe, in discussion with the author.
3,000 high school students: Dackman, Linda, in discussion with the author.

Teacher Institutes: Ibid.
"they were teachers": Falanga, Rose, in discussion with the author.

308 *"Mission Street":* Oppenheimer, "Where Are We and Where Do We Go from Here — I."
and other museums: Ibid.
became a union shop: Ansel, "The Unionization of the Exploratorium."
an extension of home: Ibid.

309 *"teach us what learning is":* Bartels, Dennis, in discussion with the author.
"feel at home": Max Otto, *Science and the Moral Life* (New York: New American Library, 1949), 138.
"like an outlet": Falanga, Rose, in discussion with the author.
"inhabited some of the exhibits": Unidentified staff member, "Séance: Interviews with Exploratorium Staff."

310 *"been working here!":* Barker, Dave, in discussion with author.
"revered nonetheless": Templeton, Michael, in discussion with the author.
"in the shadow": *New York Times* obituary for Frank Oppenheimer, February 4, 1985, Bancroft Collection, University of California, Berkeley.
social services combined: Steve Fraser, "A Lettered Numbers Man," *Los Angeles Times Book Review,* February 13, 2005.

311 *highest levels of alert:* Pavel Podvig, e-mail message to author.
programs of their own: Stephen I. Schwartz, "Warheads Aren't Forever," *Bulletin of the Atomic Scientists,* September/October 2005.
"Every single one of them": Oppenheimer, interview by Else, *Day After Trinity.*
were effectively fired: Gottfried, Kurt, in discussion with the author.
never witnessed a nuclear explosion: George Dyson, "Generation Ignorant," *Make Magazine* 7 (2006).
"care about human politics": Martin Hoffert, "Science Is, Science Isn't," *Bulletin of the Atomic Scientists* 63, no. 1 (January/February 2007).
"materials worldwide": Christopher E. Paine, "Weaponeers of Waste," Natural Resources Defense Council, April 2004. Emphasis in original. Quoted in *Bulletin of the Atomic Scientists,* March/April 2007.
"from lack of sleep": Bruce G. Blair, "Primed and Ready," *Bulletin of the Atomic Scientists,* January/February 2007.

312 *"It's not hopeless":* Freeman Dyson, interview by Lauren Redniss, *Discover,* June 2008.
much the same fate: Tom Siegfried, "Oppenheimer's Humiliation Still Haunts Science," *Dallas Morning News,* April 12, 2004.
spent his life promoting: Frank Oppenheimer, "At the end of World War II those of us who had been engaged . . . ," undated.

313 *"settle nothing worthwhile":* Ibid.
prone to generate conflict: Ibid.
"except as men make it": Otto, *Science and the Moral Life.*
"in their hearts": Keim, Liz, in discussion with the author.

316 *"hear the cane":* Lani, Shawn, in discussion with the author.

SELECTED BIBLIOGRAPHY

Note on Sources

Most of the interviews took place between 1999 and 2008. The same dates hold for most e-mails. A few quotes from early papers and articles about Frank Oppenheimer or the Exploratorium lack original sources but were cited by me in notes I assembled for a 1985 *Newsday* article, cited below. Where dates were unavailable, I sometimes give approximations. A good number of Oppenheimer's writings exist only as undated, unpublished, and sometimes partial essays, letters, articles, and ruminations. Tape-recorded conversations between myself and Frank (often noted as interviews) took place between the years 1972 and 1985; letters between us date from the same period. Unless otherwise noted in the endnotes, all papers and letters are in the possession of the author, but will eventually go to the Bancroft Library Collection or the Exploratorium. Large collections of letters are not listed individually here, but are cited in the endnotes along with the archives in which they are located.

Archives

American Institute of Physics Center for the History of Physics, Niels Bohr Library and Archives: Interview with Frank Oppenheimer by Charles Weiner, February 9 and May 21, 1973. California Institute of Technology Oral History Project, Caltech Archives: Interview with Frank Oppenheimer by Judith Goodstein, November 16, 1984. Carl A. Kroch Library, Cornell University. Federal Bureau of Investigation, Frank Oppenheimer papers released under the Freedom of Information and Protection of Privacy Act. Frank Oppenheimer Papers, Bancroft Library, University of California, Berkeley. Frank Oppenheimer Papers, University of Minnesota Archives. Robert Rathbun Wilson Collection, Fermi National Accelerator Laboratory Archives. Exploratorium Archives, including memos, progress reports, press and publicity releases, fact sheets, films and videos, catalogues, biographies, fundraising brochures, program summaries, and visitor guides.

Selected Interviews by the Author, Including Electronic Correspondence

Ansel, Joe; Bainum, Kurt; Barker, David; Bartels, Dennis; Bartlett, Albert; Bell, Jamie; Bennis, Warren; Boorstin, Jon; Brown, Ruth; Carlson, Charles; Chabay, Ilan; Crawford, Zane; Dackman, Linda; Duensing, Sally; Durr, Virginia; Else, Jon; Emerson, Ed; Falanga, Rose; Finegold, Leonard X.; Fowler, Stanley; Freichtmeir, Kurt; Friedman, Alan; Garb, Micah; Gell-Mann, Murray; Glorioso, Charles; Gordon, Bonnie; Gottfried, Kurt; Gregory, Richard; Grinnell, Sheila; Hake, Richard; Hawkins, David; Hawkins, Frances; Heckman, James J.; Hewitt, Paul; Hipschman, Ron; Hohenstein, Jack; Humphrey, Thomas; Hupert, David; Hutchinson, Brenda; Isaacs-Boorstin, Leni; Janssen, Katie; Janssen, Kristen; Johnson, Clark; Kahn, Ned; Kaiser, David; Kay, Alan; Keim, Liz; Kennedy, Donald; Kutnick, Ester; Lani, Shawn; Leigh, Jerry; Librero, Darlene; Lopez, Jorge; Marks, Gerald; Meyer, Rachel; Miller, Robert; Millero, Frank; Minor, Nicole; Morrison, Philip; Murphy, Pat; Oppenheimer, Frank; Oppenheimer, Judith; Oppenheimer, Michael; Pearce, Michael; Perlman, David; Panofsky, Wolfgang; Podvig, Pavel; Pusina, Jan; Rankin, Lynn; Renney, Mona Helen; Richards, Peter; Rodger, Nancy; Robinson, Berol; Rubin, Virginia; Sadun, Alfredo; Shaw, Larry; Semper, Robert; Schlossberg, Edward; Silverstone, Stuart; Stern, David; Sulston, Ingrid; Templeton, Michael; Tompkins, Tom; Tressel, George; Uretsky, Jack; Weiner, Charles; Wheaton, Bruce; Wilkinson, Karen; Van Kasper, F.; Zurin, Hal.

Writings by Frank Oppenheimer

Oppenheimer, Frank. "Adult Play." *Exploratorium*, February/March 1980.
———. "Aesthetics and the Right Answer." *The Humanist*, March–April 1979.
———. "Air Craft Carrier Caper," c. 1948.
———. "Albert Einstein." *Exploratorium*, April/May 1979.
———. "Amnesty," 1966.
———. Article about Mathematics for Teaching Credentials, Pagosa Springs, Colorado, 1958.
———. "Art and Science: Meaning, Tools, and Discipline." Written for catalogue of exhibit "The Expanding Visual World: A Museum of Fun." *Exploratorium*, October/November 1979.
———. "The Arts: A Decent Respect for Taste." *National Elementary Principal*, October 1977.
———. "Atom Bomb—Nightmare Weapon: Scientist Tells CIO It Must Be Controlled." *Labor Herald* 9, no. 30 (December 14, 1945).
———. "Boiling and Bubble Chambers." *Exploratorium*, Winter 1982–83.
———. "A Broader View." *Exploratorium*, 1977.
———. "Bus Essay," c. 1960.
———. "The Character of a University," 1964.
———. "Chips and Choices." *Exploratorium*, 1983–84.
———. "Color." *Exploratorium*, June/July 1977.
———. "Comments on the Unique Role of Museums and the Need for Federal Ongoing

Support for Their Operation and Expansion." Prepared for the Select Committee on Education in the course of their hearings on HR 332, the museum services bill, May 20, 1974.
——. "The Common Bonds Between Art and Science," May 1977.
——. "The Content of the Museum," 1967.
——. "Curiosity." Excerpt in *San Francisco Sunday Examiner,* September 25, 1983.
——. "Desert People," c. 1933.
——. "A Dynamic Process." *Exploratorium,* February/March 1978.
——. "Everyone Is You . . . or Me." *Technology Review* 78, no. 7 (June 1976).
——. "Exhibit Design and Conception." Paper presented to the Joint Association of Science-Technology Center/CIMUSET Conference, Monterrey, Mexico, October 24–25, 1980.
——. "Exhibit Planning." Memo, March 12, 1972.
——. "The Exploratorium as an Exemplary Institution," May 12, 1975.
——. "The Exploratorium Is a Teaching Museum." *Exploratorium,* February/March 1978.
——. "The Exploratorium and Other Ways of Teaching Physics." Paper presented at the joint meeting of the American Association of Physics Teachers and the American Physical Society in Anaheim, January 31, 1975.
——. "The Exploratorium: A Playful Museum Combines Perception and Art in Science Education." *American Journal of Physics* 40 (July 1972).
——. "A Factor of a Thousand." Version published in *Saturday Review* as "The Mathematics of Destruction," January 16, 1965.
——. "Favorite Childhood Toy—Personal Statement," December 9, 1976.
——. "The Freedom to Create," c. 1960s, most recent draft 1982.
——. "Growing Up in the Arts." *National Elementary School Principal* 56, no. 1 (September/October 1976).
——. "The Importance of the Role of Science Pedagogy in the Developing Nations." Speech, Budapest, 1965.
——. "Initiation." *Exploratorium,* August/September 1978.
——. "Interconnecting Shadows." *Exploratorium,* June/July 1978.
——. "Introduction: Diaphanous." *Exploratorium,* Fall 1984.
——. Introduction to magazine on the Science Media Conference. *Exploratorium,* 1981.
——. Introduction to "A Photo Essay by Nancy Rodger with Notes from Visiting Students." *Exploratorium,* October/November 1980.
——. "Jargon: Second Cousin Twice Removed." *Exploratorium,* 1982.
——. "The Language of Poetry and Science: Forefront Readings at the Exploratorium."
——. "Learning with an Opportunity for Choice." Speech presented at the annual meeting of the American Association for the Advancement of Science, 1980.
——. "Let the Teachers Teach and the Learners Learn."
——. "Mountain People," c. 1933.
——. "Museums: A Versatile Resource for Learning and Pleasure."
——. "Museums and Toys." *Exploratorium,* February 13, 1981.
——. "Museums for the Love of Learning—A Personal Perspective."
——. "New Admissions Policy." *Exploratorium,* April/May 1981.
——. "New Exploratorium Catalogue. Facets of Light: Colors, Images and Things That Glow in the Dark." *Exploratorium,* June/July 1980.
——. "The Numbers Game." *Exploratorium,* 1982.
——. "Paris to London," c. 1933.

———. "Persuasion." Talk given in honor of Muriel Rukeyser, Sarah Lawrence College, December 9, 1978.
———. "Persuasion, Coercion, and Overpowering," February 12, 1966.
———. "Pleasant Play." *Exploratorium,* December 1979/January 1980.
———. "The Pleasures and Personal Satisfactions of Science." Talk delivered to alumni groups at Grand Junction and Glenwood Springs, Colorado, 1964.
———. "The Poet Joins the Scientist."
———. "Political Give and Take." *Denver Post,* September 1962.
———. "Practical and Sentimental Fruits of Science." Speech, fifteenth anniversary awards dinner, *Exploratorium,* January 25, 1985.
———. "Prediction and Invention," 1968.
———. Preface to *Einstein* magazine, *Exploratorium,* 1981.
———. "A Proposal for an Alternative Course Structure," c. 1966.
———. "Proposal for a Palace of Art and Science," 1967.
———. "Public Understanding of Science." Lecture, Advisory Committee, Office of Science and Society, National Science Foundation, March 6, 1980.
———. "Public Welfare in the Atomic Age: The Support of Science." Lecture, Minnesota Bankers Association, June 1948.
———. "Recovery," c. 1947.
———. "The Role of Science Museums." Talk delivered at the Smithsonian Institution Conference on Museums and Education, University of Vermont, August 1966, excerpt in *Museums and Education,* edited by Eric Larrabee. Washington, D.C.: Smithsonian Institution Press, 1968.
———. "Science and the Ethics of Coercion," c. 1968.
———. "Science and Fear: A Discussion of Some Fruits of Scientific Understanding." *Centennial Review* 5, no. 4 (Fall 1961).
———. "Science and Immunity." *Scientific World,* 1969.
———. "Science and Invention," 1963 or 1968.
———. "Science and Peace." Lecture, Boulder, Colorado, 1946.
———. "The Sentimental Fruits of Science." Speech given on receiving the Oersted Award by the American Association of Physics Teachers, San Antonio, Texas, January 31, 1984.
———. "Simple Demonstration of the Retinal Evidence Involved in Distance Perception." *American Journal of Physics* 33, no. 12 (December 1965), 1085–88.
———. "Simple People," c. 1933.
———. "Smells: An Introduction." *Exploratorium,* Spring 1983.
———. "Stacked Deck." *Colorado Daily,* c. 1967.
———. "Stop Nuclear War: Review of Protest and Survive." Edited by E. P. Thompson and Dan Smith, Monthly Review Press.
———. "The Study of Perception as a Part of Teaching Physics." *American Journal of Physics,* July 1974.
———. "The Tail That Wags the Dog," 1949.
———. "Teaching and Learning." Robert A. Millikan Lecture, American Association of Physics Teachers, June 1973.
———. Testimony Before the House Committee on Education and Labor, Subcommittee on Post-Secondary Education. San Francisco, February 16, 1980.
———. Testimony at Hearing on Science, Mathematics, and Technology, National Committee on Excellence in Education, April 1983.
———. "Testimony with Regard to House Joint Resolution 639 Calling for White House Conference on the Humanities," January 4, 1978.

——. "The Unique Educational Role of Museums." Paper, Belmont Conference on the Opportunities for Extending Museum Contributions to Pre-College Science Education, January 1969.

——. "A War in the Shadow of the H-Bomb." *Bulletin of the Atomic Scientists,* May 1968.

——. "Watery Delights." *Exploratorium,* Summer 1982.

——. "Wave Equation," in *The Rainbow Book,* edited by Lanier F. Graham. New York: Random House, 1979.

——. "Where Are We and Where Do We Go from Here—I," June 1, 1976.

——. "Where Are We and Where Do We Go from Here—II," November 1982.

——. Paper prepared for meeting of the American Association for the Advancement of Science, Toronto, 1981.

——.Reminiscence about Robert Oppenheimer.

——. Speech, American Association of University Women, 1945 or 1946.

——. Speech, Association of School Administrators 110th annual convention, Atlanta, February 1978.

——. Speech, Exploratorium Awards Dinner, November 7, 1981.

——. Speech to graduating class of Pagosa Springs High School, 1960.

——. Speech on receipt of the American Association of Museums Distinguished Service Award, June 21, 1982.

——. Talk delivered to Berkeley Democratic Club, November 27, 1945.

——, to Arlon Elser, August 27, 1984.

——, to Bob Wilson.

——, to Mr. Binder, editor of the *Minneapolis Morning Tribune.*

——, to Bruno, November 3, no year.

——, to David Rockefeller, June 23, 1976.

——, to Dean Acheson.

——, to Dianne Feinstein, July 18, 1978.

——, to Editor, *The Chronicle,* March 26, 1978.

——, to Editor, *Colorado Daily.*

——, to Editor, *New York Times,* April 28, 1965.

——, to Ernest Lawrence.

——, to "Mr. Fleming."

——, to Fred Duncan, May 7, 1984.

——, to F. Van Kasper, April 6, 1983.

——, to "Glen."

——, to Homi Bhabba.

——, to John S. Wood.

——, to Kenneth W. Chalmers, 1958.

——, to Kevin Smith, Educational Development Corporation, February 13, 1970.

——, to Mr. Leland, 1946.

——, to Lou Branscomb, Chairman, National Science Board, International Business Machines Corporation, December 16, 1981.

——, to Mayor Thomas Currigan.

——, to M. Eugene Sundt, President of Albuquerque Gravel Products, February 18, 1970.

——, to "Miss Lee," June 6, 1958.

——, to Nat Finney, March 15, 1948.

——, to Norman Cousins, July 16, 1960.

——, to Palmer Hoyt, *Denver Post,* September 23, 1961.

——, to Peter K. Hawley, United Office and Professional Workers of America, February 14, 1946.

——, to Peter R. E. Goodchild, British Broadcasting Corporation, May 21, 1980.
——, to "Phil."
——, to Philip Abelson, September 11, 1968.
——, to President Smiley.
——, to Roy McVicker, House of Representatives, February 21, 1966.
——, to "Sam."
——, to Senator William J. Fulbright, July 3, 1967.

Other Publications and Sources

Aaland, Chris. "Fort Lewis Presents *Durango Herald* Editor with Honorary Degree and Chairman Morley Ballantine with Honorary Degree." *External Affairs,* May 21, 2004.

Albright, Thomas. "From Electric Music Boxes to Solar Energy Art: Chaotic Funkiness and Fun at Exploratorium." *San Francisco Chronicle,* October 22, 1970.

Alda, Alan. *Things I Overheard While Talking to Myself.* New York: Random House, 2007.

Allison, Samuel K., to Frank Oppenheimer, May 19, 1950.

Alperovitz, Gar. "A Dubious Advantage." *Bulletin of the Atomic Scientists,* July/August 2005.

Ansel, Joseph G., Jr. "The Unionization of the Exploratorium," in *Institutional Trauma: Major Change in Museums and Its Effect on Staff,* edited by Elaine Heumann Gurian. Washington, D.C.: American Association of Museums, 1995.

Armstrong, David. "Scouting New Worlds: Exploratorium Tries Its Hand at Partnerships Overseas and in U.S." *San Francisco Chronicle Business,* July 14, 2002.

Ballantine, Morley. "Smiling Through the Witchhunt." Probably from the *Durango Herald;* date unknown.

Barber, Benjamin. "A Failure of Democracy, Not Capitalism." *New York Times,* op-ed, July 29, 2002.

Barnett, Lincoln. *The Universe and Doctor Einstein.* New York: William Morrow & Co., 1948.

Baron, Virginia. "Nuclear Abolition Is Everybody's Business: An Interview with Jonathan Schell." *Fellowship of Reconciliation,* March/April 1999.

Bartels, Dennis M. "On-Site Science: Why Museums, Zoos, and Other Informal Classrooms Need to Be a Bigger Part of the Reform Equation." *Education Week,* September 19, 2001.

Bartlett, Albert A., to K. C. Cole, with attachments on new physics building.

——. to Director of the Exploratorium. Library of the Exploratorium, March 13, 1990.

——. "The Frank C. Walz Lecture Halls: A New Concept in the Design of Lecture Auditoria." *American Journal of Physics* 41 (November 1973): 1233–40.

——. "An Introduction to the Laboratory Manual for Physics 1140." University of Colorado, Boulder, September 1989.

Barth, John. *The Floating Opera.* New York: Doubleday & Co., 1956.

Bernstein, Jeremy. *Oppenheimer: Portrait of an Enigma.* Chicago: Ivan R. Dee, 2004.

Bielski, Vince. "Oppenheimer's Ghost: Will the Exploratorium's Quest for Commercial Success Kill the Vision that Frank Oppenheimer Had in Mind When He Founded the Museum?" *San Francisco Weekly,* February 17, 1993.

Bird, Kai, and Martin J. Sherwin. *American Prometheus: The Triumph and Tragedy of J. Robert Oppenheimer.* New York: Alfred A. Knopf, 2005.

Black, Algernon D., to Frank Oppenheimer, December 23, 1982. New York Society for Ethical Culture.

Blakeslee, Sandra. "Oppenheimer Celebration Examines the Myth and the Man." *New York Times,* June 29, 2004.

Blair, Bruce G. "Primed and Ready." *Bulletin of the Atomic Scientists,* January/February 2007.

Boorstin, Jon. *Exploratorium.* Documentary. Jon Boorstin Productions, 1974.

Born, Max. *Natural Philosophy of Cause and Chance.* Oxford, UK: Clarendon Press, 1969.

Branscomb, Lewis M. Letter on Criteria for NSB Vannevar Bush Award, November 17, 1980.

Broad, William J. "Book Contends Chief of A-Bomb Team Was Once a Communist." *New York Times,* September 8, 2002.

——. "Facing a Second Nuclear Age." *New York Times,* August 3, 2003.

Byers, Nina. "Physicists and the 1945 Decision to Drop the Bomb." Lecture, UCLA Department of Physics and Astronomy, October 13, 2002.

Cassidy, David C. *J. Robert Oppenheimer and the American Century.* New York: Pi Press, 2005.

Ciccotti, Susan. "The Exploratorium's China Trade." *Museum News,* March/April 2001.

Cirincione, Joseph. "Lessons Lost." *Bulletin of the Atomic Scientists,* November/December 2005.

Cohn, Victor. "Noted Physicist Must Plead to Teach." *Minneapolis Morning Tribune,* June 26, 1958.

Cole, K. C. "Biography, Dr. Frank Oppenheimer." Association of Science-Technology meeting, no date.

——. "An Exploratorium for Your Child's Sensorium." *Saturday Review,* November 1972.

——. *Facets of Light: Colors, Images and Things That Glow in the Dark.* San Francisco: The Exploratorium, 1980.

——. "The Rewards of a Most Unusual Friendship." *Newsday,* June 2, 1985.

——. "San Francisco's Scientific Fun House." *New York Times,* July 9, 1978.

——. "The Sentimental Fruits of Science." *Discover,* May 1983.

——. *Sympathetic Vibrations: Reflections on Physics as a Way of Life.* New York: William Morrow, 1985.

——. "Things Your Teacher Never Told You about Science." *Newsday,* March 23, 1986.

——. "Vision: In the Eye of the Beholder." *Exploratorium,* 1978.

Congress of the United States. Hearings Before the Joint Committee on Atomic Energy. Ninety-first Congress, First Session, on General, Physical Research Program, Space Nuclear Program, and Plowshare, April 17 and 18, 1969.

Corcoran, Martha. "Goggles Required: Watch Out for Cork Missiles and Flying Tennis Balls As Educators Duke It Out for the Coveted Title of 'Iron Science Teacher.'" *Teacher,* October 2000.

Cunningham, Brent. "Across the Great Divide: Class." *Columbia Journalism Review,* May/June 2004.

Dackman, Linda. "Invisible Aesthetic: A Somewhat Humorous, Slightly Profound Interview with Frank Oppenheimer." *Museum International* 38, no. 2 (1986).

Dao, James C. "Senate Panel Votes to Lift Ban on Small Nuclear Arms." *New York Times,* May 10, 2003.

Deitch, Kenneth M. *The Manhattan Project: A Secret Wartime Mission.* Perspectives on History Series. Discovery Enterprises, Inc., 1995.

Delacote, Goery. "Putting Science in the Hands of the Public." *Science* 280 (June 26, 1998).

Doss, Margot Patterson. "A Palace Full of Wonders." *San Francisco Sunday Examiner,* June 21, 1970.

Dyson, Freeman. *Disturbing the Universe.* New York: Harper and Row, 1979.

———. Interview by Lauren Redniss, *Discover*, June 2008.

Dyson, George. "Generation Ignorant." *Make Magazine* 7 (2006).

Edwards, Rob. "The A-Bomb: 60 Years On, Is the World Any Safer?" *New Scientist,* July 16, 2005.

Effects of the A-Bomb and Interview with Dr. Ryuso Tanaka, President of Hiroshima City University, May 16, 1995. http://history.sandiego.edu/gen/st/~lovenson/Theeffects.html.

Einstein exhibit at the Museum of Natural History, New York.

Else, Jon. *The Day After Trinity.* Documentary. 1981.

———. Transcripts of interviews with Frank Oppenheimer conducted for the documentaries *The Day After Trinity* and *Palace of Delights.*

———, to John Mansfield, June 11, 1981.

"Exploratorium: A Museum of Science, Art and Human Perception." January 2004.

Exploratorium. Special Memorial Issue, March 1985.

Ezarik, Melissa. "Iron Science Teacher: Educators Battle It Out for the Glory and Take Home Wealth of Teaching Know-how." *District Administration,* March 2002.

Ferguson, Charles D., and Lisa Obrentz. "Make It or Break It: The Weapons Labs Built the Bomb. Now They're Tasked with Finding Ways to Get Rid of It. Trouble Is, Old Habits Die Hard." *Bulletin of the Atomic Scientists,* March/April 2007.

Flanagan, Dennis, to K. C. Cole, November 5, 1990.

———, to *Exploratorium,* Spring 1981.

Folger, Tim. "Physicists Who Built the Bomb, and Why They Hated It." *Discover,* April 2002.

Ford, Kenneth. "Introduction of Frank Oppenheimer as recipient of the 1973 Millikan Lecture Award of the American Association of Physics Teachers." American Association of Physics Teachers.

Fraser, Steve. "A Lettered Numbers Man." *Los Angeles Times Book Review,* February 13, 2005.

Gamow, Barbara. "Evolution of a Palace." Exploratorium brochure.

Garcia, Ken. "S.F. Jewel Shines Bright at Age 30." *San Francisco Chronicle,* September 11, 1999.

Gardner, Howard. *Creating Minds: An Anatomy of Creativity Seen Through the Lives of Freud, Einstein, Picasso, Stravinsky, Eliot, Graham, and Gandhi.* New York: Basic Books, 1993.

Glaser, Robert J., to *Exploratorium,* Spring 1981.

Gold, Stephanie. "Inside Oppenheimer's Brain." *American Way,* August 1996.

Goldberger, Marvin L. to *Exploratorium,* Spring 1981.

Goldblatt, Louis, to William Coblenz, September 26, 1968.

Goodchild, Peter J. *Robert Oppenheimer: Shatterer of Worlds.* New York: Fromm International, 1985, 1980.

Gregory, Richard. *Odd Perceptions.* New York: Routledge, 1989.

Grinell, Sheila. "Starting the Exploratorium: A Personal Recollection." *American Scientist,* May–June 1988.

Hales, Dianne. "Frank Oppenheimer: Builder of Merlin's Museum." *SciQuest*, April 1981.

Hein, Hilda. *The Exploratorium: The Museum as a Laboratory.* Washington, D.C.: Smithsonian Institution Press, 1990.

Heisenberg, Werner. "Role of Modern Physics in Human Thinking," in *Physics and Philosophy: The Revolution in Modern Science.* New York: Harper & Row, 1958.

Herken, Greg. *The Brotherhood of the Bomb: The Tangled Lives and Loyalties of Robert Oppenheimer, Ernest Lawrence, and Edward Teller.* New York: Henry Holt, 2002.

Hewitt, Paul. *Conceptual Physics for Parents and Teachers.* Newbury, Mass.: Focus Publishing, 1998.

"Historical Walking Tour of Downtown Los Alamos." Brochure.

Hoffert, Martin. "Science Is, Science Isn't." *Bulletin of the Atomic Scientists,* January/February 2007.

Hoffmann, Ronald. "Ethics Growing Out of Science?" ("Honesty to the Singular Object") *Language, Lies, and Ethics*, edited by M. A. Safir. Frankfurt: Suhrkamp, 2009.

Hollinger, David A. "The Hunt for Oppenheimer: Why the Man Who Led the Effort to Build the American Atom Bomb Was Pushed Out of Government Service." *New York Times,* September 15, 2002.

———. "Life in a Force Field." *American Scientist* 94 (January/February 2006).

Homer-Dixon, Thomas. "Approaching Midnight; Doomsday Reconsidered: Speed Bump." *Bulletin of the Atomic Scientists,* January/February 2007.

House Committee on Un-American Activities. Hearings Regarding Communist Infiltration of Radiation Laboratory and Atomic Bomb Project, University of California, Berkeley, June 14, 1949.

"It Is Five Minutes to Midnight." *Bulletin of the Atomic Scientists,* January/February 2007.

Johnson, Haynes. *The Age of Anxiety: McCarthyism to Terrorism.* Orlando, Fla.: Harcourt, 2005.

Kabir, P. K. Taped interview of Frank Oppenheimer, May 30, 1974.

Kaiser, David. "The Atomic Secret in Red Hands? American Suspicions of Theoretical Physicists During the Early Cold War." *Representations* 90 (Spring 2005).

———. "Nuclear Democracy: Political Engagement, Pedagogical Reform, and Particle Physics in Postwar America." History of Science Society, 2002.

———. "The Postwar Suburbanization of American Physics." *American Quarterly* 56 (December 2004).

Keim, Liz, to Frank Oppenheimer, February 1982.

Kennedy, Donald. "Twilight for the Enlightment." *Science* 308 (April 8, 2005).

———, to *Exploratorium,* Spring 1981.

Kennedy, Randy. "The Day the Traffic Disappeared." *New York Times Magazine,* April 20, 2003.

Kerr, Peter. "Frank Oppenheimer, Nuclear Physicist, Dies." *New York Times,* February 4, 1985.

Kidder, Rushworth M. "Museum Guides Turn On to Science: San Francisco Museum Taps Teen-agers, Even Dropouts, to Make the Exhibits 'Visitor Friendly.'" *Christian Science Monitor,* June 6, 1989.

Kongshem, Lars. "The Discoverers: Science Museums Like San Francisco's Exploratorium Have a Lot to Teach Schools about the Way Children Learn." *Executive Educator,* November 1995.

Kristof, Nicholas D. "Aiding the Enemy." *New York Times,* op-ed, April 12, 2003.
Kroeger, Brooke. "When a Dissertation Makes a Difference." *New York Times,* March 20, 2004.
Lambro, Donald. "Hog-tied Spending Bills." *Washington Times,* November 12, 2007.
Leland, Marjorie. "Exploring Exploratorium." *San Francisco Progress,* September 11, 1970.
Maddox, Sam. "The Other Oppenheimer: A Life in the Shadows."*Sunday Camera,* March 24, 1985.
Marqusee, Mike. "Patriot Acts." *The Nation,* December 13, 2004.
Mather, Kirtley F. Book review, *One World or None,* edited by Dexter Masters and Katharine Way. *American Scientist* 91 (May–June 2003), 258–59.
Mayer, Jane. "The Big Idea: A Doctrine Passes." *The New Yorker,* October 14 and 21, 2002.
Miller, Matt. "Is Persuasion Dead?" *New York Times,* June 4, 2005.
Miller, Robert. "What I Did at the Exploratorium and Why I Left." Around 1987.
Minton, Torri. "Immigrants' World Opens with Science." *San Francisco Chronicle,* February 23, 2001.
Morgan, Ted. *Reds.* New York: Random House, 2003.
Morrison, Philip. *Nothing Is Too Wonderful to Be True.* Woodbury, N.Y.: AIP Press, 1997.
———. "Rightful names; and another thing . . . ," *Bulletin of the Atomic Scientists,* May 2003.
Morrison, Philip, and Phylis Morrison. *The Ring of Truth.* New York: Random House, 1987.
Morrison, Philip, and Kosta Tsipis. *Reason Enough to Hope: America and the World of the Twenty-first Century.* Cambridge: MIT Press, 1988.
Murphy, Pat. "History of the Exploratorium and Philosophy of Dr. Frank Oppenheimer."
"Museums." *Playboy,* April 1972.
Navasky, Victor S. *Naming Names.* New York: Hill and Wang, 1980.
Newsom, Barbara Y., and Adele Z. Silver. Brochure reprinted from "The Art Museum as Educator: A Collection of Studies as Guide to Practice and Policy," 1978.
Norris, Robert S., and Hans M. Kristensen. "U.S. Nuclear Forces, 2007." *Bulletin of the Atomic Scientists,* January/February 2007.
———. "Where the Bombs Are, 2006." *Bulletin of the Atomic Scientists,* November/December 2006.
Northern California Association of Scientists to Professor R. C. Tolman, United Nations Atomic Energy Commission, June 4, 1946.
Ogawa, Rodney T., Molly Loomis, and Rhiannon Crain. "Social Construction of an Interactive Science Center: The Institutional History of the Exploratorium." Presented at the biennial meeting of the European Science Educational Research Association, Malmo, Sweden, August 24, 2007.
Oliver, Chris. "Science at Play: San Francisco's Exploratorium Sets Pace for Museums." *Honolulu Advertiser,* July 11, 1999.
Oppenheimer, Frank, and Ilan Chabay. "Testimony Presented Before the U.S. House of Representatives Subcommittee on Elementary, Secondary, and Vocational Education."
Oppenheimer, Frank, and K. C. Cole. "The Exploratorium: A Participatory Museum." *Prospects* 4, no. 1 (Spring 1974).
Oppenheimer, Frank, and K. C. Cole. "Fear." *Exploratorium,* Special Issue: Frank Oppenheimer, 1912–1985, March 1985.

Oppenheimer, Frank, and Malcolm Correll. "The Library of Experiments at the University of Colorado." *American Journal of Physics* 32, no. 3 (March 1964).

Oppenheimer, Frank, and Charles L. Critchfield to Editor, *New York Times,* October 6, 1948.

Oppenheimer, Michael. "Exposure." Collected in *I Thought My Father Was God and Other True Tales from the National History Project,* edited by Paul Auster. New York: Henry Holt, 2001.

Oppenheimer, Robert J. *Atom and Void: Essays on Science and Community.* Princeton, N.J.: Princeton University Press, 1989.

——. *Science and the Common Understanding.* New York: Simon and Schuster, 1953.

Otto, Max. *Science and the Moral Life.* New York: New American Library, 1949.

Overbye, Dennis. "From Companion's Lost Diary, a Portrait of Einstein in Old Age." *New York Times,* April 24, 2004.

Paine, Christopher E. "Weaponeers of Waste." National Resources Defense Council, April 2004.

Parfrey, Jonathan, Robert W. Kingston, and Julie Atherton. "New Nuclear Weapons Would Endanger Us All." *Los Angeles Times,* May 13, 2003.

Perlman, David. "Palace Goes Mod: An Exploratorium—Kids See and Feel Science." *San Francisco Chronicle,* July 15, 1970.

——. "Scientist Frank Oppenheimer Dies at 72." *San Francisco Chronicle,* February 4, 1985.

——. "Total Solar Eclipse to Be Visible from Aruba—or S.F. Museum." *San Francisco Chronicle,* February 20, 1998.

Podvig, Pavel. "If It's Broke, Don't Fix It." *Bulletin of the Atomic Scientists,* July/August 2005.

Preuss, Paul. "On the Blacklist." *Science 83,* June 1983.

Rabinowitch, Alexander. "Founder and Father." *Bulletin of the Atomic Scientists,* January/February 2005.

Radack, Jesselyn. "Whistleblowing: My Story." *The Nation,* July 4, 2005.

Randall, John Herman, Jr. *Newton's Philosophy of Nature: Selections from His Writings.* New York: Hafner Library of Classics, 1953.

Rees, Martin J. "Approaching Midnight. Doomsday Reconsidered: Grounds for Optimism." *Bulletin of the Atomic Scientists,* January/February 2007.

——. *Our Final Hour: A Scientist's Warning: How Terror, Error, and Environmental Disaster Threaten Humankind's Future in this Century—on Earth and Beyond.* New York: Basic Books, 2003.

Regan, Jim. "Exploratorium's 'Science of Music.'" *Christian Science Monitor,* October 28, 2004.

Reichl, Ruth. "Dreams of a Mad Scientist." *New West,* July 17, 1978.

Rhodes, Richard. *The Making of the Atomic Bomb.* New York: Touchstone, 1988.

Richter, Paul. "Bush Is Seeking Newer, Smaller, Nuclear Bombs." *Los Angeles Times,* May 13, 2003.

Rogers, Michael. "Playgrounds for the Mind: In America's 200 Science Museums, Kids of All Ages Get to Frolic—and Learn." *Newsweek,* July 25, 1989.

——. "Playing with Science." *Look,* April 2, 1979.

Rosenbaum, David E. "In the Fulbright Mold, Without the Power." *New York Times,* May 3, 2004.

Rotblat, Joseph. "A Fateful Decision." *Bulletin of the Atomic Scientists,* November/December 2005.

——. "The 50-Year Shadow." *New York Times,* May 17, 2005.

Rothstein, Linda, Catherine Auer, and Jonas Siegel. "Rethinking Doomsday: Loose Nukes, Nanobots, Smallpox, Oh My!" *Bulletin of the Atomic Scientists,* November/December 2004.

Rukeyser, Muriel. "The Sun-Artist." *Collected Poems of Muriel Rukeyser,* edited by Janet Kaufman and Ann Herzog. Pittsburgh: University of Pittsburgh Press, 2005.

Salmond, John A. *The Conscience of a Lawyer: Clifford J. Durr and American Civil Liberties, 1899–1975.* Tuscaloosa: University of Alabama Press, 1990.

Savage, Milton. "Frank Oppenheimer." *California Confederation of the Arts Newsletter* 1, no. 1 (Fall 1985).

Schell, Jonathan. "The Gift of Time: The Case for Abolishing Nuclear Weapons." *The Nation,* November 29, 2001.

———. "Letter from Ground Zero: Niceties." *The Nation,* November 26, 2001.

Schevitz, Tanya. "Museums Mad About Science." *San Francisco Chronicle,* January 27, 1998.

Schollmeyer, Josh. "Minority Report." *Bulletin of the Atomic Scientists,* January/February 2005.

Schrecker, Ellen W. *No Ivory Tower: McCarthyism and the Universities.* New York: Oxford University Press, 1986.

Schwartz, John. "Thinking May Not Be All It's Thought to Be." *New York Times,* January 16, 2005.

Schwartz, Stephen I. "Warheads Aren't Forever." *Bulletin of the Atomic Scientists,* September/October 2005.

Schweber, S. S. *In the Shadow of the Bomb: Bethe, Oppenheimer, and the Moral Responsibility of the Scientist.* Princeton, N.J.: Princeton University Press, 2000.

Selvin, Joel. "Seeing at the Exploratorium Investigates How Vision Works." *San Francisco Chronicle,* July 6, 2002.

Semper, Robert, and Judy Diamond. "Use of Interactive Exhibits in College Physics Teaching." *American Journal of Physics* 50, no.5 (May 1982).

Serber, Robert. *Peace and War.* New York: Columbia University Press, 1998.

Shipley, R. B., to Frank Oppenheimer.

Siegfried, Tom. "Oppenheimer's Humiliation Still Haunts Science." *Dallas Morning News,* April 12, 2004.

Smith, Alice Kimball, and Charles Weiner. *Robert Oppenheimer: Letters and Recollections.* Cambridge: Harvard University Press, 1980.

Starr, Kenneth. "Exploration and Culture: Oppenheimer Receives Distinguished Service Award." *Museum News,* November/December 1982.

Stern, Philip M. *The Oppenheimer Case: Security on Trial.* New York: Harper and Row, 1969.

Strauss, Mark. "Essence of a Decision." *Bulletin of the Atomic Scientists,* July/August 2005.

———. "A 60-Year Conspiracy." *Bulletin of the Atomic Scientists,* November/December 2005.

Sweet, William. "Exploratorium Influences Science Museums New and Old." *Physics Today,* March 1987.

Szasz, Ferenc Morton. *The Day the Sun Rose Twice: The Story of the Trinity Site Nuclear Explosion, July 16, 1945.* Albuquerque: University of New Mexico Press, 1984.

Taleb, Nassim Nicholas. "Scaring Us Senseless." *New York Times,* July 24, 2005.

Torassa, Ulysses. "Exploratorium Blazes Trail Starting from Scratch." *Living,* May 12, 1996.

Trinity site brochure, July 16, 1945.

U.S. Atomic Energy Commission. "The Avalon Project at the Yale Law School: Findings and Recommendations of the Personal Security Board in the Matter of Dr. J. Robert Oppenheimer," May 27, 1954.

Verrengia, Joe. "20th Century Forever Altered by Atomic Bomb." *Daily Herald*, December 28, 1999.

Wang, Jessica. *American Science in an Age of Anxiety: Scientists, Anti-communism, and the Cold War.* Chapel Hill: University of North Carolina Press, 1999.

Wechsler, Judith, ed. *On Aesthetics in Science.* Cambridge: MIT Press, 1978.

Weinberg, Steven. *Dreams of a Final Theory: The Scientist's Search for the Ultimate Laws of Nature.* New York: Vintage, 1994.

Weisskopf, Victor, to Provost Oswald Tippo, University of Colorado, November 2, 1960.

——, to *Exploratorium*, Spring 1981.

Werner, Robert J., to Frank Oppenheimer, March 12, 1971.

Wilson, Jane S., and Charlotte Serber. *Standing By and Making Do: Women of Wartime Los Alamos.* Los Alamos, N.M.: Los Alamos Historical Society, 1997.

Wilson, Robert, to *Exploratorium*, Spring 1981.

——, to K. C. Cole, July 10, 1985.

Wilson, Robert R. (Director, Laboratory of Nuclear Studies, Cornell University), to Frank Oppenheimer, September 9, 1959.

Wines, Michael. "In a Lithuanian Prison, the Beauty Is Unconfined." *New York Times*, November 26, 2002.

INDEX